ROMANTIC SCIENCE AND THE EXPERIENCE OF SELF

Other people put their faces on, one after the other,
with uncanny rapidity and wear them out.

Rainer Maria Rilke,
The Notebooks of Malte Laurids Brigge (1910)

Wherever a person develops their abilities, that person is an artist.

Joseph Bevys,
'Mysteries Happen in the Main Railway Station' (1984)

Romantic Science and the Experience of Self

Transatlantic Crosscurrents from William James to Oliver Sacks

Martin Halliwell

Studies in European Cultural Transition

Volume Two

General Editors: Martin Stannard and Greg Walker

Ashgate

Aldershot • Brookfield USA • Singapore • Sydney

Published by
Ashgate Publishing Limited
Gower House
Croft Road
Aldershot
Hants GU11 3HR
England

Ashgate Publishing Company
Old Post Road
Brookfield
Vermont 05036-9704
USA

Ashgate website: http://www.ashgate.com

British Library Cataloguing-in-Publication data

Romantic science and the experience of self: transatlantic crosscurrents from
    William James to Oliver Sacks. – (Studies in European cultural transition;
    v. 2)
    1.Science – History
    I.Halliwell, Martin
    509

US Library of Congress Cataloging-in-Publication Data

Romantic science and the experience of self: transatlantic crosscurrents from
    William James to Oliver Sacks / edited by Martin Halliwell.
    (Studies in European cultural transition; v. 2)
    Includes bibliographical references and index.
    1. Self – History – 20th century. 2. Psychology – History – 20th century.
    3.Romanticism – Influence. I. Halliwell, Martin. II. Series.
    BF697R6375 1999
    150'.9—dc21   99–36375
                        CIP

ISBN 1 84014 626 5

Typeset by Pat FitzGerald and printed on acid-free paper and bound in Great Britain by MPG Books Ltd, Bodmin, Cornwall

Contents

General Editors' Preface

The European dimension of research in the humanities has come into sharp focus over recent years, producing scholarship which ranges across disciplines and national boundaries. Until now there has been no major channel for such work. This series aims to provide one, and to unite the fields of cultural studies and traditional scholarship. It will publish the most exciting new writing in areas such as European history and literature, art history, archaeology, language and translation studies, political, cultural and gay studies, music, psychology, sociology and philosophy. The emphasis will be explicitly European and interdisciplinary, concentrating attention on the relativity of cultural perspectives, with a particular interest in issues of cultural transition.

Martin Stannard
Greg Walker

University of Leicester

Preface

In this book I construct a tradition of romantic science by indicating points of theoretical interconnection in the work of five influential figures in twentieth-century transatlantic intellectual history: the American philosopher and psychologist William James; the Viennese psychoanalyst Otto Rank; the Swiss psychiatrist Ludwig Binswanger; the Danish/German child psychologist Erik Erikson; and the British neurologist Oliver Sacks. Each of these thinkers is seen to be in antagonistic relation to the orthodoxies and presuppositions of what the philosopher Thomas Kuhn calls 'normal science'. As such, the modern tradition of romantic science attempts to regalvanize the spirit of late eighteenth-century and early nineteenth-century German romanticism in which there was no strict division between art and natural science. Early romantic science, as epitomized in the work of Johann von Goethe and Novalis, was short-lived and characterized by the kind of metaphysical speculation which could not survive with the rise of strictly empirical science in the nineteenth century. However, by the end of the nineteenth century the moral, positivistic and epistemological certainties of Victorian England, Wilhelmine Germany and post-bellum America had begun to crumble, giving rise to a reassessment of romantic thought and its applicability to the development of a more flexible scientific method which William James was to call radical empiricism.

Although James was sceptical about the worst excesses of romanticism, his work provided a platform for twentieth-century romantic scientists who wished to strike a balance between the rigour of empirical science and the imaginative and therapeutic possibilities of the creative arts. The growth of psychoanalysis in Europe and America in the early twentieth century had a great impact on the reorientation of romantic science, especially its concern with the interpersonal dynamics of analysis and therapy. Each of the romantic scientists studied here has a practical and theoretical interest in the experience of individuals: thus the object of scientific study is replaced by the living subject of psychoanalytic study. However, neither Otto Rank nor Ludwig Binswanger nor Erik Erikson wished to adopt psychoanalytic thought wholesale: each reacted against the constraining presuppositions which they argued Freud had absorbed from nineteenth-century natural science. As such, Rank and Binswanger outlined rival analytic models in the attempt to go beyond the limits and constraints of normal psychology. Modern romantic science thus begins to use narrative techniques to expand the often impoverished repertoire of scientific language and to provide a vehicle for expressing the patient's experience. Rank's interest in myth, Binswanger's case-studies, Erikson's psychohistories and Sacks' clinical tales are all attempts to deal with the particularities of an individual's condition, each subsequent thinker displaying an increased awareness of the dynamics and subtleties of narrative construction. Narrative models are rarely adopted uncritically: often the narrative method is fragmentary, partial and provisional as the tensions between the romantic scientist's voice and the patient's voice become evident. Romantic science is essentially an idealistic and humanistic tradition which tries to reconnect the narrow sphere of clinical observation with the broader realities of lived experience. The reformulation of the self as an active and creative artificer is the locus where the concerns of romantic science converge with a number of emerging trends in critical theory which have reconsidered the theoretical implications of selfhood.

The five thinkers studied here have been chosen carefully for three broad reasons. Firstly, their diverse cultural backgrounds suggests a shared sensibility rather than a specific national tradition; secondly, their thought successfully fuses European and American intellectual traditions; and, thirdly, each attempts to bridge the life of the mind with the realities of bodily experience. Although there are discernible lines of influence running through the work of the five figures, each reorientates romantic science to suit his own professional needs. In many ways, romantic science is an open tradition which can be adopted for any practice which fuses scientific with aesthetic understanding and seeks to deal with the experience of individuals. Romantic science also carries an ethical undercurrent which directs the inquirer to the sensitive issue of knowing the Other and checks abstract speculation with a firm commitment to develop particular modes of active communication. There are a number of other thinkers (see Introduction) who might have been included in this book, but a study of James, Rank, Binswanger, Erikson and Sacks reveals the intrinsic hybridity of romantic science, whilst indicating the creative tensions in, and the theoretical problematics of, their practice.

I would like to thank the staff and research students in the Department of American and Canadian Studies and The Postgraduate School of Critical Theory at The University of Nottingham for providing an intellectually stimulating research community between 1993 and 1996 and also members of the Faculty of Humanities and Social Sciences at De Montfort University, Leicester for encouraging me to explore aspects of this work in my teaching. Sections of this book have been presented at The University of Central England, De Montfort University, Keele University, Leeds University and The University of Nottingham. Particularly, I wish to extend my gratitude to Douglas Tallack and Richard King whose advice and practical guidance have been invaluable. I am very grateful to the British Academy for financial support and for research grants in 1995 and 1998 to consult the James and Erikson papers at the Houghton Library, Harvard University and the Otto Rank holdings at Columbia University; also to the British Association of American Studies for their grant to finance a visit to New York in 1995 to interview Oliver Sacks. Thanks also to Laraine and my family, Ben Andrews, Tim Armstrong, Joanna Clarke, Larry Friedman, Colin Harrison, Paul Hegarty, Dave Murray and Martin Stannard for their friendship, tolerance and continued support for this project.

Introduction
The Possibilities of Romantic Science

In the introduction to *Modernist Impulses in the Human Sciences, 1870–1930* (1994), Dorothy Ross makes an explicit link between 'the late-nineteenth-century cognitive move toward subjectivity and its aesthetic ramifications'.[1] Most critics of modernism deal primarily with cultural production in the early years of the twentieth century, drawing from science just enough theoretical material to account for the conceptual developments within the field of modernist aesthetics. However, the influence of the psychological sciences upon European and American modernism has been traced in greater detail in recent years in important studies by H. Stuart Hughes, Clive Bush, Judith Ryan and Tim Armstrong.[2] An exploration of the 'move towards subjectivity' has enabled these critics to assess the significant impact of the natural and human sciences upon artistic production in the early years of the twentieth century. Ross' collection exemplifies and extends this mode of study: instead of mapping science onto art, as is usually the practice, these essays explore points of aesthetic convergence from the perspective of the human sciences. The contributors to this volume argue that traditionally disparate disciplines are found to have a strong dialogic potential, deriving mainly from a radical reappraisal of what Ross calls 'the subjectivity of perception and cognition, a subjectivity that calls into question the unity of the observing subject as well as its relationship with the outside world'.[3]

The dichotomy between the arts and sciences has an important precedent in Max Weber's argument that the transition into modernity and rational capitalism is linked closely to the fragmentation of social activity into five distinct spheres: the economic, political, aesthetic, erotic and intellectual.[4] More recently, the German theorist Jürgen Habermas has refined Weber's five spheres into three (the cognitive-instrumental, the moral-practical and the aesthetic-expressive), arguing that these separate spheres of science, morality and art result from the general

[1]   Dorothy Ross (ed.), *Modernist Impulses in the Human Sciences* (Baltimore: Johns Hopkins U.P., 1994), p. 2.

[2]   See H. Stuart Hughes, *Consciousness and Society: The Reorientation of European Social Thought 1890–1930* (New York: Vintage, 1961); Clive Bush, *Halfway to Revolution* (New Haven: Yale U.P., 1991); Judith Ryan, *The Vanishing Subject: Early Psychology and Literary Modernism* (Chicago: University of Chicago Press, 1991); Tim Armstrong, *Modernism, Technology and the Body* (Cambridge: Cambridge U.P., 1998).

[3]   Ross (ed.), *Modernist Impulses in the Human Sciences*, op. cit., p. 6.

[4]   Max Weber, 'Religious Rejections of the World and Their Directions' (1915), in *From Max Weber: Essays in Sociology*, eds and trans. H.H. Gerth and C. Wright Mills (London: Routledge, 1991), pp. 323–59.

collapse of religion and metaphysics in the nineteenth century, with each sphere becoming the primary concern of 'special experts'.[5] Instead of fulfilling the Enlightenment ideal of complete 'understanding of the world and of the self', the separation of the spheres and related 'culture of expertise' has led to institutional divisions and combative oppositions, especially between the arts and sciences.[6] Mid-century writers in Britain such as C.P. Snow, Jacob Bronowski and Aldous Huxley argued within this conceptual framework by considering the increasing specialization of the sciences against the broader understanding of the arts and humanities.[7] The most famous example of such a distinction is the English writer C.P. Snow's Rede Lecture 'Two Cultures and the Scientific Revolution' (1959), in which he argues that both cultures are intolerant of the other and inclined to caricature the opposition. While literary intellectuals see blind faith in scientific progress, he claims that 'physical scientists' discern those whom they describe as 'anti-intellectuals' committed only to the 'existential moment' as totally 'lacking in foresight'.[8] Although primarily speaking of the British condition, Snow argues that such a chasm is a 'problem for the entire West'.[9] Furthermore, he claims the divide results primarily from ignorance, only to be rectified by softening of the boundaries between the disciplines and by educational reform.

A number of critical articles followed Snow's lecture, focusing particularly on his use of conceptual shorthand and crude categories. Clearly, his attempts to qualify the schematic nature of 'two cultures' are unsatisfactory, but his lecture does succeed in highlighting two important processes in Western history – the industrial and the applied scientific revolutions – which, as technological consequences of the Enlightenment faith in reason and social progress, brought about the transformation of social organization in the nineteenth century and the 'technical utilization of scientific knowledge'.[10] The related paradigm of applied science began to preclude any practice which deviated from the goals of social utility and economic efficiency and, in the words of the socialist thinker Winnifred Wygal, tended to deal with both 'steel' and 'humanity' in the same way.[11] As an example of what Thomas Kuhn has called 'normal science', this model provided the 'foundation' for the subsequent study of an objective and

---

[5]   Jürgen Habermas, 'Modernity – An Incomplete Project', in ed. Hal Foster, *Postmodern Culture* (London: Pluto, 1983), p. 9.

[6]   Ibid.

[7]   See C.P. Snow, *The Two Cultures: The Rede Lecture, 1959* (Cambridge: Cambridge U.P., 1965); Jacob Bronowski, *Science and Human Values* (London: Hutchinson, 1961); Aldous Huxley, *Literature and Science* (London: Chatto & Windus, 1963).

[8]   Snow, *The Two Cultures*, op. cit., p. 6.

[9]   Ibid., p. 2.

[10]   Max Weber, *The Protestant Ethic and the Spirit of Capitalism*, trans. Talcott Parsons (London: Routledge, 1992), pp. 24–5.

[11]   Winnifred Wygal, 'The Political Inertia of the Intellectual', *Radical Religion*, 1(3) (Spring 1936), 31.

empirically verifiable world.[12] Because practitioners of normal science claimed to tell verifiable truths about the world, with the widespread collapse of religious frameworks in the West normal science became 'the last court of appeal and sole bearer of truth'.[13] Even though a counter-tradition to normal science arose in the early twentieth century in the work of Franz Brentano, Edmund Husserl, William James, Ernst Mach, Henri Bergson and Martin Heidegger, who all focused upon phenomenological experiences unique to the individual, it was largely ignored by empiricists whose interest was in strictly verifiable data. For example, the American cognitive philosopher Daniel Dennett argues in his provocative book *Consciousness Explained* (1991), firstly, that phenomenology is purely descriptive and lacks any method which may be useful to explanatory science; and, secondly, in its 'intimate acquaintance' with subjectivity, it is wholly inaccessible to the rigorous methodology and objective measurements of scientific instruments.[14]

Responding to these debates and working in related areas of the human sciences, the five figures whose work is explored here – William James, Otto Rank, Ludwig Binswanger, Erik Erikson and Oliver Sacks – reveal the possibilities of, as well as the difficulties in, drawing upon phenomenological inquiry to broaden the range and develop the focus of scientific study. Subjectivity may not be so much of an issue in the physical and natural sciences, but, as H. Stuart Hughes indicates, the conceptual shifts in the human sciences in the late nineteenth century have emphasized the problems of discussing 'human behaviour in terms of analogies drawn from natural science'.[15] The branches of human science with which this book deals indicate the crucial importance of subjective analysis. Although deduction (the process of reasoning from *a priori* premises), central to the domains of symbolic logic and pure mathematics, is independent of description and qualification, it is possible to conceive of an overlapping territory between inductive science (moving from the collation of empirical data to reasoned conclusions) and expressive art, because they are both involved centrally in interpretation. A study of this intellectual counter-tradition may enable the critic to move beyond Snow's categories and, as Ross asserts, erase 'the sharp separation between science and art that existed under the aegis of logical positivism and aesthetic modernism in the 1950s and turns both science and art into interpretive languages'.[16] Such interpretative overlap is most readily evident in modes of inquiry which take seriously issues of selfhood, intersubjectivity and communicative understanding.

---

[12]   Thomas Kuhn, *The Structure of Scientific Revolutions* (Chicago: University of Chicago Press, 1970), p. 10.

[13]   Brice Wachterhauser (ed.), *Hermeneutics and Truth* (Evanston: Northwestern U.P., 1994), p. 34.

[14]   Daniel Dennett, *Consciousness Explained* (London: Penguin, 1993), p. 65.

[15]   Hughes, *Consciousness and Society*, op. cit., p. 37.

[16]   Ross (ed.), *Modernist Impulses in the Human Sciences*, op. cit., p. 11.

Ross stresses that 'despite the different discourses that have emerged around cognitive modernism, aesthetic modernism, and positivist science, they are linked in a common move toward human subjectivity and a common investment in the culture of modernity'.[17] In other words, by tracing the roots of specialized sciences it may be possible to rediscover a more inclusive field of inquiry which counters the 'culture of expertise' described by Habermas. With the growth and proliferation of human sciences in the twentieth century, and particularly in the discourse of the medical humanities which orient themselves towards patient care and a study of subjectivity (for example, clinical psychology, psychoanalysis and neuropsychology), dual or multiple modes of understanding can be productive in two important ways: firstly, they help to expose the presuppositions of traditional areas of study; and, secondly, by linking rigorous study of human behaviour to modes of aesthetic creativity, they encourage the development of alternative therapeutic techniques. On this account, the categories of 'aesthetic modernism'or 'cognitive modernism' should not be separated into discrete modes of inquiry, because they derive from the same cultural manifold of human activity. In order to address these issues, this book constructs a modern tradition of romantic science beginning in the late nineteenth century and also contributes to cultural debates concerning the shifting nature and ongoing implications of modernism. This study is particularly concerned with the modernist attempt to define the elusive qualities of experience: what Henry James describes as 'an immense sensibility, a kind of huge spider-web of the finest silken threads suspended in the chamber of consciousness'.[18] The concerted attempt to describe the 'immense' parameters and the intricate experience of self motivates both modernist aesthetics and the exploratory agenda of romantic science. But, while romantic science is modernist in its desire to engage with dimensions of experience which normal science usually ignores, it is 'antimodernist' (in T.J. Jackson Lears' sense of the phrase) in its resistance to the technological forces of modernization which erode the conception of an, at least partially, self-determining identity.[19]

Snow's argument assumes that the split between the two cultures is valid because they are independent fields of inquiry and, in many ways, opposing spheres of discursive activity. But his inquiry lacks adequate historical evidence to support this account. The Platonic dichotomy between philosophy (or reason)

---

[17]   Ibid., p. 15.

[18]   Henry James, 'The Art of Fiction' (1884), reprinted in Vassiliki Kolocotroni et al., eds, *Modernism: An Anthology of Sources and Documents*, (Edinburgh: Edinburgh U.P., 1998), p. 148.

[19]   Lears claims that 'antimodernism unknowingly provided part of the psychological foundation for a streamlined liberal culture' and the twentieth-century 'therapeutic ethos'; T.J. Jackson Lears, *No Place of Grace: Antimodernism and the Transformation of American Culture, 1880–1920* (Chicago: University of Chicago Press, 1984), p. 6.

as a discourse of truth opposed to fictive art is an easy argument to deploy when a critic wishes to denigrate cultural production as being inherently worthless. Throughout modern literary history a number of critics from Philip Sidney and Percy Bysshe Shelley to M.H. Abrams have argued that this claim is utterly vapid. Indeed, from a late eighteenth-century romantic perspective, the artist is seen to champion an expressive personal truth excluded from scientific practices applied to discerning causal and natural laws. Romantic art is usually characterized by 'things modified by the passions and imagination of the perceiver', in opposition to 'the unemotional and objective description characteristic of physical science'.[20] If normal science understands the world from an Archimedean point of reference, then romantic art focuses on the primacy of the perceiving and imagining subject. Thus, the challenge for romantic scientists is to discover a method of inquiry which incorporates subjective interpretation, without reverting to metaphysical speculation or dismissing creative expression as the epiphenomena of essentially material creatures.

M.H. Abrams argues that the distinction between poetry (as a metonym for art) and science should only be seen as a 'logical device' for 'isolating and defining the nature of poetic discourse', but it has often been reified in a manner which has caused 'combative opposition'.[21] This indicates another danger with Snow's approach: in his binary formulation of two cultures he loses sight of internal differences in each sphere of activity, the mapping of which may elicit important interconnections. Indeed, the debate throws into confusion generic differences between distinct areas of research and blurs the lines between specific fields of study. It is the legacy of romantic science to deny neither objective nor subjective approaches to the study of human activity and to the natural world. Instead, its practitioners work through a series of empirical *and* imaginative developments to link disparate fields of inquiry and encompass a number of theoretical and practical possibilities which complement and extend existing modes of scientific research.

Romantic science is not a project which fuses independent spheres of inquiry for their own sake: traditionally, philosophies of holism tend to generalize from a universalist point of view at the expense of the specific and contingent. Nor should it be seen as a purely reactionary counter-tradition, the sole aim of which would be to check the narrow or intolerant dimensions of normal science. Instead, if the truth claims of scientific thought are seen as no more (or no less) valid than the verisimilitude of aesthetic discourse, it becomes possible to blend disparate modes of discourse, in the language of the American post-Philosopher Richard Rorty, to strategically 'redescribe' the terms of inquiry into human behaviour. This act of hybridization would enable the discourse of romantic science to 'be

---

[20]   M.H. Abrams, *The Mirror and the Lamp* (New York: Oxford U.P., 1953), p. 299.

[21]   Ibid.

filed alongside all the others, one more vocabulary, one more set of metaphors which he thinks have a chance of being used'.[22] As Rorty comments:

> whereas the metaphorical looks irrelevant to . . . positivists, the literal looks irrelevant to Romantics. For the former think that the point of language is to represent a hidden reality which lies outside us, and the latter thinks its purpose is to express a hidden reality which lies within us.[23]

By linking these different versions of 'hidden reality' the romantic scientist can expose the weaknesses inherent in conventional empirical science. This does not mean romantic scientists simply elide perspectival differences, but that they attempt to compensate for the inadequacies in each point of view by juxtaposing a series of apparently irreconcilable perspectives. Although rarely an excessively utopian discourse, romantic science projects the emancipatory aim of redescribing selfhood in therapeutic or pragmatic ways, with which the realities of experience cannot always be squared. Despite the practical difficulties of carrying through these aims, when faced with the inadequate alternative of normal science, the five thinkers discussed here each prefer to work through the discourse of romantic science as a problematic, but potentially liberating, mode of inquiry.

Theories of Selfhood and Consciousness

Complementing the focus on a twentieth-century tradition of romantic science, this study is stimulated by two contemporary trends in current scholarship: first, a renewed interest in theories of selfhood in theoretical discourse; and, second, the study of consciousness as the prevalent focus for philosophers of cognitive science. Both trends mark a renewed interest in the parameters of 'the human' in a millennial period when the 'post-human' conditions of body modification and genetic manipulation are becoming distinct realities. Erik Erikson commented as early as 1959 that 'industrial man's attempt to identify with the machine' inevitably leads to 'the question as to what . . . is left of a human "identity"'.[24] Romantic science does not construct an essentialist model of the self which erases specific cultural and gender differences, but attempts to focus on skills and capabilities which may have been lost or never acquired by individuals suffering particular forms of mental or bodily dysfunction. Romantic science inscribes a humanistic and optimistic model of human capability and emphasizes the versatility of individuals to adapt and to develop in adverse circumstances. The

---

[22] Richard Rorty, *Contingency, Irony, and Solidarity* (Cambridge: Cambridge U.P., 1989), p. 39.
[23] Ibid., p. 19.
[24] Erik Erikson, *Insight and Responsibility* (New York: Norton, 1964), p. 105.

re-emergence of selfhood and consciousness in current debates suggests that some of the humanistic assumptions attacked by logical positivists and first-wave post-structuralists are in need of reassessment in the late twentieth century. By returning to the modernist moment in European and American thought it is possible to tap into a largely unexplored tradition of romantic science which outlines the possibilities, as well as the difficulties, of retaining a coherent definition of 'the human' in a period of increasing mechanization and intellectual specialization. Such a conception is not achieved by retreating from technology, but by exploring the borders and limits of selfhood 'either in terms of its mechanical functioning, its energy levels, or its abilities as a perceptual system'.[25]

Issues of personal identity and selfhood re-emerged on the critical agenda in the 1980s as a backlash against anti-humanist trends in the social sciences and language-based philosophy which sought to challenge previous subject-centred notions of human behaviour. The early century phenomenology of Brentano and Husserl, which promised to accurately describe the contents of consciousness, was attacked by the post-structuralist critics Jacques Derrida and Paul de Man in the late 1960s, who questioned the very existence of non-linguistic thoughts. In literary theory, positing an author's consciousness to establish stable meaning and guarantee that the author's intentions are conveyed by the transparent words of a text (evident in the phenomenologically influenced theory of the French critic Georges Poulet), has been widely discredited in favour of a model of language which can be seen to generate multiple meanings, depending on the individual reader's perspective on a text.

This theory has important consequences not just for literary criticism but for all intersubjective understanding: the implications of reading a literary text should also concern the inquirer who attempts to read the behaviour and expressions of another human being. As Poulet claims: 'critical consciousness relies, by definition, on the thinking of "another"; it finds its nourishment and its substance only therein'.[26] The implication here is that if the inquirer is sensitive to the different mind-set of other humans (as Poulet says, 'the significance of the *cogito*'), then such intersubjective inquiry will form 'an act of self-discovery'.[27] But, as post-structuralist theorists assert, consciousness is not simply a transparent and coherent container of words and meanings: verbal and behavioural languages are as mutable and contradictory as written discourses. Paul Ricoeur is one theorist who has taken seriously 'the existence of an opaque subjectivity which expresses itself through the detour of countless mediations – signs, symbols, texts and human praxis itself'.[28] Ricoeur's use of 'detour' may

---

[25]   Armstrong, *Modernism, Technology and the Body*, op. cit., pp. 4–5.
[26]   Georges Poulet, 'The Self and Other in Critical Consciousness', *Diacritics*, 2 (1972), 46.
[27]   Ibid., p. 47.
[28]   Richard Kearney, *Dialogues with Contemporary Continental Thinkers* (Manchester: Manchester U.P., 1984), p. 32.

suggest a way of bypassing language, but it actually reveals an important departure from the early romantic understanding of language as mid-world (*Zwischenwelt*), through which God's 'Book of Nature' could be read. With the attack of Darwinism on theories of natural design in the last quarter of the nineteenth century and poststructuralist criticisms of textual authority since the 1960s, Ricoeur's 'detour' implies there is no originary or transparent meaning to be discovered in or through language. Such opacity amplifies the difficulties of intersubjective communication, but may also lead towards a positing of the radical Otherness of other beings. Responding to Poulet, the American literary critic J. Hillis Miller claims:

> another mind is so alien, so impenetrable that it is never possible . . . to lift the veil which hides the other from me. This means I can never confront the other person as an immediate presence, only encounter indirect signs and traces of his passage.[29]

What seems to be a statement about the care and sensitivity needed to interpret another person's behaviour (and to reflect self-consciously on the process of interpretation), actually provides fuel for sceptics who question the reality of objects and the existence of other sentient beings. Although such a sceptical position is obviously untenable for practitioners dedicated to medical care, it must be taken seriously for theorists of the human sciences.

If one takes seriously Fredric Jameson's analysis of the erosion of the 'depth model' as a metaphor for establishing self-identity (the 'emotional ground tone' of 'intensities' serving to replace the 'hermeneutic model of inside and outside'), then so too must one address the post-structuralist attack on truth and the self (as a receptacle of truth) as metaphysical and uncritical terms.[30] In his critique of the decline of stable notions of selfhood, Jameson states that 'what replaces . . . depth models is for the most part a conception of practices, discourses, and textual play', or, in other words, 'depth is replaced by surface, or by multiple surfaces'.[31] A modernist concern for the crisis of selfhood is replaced by a permanent schizophrenic condition resulting from a 'breakdown of the signifying chain', or by a 'simulacrum' for which the original has been lost or had never existed as such.[32] Recuperating the remnants of selfhood in this conceptual territory is not only a treacherous activity but seemingly impossible, with every strategic reading undermining itself. On this view, the self is redescribed less as an empty shell and more as a corpse from which creative life cannot issue.

---

[29]   J. Hillis Miller, 'Geneva or Paris: the Recent Work of Georges Poulet', *University of Toronto Quarterly*, 39(3) (1970), 221.

[30]   Fredric Jameson, *Postmodernism, or the Cultural Logic of Late Capitalism* (New York: Verso, 1991), pp. 6, 12.

[31]   Ibid., p. 12.

[32]   Ibid., pp. 26, 6.

Jameson takes Jean Baudrillard to task as a critic who best epitomizes the recent theoretical dismantling of self – more explicitly, the embodied self – as a meaningful metaphor. A powerful example of this is to be found in Baudrillard's book, *The Ecstasy of Communication* (1987), in which the French theorist claims that 'the body as a stage, the landscape as a stage, and time as a stage are slowly disappearing'.[33] Furthermore,

> the religious, metaphysical or philosophical definition of being has given way to an operational definition in terms of the genetic code (DNA) and cerebral organization (the informational code and billions of neurons). We are in a system where there is no more soul, no more metaphor of the body – the fable of the unconscious itself has lost most of its resonance. No narrative can come to metaphorize our presence; no transcendence can play a role in our definition; our being is exhausting itself in molecular linkings and neuronic convolutions.[34]

If the metaphors of 'body' and 'soul' have lost their potency to describe what remains of selfhood for Baudrillard, then there can be 'no more individuals, but only potential mutants'.[35] This vision of mutation allows no possibility of recouping the self from a condition of exile, in which state one could maintain 'a pathetic, dramatic, critical, aesthetic distance': that is to say, one could still deal with selfhood as alienated, fugitive or diminished.[36] Instead of the modernist trope of exile, Baudrillard deploys a 'figure of metastasis' as 'a deprivation of meaning and territory, [or] lobotomy of the body.'[37] This 'figure of metastasis' works on two levels: firstly, as a condition of permanent transformation from depth to surface, or from whole to fragment (from which point neither depth nor whole can be reconstructed); and, secondly, as an expression of the interminable spreading of disease, a malignant cancer which rends the self and for which there is no treatment.

Such ideas may appear beguiling, but only undermine one version of the self understood as metaphysical presence or as a transcendent being. If one deploys the fiction of selfhood as a heuristic device, given specific content depending on the context in which it is posited or for what purpose it is appropriated, then the trope of internal disorder can be recouped as provisional agency, where disorder represents the active mutation of what has been understood to be permanent or essential characteristics. Especially in medical cases when the metaphors of disease are made literal (the cancer patient or the schizophrenic) then such a constructed notion of selfhood is extremely limited, but nevertheless serves an

---

[33] Jean Baudrillard, *The Ecstasy of Communication*, trans. Bernard and Caroline Schutze (New York: Semiotext(e), 1987), p. 19.

[34] Ibid., pp. 50–51.

[35] Ibid., p. 51.

[36] Ibid., p. 50.

[37] Ibid.

important role in a set of practices which take patient care as their precondition. The interpretation of the patient's experience, or, even better in terms of therapy, the patient's reading of him/herself (where reading is conceived as an active and creative process) provides the potential for such conceptual redescription in an environment of lived experience.

Rather than recouping the self as a transcendent marker of subjectivity, the embodied self has recently been championed as a viable heuristic construct which has served to counter the anti-humanistic theoretical trends which became dominant in the 1970s. Despite the non-foundational moves made by contemporary theorists like Richard Rorty, the self has emerged as the locus of experience in various fields of theory: the psychoanalytic work of Roy Schafer and Slavoj Žižek, the philosophy of Stanley Cavell and the narrative hermeneutics of Paul Ricoeur. Instead of forming a firm foundation for theory, the self has come to be seen as firmly embedded in discourse. To appropriate Jacques Lacan's well-known claim: the self is understood to be spoken in discourse rather than the self being the sovereign agent of enunciation. In this respect, most current theories of self-construction also consider the linguistic and narrative parameters of the self.

This return to theories of the self is foreshadowed in the work of the French theorist Michel Foucault. After eschewing the individual as the focus of attention in the 1960s, he re-established the self as a theoretical site in *The History of Sexuality* (1976-84) and his seminar on *Technologies of the Self* (1982).[38] His earlier criticisms of phenomenology were aimed at its humanistic implications and its uncritical positioning of a 'meaningful' universe around the undeconstructed figure of man. When viewed within the framework of the medical humanities, Foucault's image of the erasure of man ('a face drawn in the sand at the edge of the sea') causes severe theoretical difficulties for the basic tenets of patient care.[39] However, as Heidegger discusses in his 'Letter on Humanism' (1947), the revaluation of a humanistic tradition neither necessarily entails the rejection of humane values, nor the abandoning of medical ethics.[40] Romantic science is committed to the maintenance of humane values and attempts to redescribe selfhood in terms which undercut Foucault's early reading

---

[38]   Foucault shares with other major critical theorists a sceptical attitude to what Alan Sinfield calls the 'essentialist-humanist approach' to the individual seen as 'autonomous' and 'self-determining'; Alan Sinfield, *Faultlines: Cultural Materialism and the Politics of Dissident Reading* (Oxford: Clarendon, 1992), p. 37. My use of the term 'individual' in this book does not ignore the power structures that give rise to individualism, but is a shorthand way of describing a single human entity who, at least on the surface, is unified in action and belief (even though each of the romantic scientists discussed here problematize this unity).

[39]   Michel Foucault, *The Order of Things*, trans. unidentified collective (London: Routledge, 1986), p. 387.

[40]   Martin Heidegger, *Basic Writings*, ed. David Farrell Krell (London: Routledge & Kegan Paul, 1978), pp. 217–65.

of subjectivity as subjugation. As such, romantic science seeks to be an expansive, an understanding and a tolerant science.

Closely paralleling the renewed interest in selfhood in theoretical debates, the most hotly contested concept in American and British analytic philosophy in recent years has been a reformulation of consciousness. Such thinkers as Paul and Patricia Churchland, Thomas Nagel, Roger Penrose, Daniel Dennett, John Searle and Francis Crick have all attempted to redefine consciousness as the foundation of human cognitive activity.[41]

The general argument turns on a fairly traditional conflict between behaviourists (represented by Dennett and the Churchlands), who treat everything non-empirical as, at best, the epiphenomena of brain activity and, at worst, an erroneous understanding of cognition, and those thinkers (such as Searle and Nagel) who are sensitive to the irreducible areas of subjective experience which strict behaviourism cannot address. But Dennett is more than just a traditional behaviourist: he is a nominalist in the sense that he does not entertain final essences; he is interested in telling stories to illustrate his philosophical points; and he concurs with Rorty in his belief that metaphors are all one has to redescribe truths.[42] In his most provocative book *Consciousness Explained*, Dennett argues that although complex mental experiences cannot be reduced easily to a chemico-physical level, by following a scientific mode of interrogation one can simultaneously dispel the mystery surrounding the ineffability of mind and set out an objective explanation of cognition. He rejects the phenomenological investigation of mind because it is too personal and implies falsely that introspective inquiry is the only viable approach. For these reasons, he argues it is 'invulnerable to correction' and, therefore, inaccessible to materialistic science.[43] For Dennett, there are no 'intrinsic qualities' of experience and no self, or 'Central Meaner' (understood as a 'Cartesian theatre'), to imbue these qualities with language.[44] He claims behaviourism can offer a 'neutral' answer to the problem of verification which phenomenology poses, because it seeks an 'intersubjectively verifiable method' of analysis in its interpretation of 'vocal sounds . . . that are *apparently* amenable to a linguistic or semantic analysis'.[45] In developing such a position, Dennett adopts an 'intentional stance' towards the subject of the utterances which enables him to understand the speaker

[41]   See Paul Churchland, *Matter and Consciousness* (Cambridge, MA: MIT Press, 1984); Patricia Churchland, *Neurophilosophy* (Cambridge, MA: MIT Press, 1986); Thomas Nagel, *The View From Nowhere* (Oxford: Oxford U.P., 1986); Roger Penrose, *The Emperor's New Mind* (Oxford: Oxford U.P., 1989); Dennett, *Consciousness Explained*, op. cit.; John Searle, *The Rediscovery of Mind* (Cambridge, MA: MIT Press, 1992); Francis Crick, *The Astonishing Hypothesis* (London: Simon & Schuster, 1994).

[42]   See Rorty's essay 'Holism, Intrinsicality, and the Ambition of Transcendence', in ed. Bo Dahlborn, *Dennett and His Critics* (Oxford: Blackwell, 1995), pp. 184–202.

[43]   Dennett, *Consciousness Explained*, op. cit., p. 67.

[44]   Ibid., pp. 65, 228.

[45]   Ibid., p. 74.

as an agent capable of rationality.[46] The intentional stance explains how intersubjective exploration can uncover knowledge about the world and self, whilst circumventing the problems posed by retaining a model of consciousness.

In *The Rediscovery of Mind* (1992), John Searle argues against Dennett by positing a form of biological monism (or 'biological naturalism') as a physico-mental continuum which accounts for a private subjective perspective and bodily sensations.[47] Searle bases his theory upon the contention that 'mental phenomena are caused by neurophysiological processes in the brain and are themselves features of the brain'.[48] Because they emerge from the brain these 'special features' like 'consciousness, intentionality, subjectivity, [and] mental causation' cannot be reduced to neural processes (in the same way as liquidity is a 'higher-level emergent property of H_2O molecules' under certain conditions).[49] Searle claims consciousness is biological at root: it is a feature of the brain, but, in its 'first-person ontology', it is also a 'higher-level' mental phenomena.[50] In a number of articles and, most notably, in an exchange in *The New York Review of Books* (November 1995), Searle claims that Dennett's eliminative account cannot do justice to those special features of the brain which he describes, whereas Dennett accuses Searle of displaying a vestigial dualism. Whether either thinker has 'solved' the problem of consciousness is not the real issue here. Despite their refinements, this latest manifestation of philosophical rivalry displays a continuity with the classical philosophical split between monism (materialism or idealism) and dualism. Monism stresses the triumph of matter over mind (or vice versa), whereas dualism retains both within the philosophical frame, but leaves itself vulnerable to criticism from either of the other positions.[51] Both theories are open to criticism: in his biological monism, Searle's attempts to rescue the idea of 'the human' actually disguises a residual dualism and Dennett's materialistic behaviourism eliminates dimensions of human activity at the heart of rival models of consciousness.

In response to these debates this study considers how and in what ways the self – the conscious self *and* the embodied self – has endured and developed as a theoretical construct within the medical humanities in the twentieth century. Different variations of romantic science are represented by the five thinkers: James fuses philosophy, physiology and psychology with an interest in spiritualism and religion as meaningful human concerns; Rank attempts to situate

---

46   Ibid., p. 76.
47   Searle, *The Rediscovery of Mind*, op. cit., p. 1.
48   Ibid., p. 1.
49   Ibid., pp. 2, 34.
50   Ibid., p. 17.
51   Although the implication here is that a dualistic account of mind and matter is the only alternative position to materialism and behaviourism, as discussed in the chapter on William James, dualism of description (rather than dualism of substance) is not necessarily incommensurable with philosophical monism.

notions of the soul and creativity at the centre of his therapy; Binswanger seeks a broader anthropological study which re-establishes the self (or *Dasein*) at the centre of analytic discourse; Erikson gives equal attention to biological constitution, social forces and the development of personal identity; and Sacks outlines a vision of a double science, in which neurology is harnessed to psychology in order that he might focus on how neurological disorders appear from the patient's point of view. None of these thinkers finally solves the philosophical problematics related to consciousness and subjectivity, but they do make significant attempts to redescribe selfhood with the terms of their own practice.

In order to frame a discussion of romantic science within Foucault's description of the modern episteme (stretching roughly from the eighteenth to the mid-twentieth century and characterized by the problem of subjectivity), the following chapters trace an extended discussion of selfhood through a selective reading of five twentieth-century romantic scientists. Beginning with the ferment of experimental and theoretical activity in Germany in the late eighteenth and early nineteenth centuries, romantic science historically opens a field of investigation in the human and the natural sciences, in which systematic observation is yoked to an interest in the non-behaviouristic aspects of human activity, such as imaginative creativity and spirituality. The following section of the introduction considers how romantic science as a self-conscious movement was short-lived, becoming largely discredited by the growth of a strictly empirical science in mid-nineteenth-century Europe. However, William James' conception of radical empiricism can be seen to inaugurate a modern tradition of thought with its renewed interest in theories of selfhood and aesthetic creativity (especially with reference to therapy), which provided a forgotten romantic potential richly mined by the five thinkers.[52] All five display strong and discernible connections with early German romanticism, but also reveal their theoretical and cultural differences from the early Germanic models.

A number of other figures could arguably be placed within the tradition of romantic science: Carl Jung, Alexander Luria, Wilhelm Reich, Erich Fromm, R.D. Laing, Félix Guattari, Julia Kristeva and Adam Phillips, among others. Although claims could be made for featuring any of these in this book, the reasons for selecting the five thinkers as the focus of study are fourfold. Firstly, they are all both theorists and practitioners: Binswanger was and Sacks is a professional physician; James consulted with private patients at his home; and Rank and Erikson were lay analysts. This makes for a blend of theory directed towards the practical ends of patient care, with each dimension acting as a check on the other: theory counters technique emphasized in and for itself, and

---

[52]   William James began to formulate the term 'radical empiricism' in the 1890s, although his collection *Essays in Radical Empiricism* (Lincoln: University of Nebraska Press, 1996) was not published until 1912.

technique ensures that theoretical problems are directed towards practical ends. Secondly, the work of each is characteristic of the discursive manifold of romantic science within his own branch of the medical humanities: they focus centrally on selfhood as the vehicle for understanding and expression while retaining a firm focus on empirical reality. Thirdly, the five figures form a diverse transatlantic tradition of thought which draws on European romanticism but, either through conscious choice or enforced circumstances, seems best suited to American cultural conditions: James blended Germanic philosophy with his New England heritage of religious individualism; Rank's early education in Vienna under the tutelage of Freud contrasted to his experiences with the Paris literati and his reception in New York in the 1930s; Binswanger's existential theories found modest support in his native Zürich but were widely disseminated across the Atlantic in the 1950s; Erikson's Danish roots, German childhood and Viennese training fused with his enforced adult exile to America; and the British neurologist Oliver Sacks has lived and worked in New York since the early 1960s. Fourthly, although the five can be positioned in a coherent transatlantic tradition of romantic science, they are distinctive innovators in their own right, questioning the parameters of inherited disciplines in order to strategically redescribe selfhood outside traditional categories of understanding.

The Genre and Genealogy of Romantic Science

Romantic science is an oxymoronic term which inscribes its object of study within philosophical parameters, but also suggests sites of tension in the field of investigation. This section lays the ground for the following chapters by providing a brief overview of romantic science in Germany at the turn of the nineteenth century, in order to link romantic concerns with contemporary theories of the self. Because Oliver Sacks appropriates the label of romantic science from the Russian neuropsychologist Alexander Luria and he is the only one of the five figures to deploy the term explicitly, he is discussed here to foreground the central concerns of this study. Romantic science, firstly, characterizes a mode of writing which cannot easily be classified with reference to one or other available generic categories and, secondly, describes a particular approach to the complex theoretical problems which arise within the human sciences. As such, romantic science is both a genre of writing and a tradition of thought. Sacks and Luria are sympathetic to both of these goals, particularly in the field of neurology. However, the general comments below are also applicable to the aims and interests of James, Rank, Binswanger and Erikson, whereas more detailed commentary on Sacks has been reserved for a later chapter.

Romantic science is a particularly heterogeneous genre, encompassing an inquiry into selfhood (encompassing forms of inductive science with philosophical anthropology), aesthetic responses, phenomenological description and behavioural observation. As the American philosopher Stanley Cavell indicates, the inherent problem with describing a 'full-blown' genre is that it limits the evolution of a genre to simply recombining the discursive elements already established.[53] It is important to note that there is no such thing as a 'full-blown' form of romantic science because it does not exist in any one particular discipline of study; Cavell attests that 'late members can "add" something to the genre because there is . . . nothing one is tempted to call *the* features of a genre which all its members have in common'.[54] Nevertheless, there *is* an internal coherence to the genre of romantic science as each subsequent writer reinterprets the terms of the discourse within his specialist field of study. For Cavell, 'membership in the genre requires that if an instance (apparently) lacks a given feature, it must compensate for it, for example, by showing a further feature "instead of" the one it lacks'.[55] From this perspective, romantic science is a dynamic genre which recombines features (and adds to them) depending on its context. It closely resembles Raymond Williams' flexible historical model in which diachronic development passes through a series of emergent, dominant and residual transitions. Within the genre each emergent discourse (in which 'new meanings and values, new practices, new significances and experiences, are continually being created') would be a new dominant, but within a different field of application: 'the new feature introduced by the new member will, in turn, contribute to a description of the genre as a whole'.[56] For example, the psychoanalytic discourse evident in the work of James and Sacks in emergent and residual forms respectively (at least in terms of the history of psychoanalysis) is evident as a dominant discourse in the work of Rank, Binswanger and Erikson (although they each depart from the tenets of Freudian psychoanalysis in different ways). In short, it is important to stress that there is no one form of writing which epitomizes romantic science; it describes an open discursive field which enables thinkers to deploy a combination of fragmented discourses to attend to a particular set of theoretical problems.

Sacks diverges from classical neurology because he claims it does not deal with the subjective states of mind which result from neuronal firings. Defined as a physical science which examines the matter of the brain and nervous system, neurology uses language only as it needs to linguistically represent anatomical detail. Whereas precise referential language renders the blurred margins of hermeneutics negligible, Sacks' vision of neuropsychology emphasizes both the critical and interpretative roles of the inquirer. When subjectivity becomes an

---

[53] Stanley Cavell, *Pursuits of Happiness* (Cambridge, MA: Harvard U.P., 1981), p. 28.

[54] Ibid.

[55] Ibid., p. 29.

[56] Raymond Williams, *Problems in Materialism and Culture* (London: Verso, 1973), p. 41; Cavell, *Pursuits of Happiness*, op. cit., p. 29.

issue, referential language loses its precision and neurology, in its attempt to deal with non-material reality, becomes descriptive. Luria argues that it is possible for neurological science to humanize itself by broadening its field of study to incorporate and focus attention on the disorderly experiences of patients. By attending closely to the personal experiences of patients, especially those suffering from neurological disorders, the inquirer may glean more information than can be gathered from a neurological model constructed *in vacuo*. Indeed, as Cavell makes explicit in his discussion of genre: 'an interest in an object is to take an interest in one's experience of the object.'[57] In other words, the subjective experiences which neurology ignores become a central area of exploration for Luria and Sacks.[58]

The collation and description of subjective experiences, then, provide the basis of rigorous inquiry into disorders which neurology is unable to explain fully. Luria rejects methods which crucially rely on the measurement of instruments to emphasize the importance of the critical and interpreting mind:

> I am inclined to reject strongly an approach in which these auxiliary aids become the central method and in which their role as servant to clinical thought is reversed so that clinical reasoning follows instrumental data as a slave follows its master.[59]

The sympathies of Luria and Sacks lie firmly on the phenomenological fork of the divide: for example, Sacks discerns a separation between 'the apparent poverty of scientific formulation and the manifest richness of phenomenal experience', with the 'poverty' of science resulting from the methodological need to bracket, reduce or eliminate the 'richness' of lived experience.[60] In this view, any truth (a valid assertion about how the brain functions) elicited from a strictly scientific method is always narrow or partial, because it fails to recognize experiences which do not fit neatly within the framework of a single mode of inquiry. Behaviourism, the science which Dennett practises and promotes, is treated cautiously by Sacks because he sees it to be limited in its dismissal of private mental states.

---

[57]   Cavell, *Pursuits of Happiness*, op. cit., p. 7.

[58]   The other romantic scientists also react to forms of scientific inquiry which do not attend to the manifold of experience: James reacts to late nineteenth-century behaviourism which deals with physiology more than psychology; Rank and Binswanger take issue with Freudian psychoanalysis because they claim it undervalues the spiritual and aesthetic dimensions of consciousness; and Erikson argues against restricted and one-sided branches of psychology.

[59]   Alexander Luria, *The Making of Mind: A Personal Account of Soviet Psychology*, eds Michael and Sheila Cole (Cambridge MA: Harvard U.P., 1977), p. 177.

[60]   Oliver Sacks, 'Neurology and the Soul', *New York Review of Books*, 37(20) (22 November 1990), 44.

While Luria supports the re-introduction of romantic issues into the field of empirical science, he is quick to indicate that romantic science pursued to its extreme lacks logic and rigour and thereby prevents the practitioner from reaching 'firm formulations'.[61] On this issue, he admits that 'sometimes logical step-by-step analysis escapes romantic scholars, and on occasion, they let artistic preferences and intuitions take over. Frequently their descriptions not only precede explanation but replace it.'[62] Luria's ideal is to discover a methodological balance, if not a synthesis, where close observation facilitates a causal explanation of phenomena without eliminating the 'manifold richness' of phenomenology: for him, the greater the number of perspectives the more likely is the possibility of understanding.[63] Sacks also endorses this multi-perspectival method, although critics have argued that his visionary aims (typical of the other romantic scientists) are never fully realized in his written work or in his patient care. Nevertheless, in order to provide a context for these ideas it is useful to consider the ramifications of the term 'romantic science', both in its specific historical context and as a mutable generic concept.

Towards the beginning of his review article, 'Neurology and the Soul' (1990), Sacks quotes from the first part of Goethe's *Faust* (1808), in which the German poet celebrates the 'colors of life' in contrast to the drab greyness of theory.[64] This quotation closely echoes one of Luria's citations and aligns modern romantic science with the forms of inquiry which flourished in Germany between the late eighteenth and mid-nineteenth centuries.[65] In this article Sacks reviews his previous work, outlines other thinkers who share his sympathies (the American neuroscientist Gerald Edelman and the medic Israel Rosenfield) and acknowledges his debt to particular currents of romantic thought. It is important not to lose sight of the relevance of the twentieth-century renaissance of romantic science for contemporary critical debate by too easily eliding Sacks' position with early romanticism. In order to check this impulse, only issues which illuminate philosophical similarities between early and modern manifestations of romantic science are discussed here. As a result, the complicated genealogy of European and American romanticism is incompletely rendered.

In an essay on 'Romanticism in Germany', Dietrich von Engelhardt argues that, unlike other European countries where romanticism was prevalent, in German culture one is able to 'speak of Romantic science and medicine side by side with literature and the other arts'.[66] Here science refers to 'a metaphysical

---

[61]   Luria, *The Making of Mind*, op. cit., p. 175.
[62]   Ibid.
[63]   Ibid., p. 178.
[64]   Sacks, 'Neurology and the Soul', op. cit., p. 44.
[65]   Luria, *The Making of Mind*, op. cit., p. 174.
[66]   Dietrich von Engelhart, 'Romanticism in Germany', in Roy Porter and Mikulas Teich, eds, *Romanticism in National Context* (Cambridge: Cambridge U.P., 1988), p. 111.

form of scientific research' which is essentially speculative, but harnessed to an empirical approach to natural phenomena. Although many romantic scientists were influenced by the *Naturphilosophie* of Schelling and Hegel and their attempt 'to construct all natural sciences from *a priori* speculation', they were committed to the practical application of speculative knowledge.[67] There is no one essential form of romantic science, partly because many branches of the natural sciences were represented, among them Justus Liebig in chemistry, Alexander von Humboldt in universal science, Johann Ritter and the Dane Hans Oersted in physics, Henrik Steffens in geology and Carl Carus and Gotthilf Schubert in medicine, and partly because no practitioner claimed his own discipline was categorically distinct from the others. As Alfred North Whitehead indicates, it is important to note that 'all the sciences dealing with life were still in an elementary observational stage, in which classification and direct description were dominant'.[68] Rather than being viewed as mutually combative, the various dimensions of natural science were seen to be interconnected and subjected to a unitary metaphysics of nature.

The *Naturphilosophen* argued that mind and matter had a dialectical relationship and, following Kant, they countered materialist philosophy by emphasizing the importance of subjectivity. By challenging the static model of Newtonian mechanics they attempted to understand the natural and human world as a dynamic process.[69] Accordingly, the spatial and atomistic metaphors of the French materialist La Mettrie in *L'Homme machine* (1747) were replaced by organic and vitalistic descriptions of nature and of the forces which hold the universe together. A single vision, extrapolated from the perception of the 'outward' seeing eye, was countered by a multiple vision in which the 'inward' creative eye perceives the manifold aspects of Nature. Such a distinction is analogous to Schelling's theory of an active mind which challenged the dominant Lockean model of mind as passive receptor. As Andrew Bowie indicates in his reevaluation of Schelling's place in German idealist philosophy: 'what is able to know itself must be more than *what* it knows'.[70] Schelling does not claim that the mind represents the mythical foundations of the Absolute; rather, the inquiring mind, while partially 'determined by something posited outside itself',

---

[67]   T.A.M. Snelders, 'Romanticism and *Naturphilosophie* and the Inorganic Natural Science, 1797–1840: An Introductory Survey', *Studies in Romanticism*, 9(4) (1970), 195.

[68]   Alfred North Whitehead, *Science and the Modern World* (Cambridge: Cambridge U.P., 1929), p. 79.

[69]   The attacks levelled against Newton often lose sight of the fact that he actually conceived of a universe made up of matter *and* spirit. Not until the eighteenth century did scientific materialism reduce this dualistic account to one fundamental substance which was understood with recourse to material laws.

[70]   Andrew Bowie, *Schelling and Modern European Philosophy* (London: Routledge, 1993), p. 23.

constitutes something which cannot be reduced to the level of matter.[71] This philosophy was popular among early romantic scientists, enabling them to consider 'the totality of all being' without losing sight of their 'experience of the object world'.[72] Moreover, the influence of Schelling's theories spawned a common belief in the mutual reciprocity of self-knowledge and knowledge of nature: the mind, as a part of nature, is structured by the same laws and is not representative of a separate and ideal sphere of existence. To understand human beings is to understand nature, and vice versa: 'Nature attains perfection in living organisms, where the world of physical phenomena overflows into that of the mind'.[73] Andrew Bowie frames this understanding in terms of Schelling's description of 'existence' or '*Daseyn*', directly linking his work to Heidegger's *Being and Time* (1927) and providing a philosophical source for the development of Ludwig Binswanger's existentialist version of romantic science.[74]

Schelling's influence is important for any consideration of romanticism, but romantic science represents a slight shift away from his idealistic tendencies, together with a retreat from the Dionysian exuberance of early German romantic poetry. It is not that the early practitioners of romantic science rejected idealism *per se*, but they saw intuition and feeling as complementary to an understanding of nature achieved through the faculty of reason. Consequently, it is possible to show a closer connection between the German natural scientists and 'the pre-romantic romantics', figures like Herder, Goethe and Novalis.[75] As 'men of sentiment' they displayed a relish for forms of classicism together with a regard for an emotional response for and in the world, and so stand as historical mediators between the intellectualism of the *Aufklärung* and the elevation of feeling displayed in romantic poetry. Natural scientists were wary of applying dogmatic reason too readily to their studies; they detected the stultifying alliance between a rationalist method and what they interpreted as the rigid laws of Newtonian mechanics. Reason was used only as an organizing tool to help as they collected empirical data and did not represent the basis of their inquiry. T.A.M. Snelders distinguishes one major difference between the *Romantik* and the *Aufklärung* approaches to scientific knowledge:

---

[71]   Ibid, p. 24.

[72]   Snelders, 'Romanticism and Naturphilosophie', op. cit., p. 197; Bowie, *Schelling and Modern European Philosophy*, op. cit., p. 24.

[73]   Engelhart, 'Romanticism in Germany', op. cit., p. 113.

[74]   Bowie, *Schelling and Modern European Philosophy*, op. cit., p. 24. Rollo May, an American popularizer of Binswanger's work in the late 1950s and 1960s, claims that Schelling's lectures given at Berlin in 1841 (in which he attacks Hegel's totalizing rationalist system) marks one of the earliest significant moments in the founding of existentialism as a dominant European philosophy; Rollo May, *The Discovery of Being* (New York: Norton, 1986), p. 54.

[75]   Ralph Tymms, *German Romantic Literature* (London: Methuen, 1955), p. 9.

the Newtonian mechanical-atomistic explanation of all natural phenomena was supplanted by a dynamical and organic concept, with a concomitant substitution, at the extreme, of sentiment for the critical mind. One tried to unriddle phenomena in the natural sciences, the enigma of life and disease, by intuition instead of by experiment.[76]

Intuition and creative imagination came to be valued as powerful faculties for unveiling the enigmas of Nature which lay outside the scope of scientific instrumentation (represented by William Blake's engraving of Urizen dividing up the material world with his calipers) and apparent only to the inquiring mind.

In their combined interest in science and literature, Goethe and Novalis perhaps best represent the synthesizing potential of romantic science, although their reputations depend primarily on the quality of their art. For example, the notebooks of Novalis (prefiguring Otto Rank's daybooks) contain scientific musings intermingled with epigraphs, aphorisms and poetic fragments.[77] These notebooks epitomize the notion that knowledge of the mind also provides knowledge of nature as macrocosm: 'the world, the human world, is just as manifold as man is'.[78] The fragment form, found elsewhere in the poetry of Blake and Coleridge, also displays the romantic dislike for grand systems. The universal systems favoured in the Enlightenment were seen by the romantics as 'prisons of the spirit'; in a drive to generality, the system tends to neglect the particular and unique.[79] To evoke Blake's image, instead of being able to discern eternity in a grain of sand, the single grain is neglected for the sake of conceptualizing eternity. The fragment proved to have greater significance than merely its function as a poetic form: 'articles by scientists and medical practitioners, like literary works, often appear in unsystematic, fragmentary, aphoristic, even poetic, form. This form is chosen deliberately, as it is meant to mirror what can be understood of nature'.[80] It was thought that the 'mysteries' of nature could only be understood at moments of intuitive insight and recorded in brief jottings and unconnected fragments. While modern thinkers may wish to dispense with myths of romantic genius, the fragment has been retained as an important element in the heterogeneous written genre of romantic science.

The early flowering of romanticism had much to offer the natural and human sciences, but it contains inherent weaknesses which bear upon the twentieth-century renaissance of romantic science. For example, the transcendental belief

---

[76]   Snelders, 'Romanticism and *Naturphilosophie*', op. cit., p. 194.

[77]   One example of Novalis' technique of combining dual modes of understanding through analogy is evident in a fragment from his *Encyclopedia*: 'Physics is nothing but the teaching of the imagination'; Novalis, *Pollen and Fragments*, trans. Arthur Versluis (New York: Phane Press, 1989), p. 93.

[78]   Ibid., p. 96.

[79]   Isaiah Berlin, *Against the Current: Essays in the History of Ideas* (London: Hogarth, 1979), p. 9.

[80]   Engelhart, 'Romanticism in America', op. cit., p. 112.

that human insights are somehow related to an organic and natural unity is often a matter of wonder and speculation rather than an empirical questioning of metaphysical premises. Although the early romantic scientists did not seek grand explanations, this sense of a wordless communion with nature had no direct applicability to science. Instead, romantic rapture must be transposed into a description of phenomena which can be recorded and analysed. The philosophical implications of the transposition from mental experience to linguistic description can be seen from two distinct, yet interconnected, perspectives.

Firstly, a wordless and essentially incommunicable vision can be shown to be distorted, or even debased, by transferring it to a linguistic medium.[81] The purity of silent communion is frustrated and dismantled by the self-conscious recalling and transcribing of it. However, by making use of symbolic language, a discourse which is suggestive rather than indicative of actual experience, the romantic writer can avoid over-intellectualizing. As such, the impressionistic romantic enterprise runs tangential to a scientific mode of interpretation which demands precision of description. The second, broadly post-structuralist, perspective would cast doubt over whether there is *any* raw experience prior to the language used to describe it. This corresponds to J. Hillis Miller's criticism of Poulet's moment of 'double consciousness' in which one can self-consciously know the *cogito* of the other.[82] For Miller, language should be recognized as 'the instrument by which the mind explores its own depths' without ever reaching epistemological bedrock.[83] Such an understanding moves away from phenomenological revelation towards an analysis of the discourses which constitute both experience itself and the subject of experience. As Paul de Man remarks in his Gauss Lecture of 1967, it is impossible to make 'the actual expression coincide with what has to be expressed, of making the actual sign coincide with what it signifies'.[84] In other words, there is no self-present sign which represents the 'origin or constitutive focus that is ontologically prior' to experience.[85] De Man's critique of romanticism dovetails with Jacques Derrida's interrogation of Husserlian phenomenology, in which Derrida argues that there is no unmediated phenomenological meaning (conceived by Husserl as pure ideality or self-presence) outside the realm of signification. This idea not only throws a sceptical shadow over the primacy of mental experience, but also undermines the romantic notion of transcendental consciousness by questioning the status of the subject as the locus

---

[81] For example, Coleridge speaks of the treacherously ambiguous 'secondary imagination' as a dissolving, diffusing and dissipating process; Samuel Taylor Coleridge, *Biographia Literaria* (London: Everyman, 1984), p. 167.

[82] Miller, 'Geneva or Paris', op. cit., p. 216.

[83] Ibid., p. 225.

[84] Paul de Man, *Romanticism and Contemporary Criticism* (Baltimore: Johns Hopkins U.P., 1993), p. 12.

[85] Ibid., p. 6.

of the experience. To add to this, de Man questions the 'forms of romantic deceit' which claim 'self-autonomy . . . as a philosophical truth about the nature of human existence' and 'the work of art as a self-engendered world of the subject's own making'.[86]

Although there have been a number of challenges to the romantic myth of divine creator, these alone are insufficient to dismiss the project of romantic science as being founded on entirely misguided principles. Romantic science appropriates a diluted romanticism rather than an excessive idealism; although some scientists have been influenced by strains of romantic thought, their need to follow a scientific method will necessarily distance them from the tenets of aesthetic romanticism. Moreover, the apparently irreconcilable tension between art and science is primarily due to a reification and an exaggerated opposition between the terms. As de Man argues, the romantic influence upon science should not be 'measured by the contribution [it] makes to the elaboration of a cogent historical outline' of romanticism.[87] By granting priority to subjectivity and psychological complexity, romantic science duplicates the counter-Enlightenment spirit of first-generation romantics, but without their tendency to repudiate an inquiry based solely on reason. In the words of Gerald Edelman, the challenge for romantic scientists is to put 'the mind back into nature' without resorting to an out-moded dualism of substance and the philosophical problems that would entail.[88] The importance of critical theory for this study is in its attention to language and narrative, rather than in its dealing with an ineffable realm of mentality. As will become evident, narrative plays an increasingly pivotal role in the romantic-scientific account of mental and bodily life.

Romantic science does not just seek to bridge the opposition between the arts and sciences, but also attempts to reconnect with the moral-practical sphere outlined by Habermas. In order to understand the moral agenda of romantic science it is useful to return to Stanley Cavell, whose work combines a reconsideration of romanticism with a post-structuralist view of language. Cavell aims to reinstate a morally orientated romanticism in the face of the scepticism which post-structuralist insights often engender. Mirroring Derrida's view of language (and, to a certain extent, Lacan's view of the self), Cavell argues that language, instead of providing clarity and precision, may actually reveal a lack of sense. What he calls the 'questionable' search for the self will always lead one away from unquestioning security towards a proliferation of meaning.[89] If

---

[86]   Ibid., p. 6.

[87]   Ibid., p. 95.

[88]   Gerald Edelman, *Bright Air, Brilliant Fire* (New York: Penguin, 1994), p. 9.

[89]   In similar vein, Richard Poirier writes: 'the democratic impulse shared by Emersonian pragmatists also involved a recognition that language, if it is to represent the flow of individual experience, ceases to be an instrument of clarification or of clarity and, instead, becomes the instrument of a saving uncertainty and vagueness'; Richard Poirier, *Poetry and Pragmatism* (London: Faber & Faber, 1992), pp. 3–4.

meaning is viewed as radically unstable, the inquirer may be sceptical about the possibility of knowing the world and other beings in it, or, at the very least, may be sceptical about saying anything significant about them. Cavell sees the futility of attempting to claim 'a final philosophical victory' over scepticism and acknowledges that because one must act as if the world were inhabited by objects and conversant peoples then, to all intents and purposes, they do exist.[90] As the first chapter discusses, there is much in this claim indebted to James' view of pragmatism, but Cavell argues in terms of needing to 'accept' or to 'receive' the world as it is, rather than reshaping it as we might wish it to be.[91] This does not mean one cannot 'redescribe' the world (in Rorty's sense of the phrase), only that the attempt to engage with a world of others should not be abandoned for solipsistic flights of imagination.

Crucial for a consideration of romantic science (and echoing Poulet as much as Heidegger), this kind of acknowledgement of others turns out to be an awareness of the self's place in the world.[92] To this end, Cavell claims that the 'quest' of romantic writing is 'the recovery of the self, as from an illness'.[93] In the work of Rank and Sacks metaphors of 'rebirth' and 'awakening' suggest that recovery of the self is often recovery from a condition characterized by loss of meaning, brought about through debilitating illness in which aspects of 'the human' are stripped away.[94] Recovery from such a diminished state often entails the awakening of self to an acknowledgement of itself as an agent among others in the world. Rebirth and awakening do not merely describe moments of romantic epiphany, but entail a moral commitment to live seriously with the threat of scepticism and self-doubt. Cavell indicates that it is possible to retain the existential rhetoric of 'becoming' and 'self-authoring' without swallowing the 'discredited romantic picture of the author or artist as incomprehensibly original'.[95]

The relevance of Cavell's thought to the theoretical-therapeutic perspective of romantic science is that, in both, the hermeneutic activity begins with 'a loss of self-knowledge; of being, so to speak, at a loss'.[96] To appropriate a cliché of popular psychology (and to go beyond that cliché), such a commitment entails the

---

[90] Stanley Cavell, This New Yet Unapproachable America (Albuquerque: Living Batch, 1989), p. 38.

[91] Ibid., p. 133.

[92] This kind of acknowledgement is particular apposite in a discussion of the human sciences when the focus of study is the nature of human existence. As Hans-Georg Gadamer succinctly states: 'Knowledge in the human sciences always has something of self-knowledge about it'; Wachterhauser, Hermeneutics and Truth, op. cit., p. 29.

[93] Stanley Cavell, The Senses of Walden: An Expanded Edition (Chicago: University of Chicago Press, 1992), p. 80.

[94] Stanley Cavell, Conditions Handsome and Unhandsome (Chicago: University of Chicago Press, 1990), p. 57.

[95] Ibid., pp. 111, 110.

[96] Cavell, This New and Unapproachable America, op. cit., p. 36.

self's learning not only to live with loss, but also with the refusal to erase or forget that loss. Even if one cannot speak of the self as 'I' in a fully conscious sense, one may be able to redescribe the active bodily performance of the self. One of the aims of romantic science is to develop techniques which encourage individuals suffering from loss to reconstruct the embodied self as a meaningful and active vehicle of agency. But this possibility should be accompanied by a commitment to forget neither the provisionality nor the fragility of selfhood. Self-knowledge is never indubitable, but the possibility of knowing is a risk which Cavell invites the individual to take. The human scientist who refutes such an understanding may 'profit in gaining the whole world' at the risk of 'losing one's soul'.[97] As William James claims in his discussion of Pascal's wager, the gains of risking an encounter with the self might be infinite, whereas refusing to take the risk will certainly lead to 'finite loss'.[98] Both James and Cavell stress that one must take this risk, but one must also retain the possibility of being challenged or proved wrong. The reason why Cavell takes scepticism so seriously is that these relationships are always liable to collapse from fragility and the fear of doubt. Similarly, the reason why the human sciences need to be tolerant of 'the human' is because, for Cavell, 'being human' is tantamount to being fallible. In terms of the therapeutic dimension of romantic science, the 'moral urgency' to understand the self and others is not a moralizing tendency, rather it is the possibility of 'acting beyond the self and making oneself intelligible to those beyond it'.[99]

Each of the five principal thinkers in this book treats seriously the possibility of knowing the self through particular categories of understanding. However, it is the narrative potentiality of self as active artificer (as Cavell outlines it) or creator (as Rank discusses) to which each turns in order to indicate areas in which aesthetic modes of expression can complement the conventional scope of the human sciences. As such, a movement away from epistemology towards a hermeneutic and pragmatic practice can be traced through modern romantic science, inflected by both an expressive and a moral romanticism. Romantic scientists not only seek to dissolve the perceived tension between empirical science and expressive art, but to shift attention away from the strictly clinical experience of self to engage with the experience of cultural life. To this end the American psychoanalytic critic Philip Rieff echoes Habermas' desire to reconnect the three spheres of existence across a cultural continuum: 'culture is another name for a design of motives directing the self outward, toward those communal purposes in which alone the self can be realized and satisfied'.[100]

---

[97]   Cavell, *Conditions Handsome and Unhandsome*, op. cit., p. 26.

[98]   William James, *The Will to Believe & Human Immortality* (New York: Dover, 1956), p. 5.

[99]   Cavell, *Conditions Handsome and Unhandsome*, op. cit., p. 46.

[100]   Philip Rieff, *The Triumph of the Therapeutic: Uses of Faith after Freud* (Chicago: University of Chicago Press, 1987), p. 4.

While James, Rank and Binswanger go some way to dissolve the boundaries between theory and the practical application of romantic science by linking their conception of the self with broader cultural concerns, Erikson and Sacks most successfully combine a theorization of selfhood and the development of a set of viable therapeutic techniques within the framework of lived experience. Accordingly, the early chapters trace the movement from philosophical conceptions of self to considerations of a narratively constructed self, whereas the later chapters on Erikson and Sacks reflect their concerted attempts to synthesize theories of self, narrative and cultural experience.

Chapter 1

William James: The Pragmatic Romantic

William James (1831–1910) was professionally a philosopher, scientist and university lecturer, in contrast to the four thinkers discussed in the following chapters whose theories feed directly into their therapeutic practices. During his time at Harvard University between 1872 and his death, James staked his credentials as a late-Victorian polymath, his attention shifting from medicine, anatomy, physiology and psychology to a developing interest in philosophy and religion. James also brought to his studies an aesthetic sensibility which derived from his early twenties when he considered painting as a career, together with the influence of his famous literary brother, Henry. The use of hybrid and multiple discourse characterizes James as a thinker who fuses established genres of writing in order to broaden his field of investigation. Underpinning these various discourses is a central preoccupation with the problem of selfhood, which aligns his thought closely to the tenets of romantic science. Although there may seem to be no immediate connection between James' early theoretical work and the application of these ideas to individual experience, the critics Donald Meyer and Eugene Taylor claim his interest in 'supernormal' mental states fed directly into the therapeutic sessions which he began on a private basis during the 1890s and later developed into his action-oriented theory of pragmatism.[1]

This chapter demonstrates that this therapeutic strain is a vital dimension of James' romantic science and should not be isolated from his theoretical interest in the active and experiencing self. His thought appropriates a number of identifiable European romantic ideas within the framework of his distinctly New England philosophy, while his theory of the self as an active agent emerges directly from his early attacks on determinism and behavourism, prevalent in Anglo-American thought in the late nineteenth century. Accordingly, this chapter charts the development of James' work away from natural science and epistemology towards his theory of pragmatism, religious experience and therapeutic philosophy by focusing on his two principal works, *The Principles of Psychology* (1890, hereafter *Principles*) and *Varieties of Religious Experience* (1902, hereafter *Varieties*), which are discussed in detail, together with a number of his influential essays.

---

[1]   Eugene Taylor, *William James on Exceptional Mental States: The 1896 Lowell Lectures* (Amherst: University of Massachusetts Press, 1984), p. 150.

From Physiology to Ethics

In 1872 James was appointed instructor of physiology and anatomy at Harvard, where he devised a groundbreaking course on experimental psychology in 1875, 'The Relations between Physiology and Psychology'. James' experimental psychology was a hybrid of European and American ideas, deriving from his own interest in personal experience and linked to the French experimental physiology of Claude Bernard, the experiential psychology of the German Wilhelm Wundt and the work of the American thinkers Chauncey Wright and Charles Sanders Peirce on evolutionary theory and the philosophy of science. Stimulated by his students' interest in the course, James was contracted in 1878 to write a book for the American Science Series synthesizing the major trends of late nineteenth-century psychology. He hoped to finish the study within two years, but it was not published until 1890, when it appeared as the two-volume *The Principles of Psychology*. Throughout *Principles* James stressed a desire to work within the parameters of natural science and avoid the theoretical obscurity of metaphysics, which had become largely discredited as a branch of philosophy in the nineteenth century, mainly because of its immunity to empirical testing.[2] As such, he viewed psychology as a division of natural science which he wished to develop with reference to personal experience. However, he was well aware that to explicate the central principles of psychology without brushing against metaphysical questions would be an arduous, if not an impossible, task.

In 1892 James responded to the critic George Trumbell Ladd's article, 'Psychology as So-called Natural Science', published in the *Philosophical Review*, in which he takes issue with James' lack of consistency in *Principles* and detects instances when he encroaches on metaphysical terrain.[3] In response to Ladd, James argued that if 'psychology is ever to conform to . . . the other natural sciences, it must renounce certain ultimate solutions',[4] beyond which it cannot hope to step: 'psychology when she has ascertained the empirical correlation of the various sorts of thought or feeling with definite conditions of the brain can go no farther – can go no farther, that is, as a natural science. If she goes farther she becomes metaphysical'.[5] In his attempt to resist metaphysical speculation, James did not hesitate to indicate areas where psychology 'can go no farther'.[6]

---

[2]   In an 1892 article in *Philosophical Review*, 'A Plea for Psychology as a "Natural Science"' James commented: 'I wished, by treating psychology like a natural science, to help her to become one'; William James, *Collected Essays and Reviews* (New York: Russell & Russell, 1969), p. 317.

[3]   George Trumbell Ladd, 'Psychology as So-called Natural Science', *Philosophical Review*, 1 (1892), 24–53.

[4]   James, *Collected Essays and Reviews*, op. cit., p. 317.

[5]   James, *The Principles of Psychology*, vol. 1 (New York: Dover, 1960), p. vi.

[6]   Ibid.

The opening pages of *Principles* illuminate one of James' central theoretical dilemmas: how to outline the central tenets of psychology while avoiding metaphysical speculation. He outlines a 'positivistic' model of the self in order to resist two rival metaphysically informed philosophies – the associationist and spiritualist theories – but he stresses points where natural science opens into 'queries which only a metaphysics alive to the weight of her task can hope successfully to deal with'.[7] This comment proves crucial for understanding James' work as a example of romantic science, stimulating him to claim: 'the reader will in vain seek for any closed system in the book' and 'the best mark of health that a science can show is this unfinished-seeming front'.[8] While *Principles* extends to over 1350 pages and ostensibly takes its place alongside Lyell, Darwin, Comte and Spencer as a product of nineteenth-century systematic rigour in empirical science, it opens out into a modernist text which explores queries and uncertainties of epistemology. As such, *Principles* can be viewed as a pivotal work, interpreted either as a late-Victorian example of rigorous empiricism or as a more troubled expression of modernity. In many ways, James' work stands free from the Germanic shadow of early romantic science, but it is the historical and cultural distance from that earlier moment which enabled him to develop and redirect the initiatives of romantic science.

In the first chapter of *Principles*, 'The Scope of Psychology', James expands upon his concerns about the inadequacy of spiritualist and associationist modes of psychological explanation, the first which he associates broadly with Continental philosophy and the second with an Anglo-American tradition of empirical thought. Firstly, he argues that the bewildering 'variety and complexity' of psychological phenomena have been historically unified through the theologically informed theory of 'the personal Soul'.[9] This spiritualist philosophy is devoted to abstract and ahistorical principles, characterized by the Cartesian theory of mind and Kantian idealism, the latter which found an American outlet in New England Transcendentalism and, more immediately for James, in the work of the Harvard philosopher Josiah Royce. The rival associationist theory (a 'psychology without a soul') proceeds by examining 'common elements in the divers mental facts rather than a common agent behind them'.[10] Championed by John Locke and the British empiricists, associationism explains phenomena by reference to 'the various forms of arrangement of these elements'.[11] Unlike spiritualism, associationists claim that the self does not pre-exist perception and emotion, but emerges *a posteriori* 'as their last and most complicated fruit'.[12] James is wary of the spiritualist theory insofar as it posits

---

7   Ibid., p. vii.
8   Ibid.
9   Ibid., p. 1.
10   Ibid.
11   Ibid.
12   Ibid., p. 2.

essences or faculties like 'Cognition' or 'Memory' as the 'absolute properties of the soul'.[13] Because neither faculty is empirically verifiable, he rejects the mysterious Cartesian interface between mind and body together with the metaphysical unity which Royce asserts. Explicitly reacting to Royce's notion of an all-encompassing principle by which every empirical contradiction can be resolved, James argues that essences are unverifiable because they exist prior to experience. However, he does not reject the spiritualist's theory wholesale, gravitating towards a conception of mind which can be inferred through an analysis of its 'indubitable expressions'.[14]

James' theory of an active mind is crucial for understanding his break from Spencerian psychology, in which environmental 'conditions' are seen to determine consciousness.[15] James taught Herbert Spencer's psychological treatise *First Principles of Psychology* (1864) at Harvard, praising him for advancing psychology beyond 'the old-fashioned 'rational psychology,' which treated the soul as a detached existent, sufficient unto itself'.[16] But, in his 1884 essay, 'The Dilemma of Determinism', James argues that Spencer's behaviourism is too mechanical and deterministic and, in an 1878 essay, he asserts that survival should be understood as 'only one out of many interests'.[17] Thus, when he claims in *Principles* that 'minds inhabit environments which act on them and on which they in turn react', he implies that the conscious self is partially determined by the world, but retains an ability to interact socially and to make ideal choices.[18]

James also rejects associationism for what he deems to be a residue of the spiritualist theory: the 'fantastic laws of clinging' by which ideas are associated and arrange themselves in 'an endless carpet . . . like dominoes in ceaseless change, or the bits of glass in a kaleidoscope'.[19] Although he agrees with the idea that the self arises from lived experience, he is dubious about the position of sceptics who reject the idea of a knowing self, but then cannot explain how the ideas that constitute a particular memory are arranged and configured. While he agrees that the object-world is experienced primarily through the senses and should not be conceived from a perspective which pre-exists sensation, he claims:

---

[13]   Ibid., p. 2.
[14]   Ibid., p. 11.
[15]   Ibid., p. 6.
[16]   Ralph Barton Perry, *The Thought and Character of William James*, vol. 1 (Boston: Little, Brown, 1935), p. 482. As the founder of Social Darwinism, Spencer tended to disregard emotions and beliefs in favour of the survival instinct. William Graham Sumner was a fervent American advocate of Spencer's ideas, pronouncing survival as the sovereign impulse in life. James, among others like Lester Ward, reacted to Sumner's determinism by arguing that humans can meaningfully influence their environment.
[17]   James, *Collected Essays and Reviews*, op. cit., p. 43.
[18]   James, *Principles*, vol. 1 , op. cit., p. 6.
[19]   Ibid.

> The bare existence of a past fact is no ground for our remembering it. Unless we have seen it, or somehow undergone it, we shall never know of its having been. The experiences of the body are thus one of the conditions of the faculty of memory being what it is.[20]

It is this disregard for physiological reality which renders both rival theories untenable for James. Instead of relying upon a wholly determining environment or 'the fantastic laws of clinging', James focuses upon bodily 'conditions' through which the self develops within, and in negotiation with, natural and social environments. Without fully refuting either theory, James suggests that he and the reader must assume the 'coexistence' of thoughts and brain-states as the 'ultimate laws of our science'.[21] He devises a philosophy of empirical parallelism which entails a causal reciprocity between mental and physical realms. Although ultimately this type of parallelism can be reduced to a dualistic level, in his claim that 'no mental modification ever occurs which is not accompanied or followed by a bodily change' James conceives of a 'psycho-physic formula' which goes some way to avoid the metaphysical problems implicit in the two rival theories.[22]

By speaking in terms of empirical parallelism James avoids surrendering to a purely physiological description, to the utter contingency of the mental and physical or to a theology of soul. He does not dismiss the problems implicit in his own theory, but clears the philosophical ground to explore questions of selfhood in more detail. Indeed, he seems to understand the futility of searching for indubitable foundations when he humbly suggests that nature 'has mixed us of clay and flame, of brain and mind, that the two things hang indubitably together and determine each other's being, but how or why, no mortal may ever know.'[23] By beginning *Principles* with a description of physiology, James is able to direct his attention towards psychological processes without reducing his study to crude materialism. Moreover, he claims that conceiving 'the chain of events amongst the cells and fibres as complete in itself' would be 'an unreal abstraction', while speaking solely in terms of Roycean 'ideas' would ignore the fundamental importance of their organic cause.[24]

Broadly, he follows an evolutionary argument in which the more complicated cerebral mechanisms are evident in higher organisms whose hemispheres have developed to a greater degree than those lower down the evolutionary tree. He detects an organism's evolution to have taken two directions: 'the lower centres passing downwards into more hesitating automatism and the higher ones upwards into larger intellectuality'.[25] Intelligent action, in turn, is characterized by three

---

[20]   James, *Principles*, vol. 1, op. cit., p. 4.

[21]   Ibid., p. vii.

[22]   Ibid., p. 5.

[23]   Ibid., p. 182.

[24]   Ibid., p. 24.

[25]   Ibid., p. 79.

qualities: sentience; the ability to discriminate; and the ability to project goals towards which action can be directed. He argues that as the physiological seat of the reflex evolves from the spinal cord and the lower brain to the centres of the cerebral cortex, so these abilities are made possible by the 'passage of functions forward to the ever-enlarging hemispheres'.[26] He also asserts that basal reflexes, which in lower creatures are conditioned entirely as fixed responses to sensory stimuli (for example, the frog's nervous system), have evolved into cerebral tendencies which are 'modifiable by education' in higher organisms.[27] Whilst this argument owes much to Spencer and the American Darwinists, James moves strategically to argue that it is possible to speak of 'cerebral reflexes' in higher creatures as cortical transactions, in order to arrive at 'psychological truth . . . without entangling ourselves on a dubious anatomy and physiology'.[28] Importantly, these psychological truths are not ineffable ideas severed from their organic roots, but aptitudes and abilities which emerge as the fruits of experience. In order to develop this argument, James turns to his consideration of the 'aptitude of the brain for acquiring habits'.[29]

In his initial analysis of the habits of living creatures he adopts the perspective of a behaviourist. By observing a variety of organisms under particular environmental conditions the behaviourist can perceive and record certain repetitive tendencies in creatures following similar patterns of response. James outlines two broad categories of habit: instincts, which are fixed and appear to be innate, and habits, which are 'the result of education'.[30] This division comes within range of the classical philosophical debate on determinism and free will: on the one hand, responses which follow the predetermined laws of cause and effect and, on the other, those which are 'variable' and can be adapted and modified 'to suit the exigencies of the case'.[31] James argues that more complex organisms possess a wide repertoire of flexible habits to counteract the forces of necessity, enabling them to adapt to changes in environment without complete transformation or total dissolution of form. Thus, James posits the idea of 'plasticity' which is present in organisms who possess 'a structure weak enough to yield to an influence, but strong enough not to yield all at once'.[32]

By retaining a notion of free will, James echoes the English polymath John Stuart Mill's argument that free will can be maintained in the face of determinism: 'our will, by influencing some of our circumstances, can modify

---

26 Ibid., p. 79.
27 Ibid., p. 80.
28 Ibid.
29 Ibid., p. 103.
30 Ibid., p. 104.
31 Ibid.
32 Ibid., p. 105.

our future habits or capacities of willing'.[33] In what was perhaps his most influential book upon Victorian thought, *A System of Logic* (1843), Mill argues that the philosopher of necessity, who claims that character is wholly determined by circumstance, makes a critical error: a human 'has, to a certain extent, a power to alter his character. Its being, in the ultimate resort, formed for him, is not inconsistent with its being, in part, formed by him as one of the intermediate agents'.[34] Mill's argument for compatibility does not deny causal necessity, but instead reserves a philosophical space for self-determinism. By extending Mill's argument into the domain of biology, James counters Spencer's passive account of mind by claiming that individuals can act upon an environment at the same time that they are moulded by it. However, both thinkers are vulnerable to the criticism that 'experience' (an individual's sensual and perceptive interaction with the world) remains the sovereign factor for determining action. But it is the romantic scientist's claim that experience is deeply embedded in bodily activity which enables James to distance himself from a metaphysical conception of experience and to dissolve the strict division between self and world.

In the same chapter, James moves subtly from a consideration of habit as a physiological principle to the learning of skills through training and discipline. He argues that this developmental process occurs throughout the neural systems of higher organisms, but only humans can develop complex intellectual activities. Drawing heavily on the British physiologist William Carpenter's study, *Principles of Mental Physiology* (1874), James asserts two critical points: firstly, the channelling of good habits can be acquired through discipline and effort and, secondly, as skills are learnt and refined the effort expended in the accomplishment of them decreases. In a skilled activity the chain of events which constitutes the action can commence by 'a single instantaneous "cue" '.[35] Rather than developing through a series of laborious stages, each 'muscular contraction' instigates the appropriate contraction of the next in an automatic sequence, comparable to an involuntary wave of muscular peristalsis.[36] As skills are learnt and mastered, the higher regions of the brain are set free from the process of learning to engage with other tasks. Acquired habits may seem like involuntary actions, but, unlike reflexes, there is a realm of consciousness below the level of direct attention which can regulate activities if learnt habits 'go wrong'.[37] James also introduces what he deems to be the 'ethical implications of the law of habit': by acquiring good habits and 'useful actions' at an early age individuals can

---

[33]   John Stuart Mill, *Autobiography and Literary Essays* (Toronto: University of Toronto Press, 1981), p. 169.

[34]   John Stuart Mill and Jeremy Bentham, *Utilitarianism and Other Essays* (London: Penguin, 1987), p. 117.

[35]   Ibid., pp. 115–16.

[36]   Ibid., p. 116.

[37]   Ibid., p. 118.

reserve intellectual and muscular energy to pursue other ends, in order 'to make our nervous system our ally instead of our enemy'.[38]

There are two crucial features of this account of habit that become central expressions of James' romantic science: firstly, the implication of an underlying moral pattern and, secondly, the principle of action. To examine these features, it is helpful to compare James' theory of habit with the work of two key modernist figures who developed post-romantic ideas in the late nineteenth century: the German philosopher, Friedrich Nietzsche, and the English aesthete, Walter Pater.

In *Human, All Too Human* (1878), Nietzsche shifted away from his early romantic associations with Richard Wagner and his philosophical allegiance with Schopenhauer to address 'the history of moral feelings', which led into his later rejection of Christian morality and formulation of will-to-power.[39] Here he considers morality to be closely connected to a veneration for, and preservation of, custom: 'to be moral, correct, ethical means to obey an age-old law or tradition'.[40] Goodness is therefore a value which rests on the weight of its inheritance, whilst 'evil is to be "not moral" (immoral), to practice bad habits, go against tradition'.[41] Nietzsche is critical of the conservative belief that 'because one feels good with one custom' it becomes the 'only possibility by which one can feel good' and he detects that some societies tolerate and preserve difficult and 'burdensome' customs because they seem to be 'highly useful' in the pursuit of more important ends.[42] Following Emerson's comments on passive conformity, Nietzsche later developed this position to argue that often slave mentalities are preserved through a desensitization to the constraints of custom and the naturalization of traditional morality.

However, far from rejecting habits wholesale, in *The Gay Science* (1882) Nietzsche claims: 'I love brief habits . . . and consider them an inestimable means for getting to know many things and states.'[43] From this perspective, habits which facilitate the development of an active self are seen to be both useful and enjoyable; but, he claims: 'enduring habits I hate. I feel as if a tyrant had come near me and as if the air I breathe had thickened when events take such a turn that it appears that they will inevitably give rise to enduring habits.'[44] Rather than providing personal enlightenment, he argues 'enduring habits' lead to suffocating bondage and repetition of sameness. By perpetuating habit, the goodness inherent

---

[38]   Ibid., p. 120.

[39]   Friedrich Nietzsche, *Human All Too Human*, trans. Marion Faber and Stephen Lehmann (London: Penguin, 1994), p. 39.

[40]   Ibid., p. 66.

[41]   Ibid.

[42]   Ibid, p. 67.

[43]   Nietzsche, *The Gay Science*, trans. Walter Kaufmann (New York: Random House, 1974), p. 236.

[44]   Ibid., p. 237.

in 'brief habits' (what is good for the self at a particular moment) collapses into what custom or law decrees to be good.

James' theory of habit is clearly different from the Nietzschean view. Early in *Principles* James outlines a patriarchal moral hierarchy which moves from drunkards and tramps, through 'the bachelor', 'the father', 'the patriot' and upwards to the heights of the 'philosopher and saint whose cares are for humanity and eternity'.[45] The formation of good habits is thought to influence and to contribute to the perpetuation of shared social values. However, there are other aspects of James' thought which push him much closer to Nietzsche. On one level, James appears to espouse a personal morality which adheres closely to public values, but he also claims the 'abrupt acquisition' of new habits is the 'best way' of acting upon resolution.[46] Furthermore, echoing Nietzsche's view on the extreme modes of life, in which the individual either seeks release of powerful creative energy or desires worldly abstinence through solitude, James claims that the 'best way' in which to keep 'the faculty of effort alive' is either by practising 'a little gratuitous exercise every day' or through 'a sharp period of suffering, and then a free time'.[47] Thus, he argues that public and private morality cannot be easily elided and individuals should find techniques for disrupting the passive absorption of naturalized values. Reflecting his refusal to simply accept Spencer's fatalistic account of self, James supplements his behaviourist study with an introspective method of self-analysis and therapeutic techniques for resisting stasis through bodily activity. Following Nietzsche, he implies that the acquisition of 'good' habits should be seen as enabling techniques for self-creation ('to make one's self over again'), rather than a fixed pattern which restricts individual liberty.[48]

James' principle of action is one of the distinctive marks of nineteenth-century Anglo-American thought in his work. Against the 'sentimentalist and dreamer, who spends his life in a weltering sea of sensibility and emotion, but who never does a manly concrete deed', James encourages active decision-making: 'we must take care to launch ourselves with as strong and decided an initiative as possible'.[49] In this way, he positions himself theoretically (although not ideologically) alongside the likes of Thomas Carlyle (an advocate of hero-worship) and the American President Theodore Roosevelt, against the European aesthetes emerging towards the end of the nineteenth century.[50]

---

[45]   James, *Principles*, vol. 1, op. cit., p. 23.

[46]   Ibid., p. 124.

[47]   Ibid., p. 124, 126.

[48]   Ibid.

[49]   Ibid., p. 125.

[50]   Kim Townsend convincingly argues that the model of masculine activity which both James and Roosevelt promote derives from an ideal of manliness prevalent at Harvard University in the 1890s; Kim Townsend, *Manhood at Harvard: William James and Others* (Cambridge, MA: Harvard U.P., 1996).

The critic Philip Fisher uses Walter Pater's famous concluding remarks to his art-historical study *The Renaissance* (1873) to exemplify the position against which James reacts. Pater claims that 'in a sense it might even be said that our failure is to form habits . . . after all, habit is relative to a stereotyped world, and meantime it is only the roughness of the eye that makes any two persons, things, situations look alike',[51] a comment which Fisher interprets to mean that 'perception, not action' is made 'the center of the self'.[52] However, whereas James asserts that strenuous labour should be privileged over the sensuous experiences of Pater's 'hard, gem-like flame', the soft organic metaphors he favours in *Principles* lessen the distance between an autonomous self and the kind of interactive individuation which the aesthete seeks. In a later chapter, James explicitly speaks in painterly vocabulary: 'interest alone gives accent and emphasis, light and shade, background and foreground – intelligible perspective, in a word'.[53]

In contrast to Pater's post-Kantian expression of disinterested pleasure, James adheres to the neoclassical notion that art should be instructive and morally edifying. James would seem to reject the doctrine of *l'art pour l'art* because 'one becomes filled with emotions which habitually pass without prompting to any deed, and so the inertly sentimental condition is kept up. The remedy would be, never to suffer one's self to have an emotion at a concert, without expressing it afterward in some active way.'[54] James' two examples of action, 'speaking genially to one's aunt, or giving up one's seat in a horse-car', indicate that art and music should serve to invoke compassion and altruism.[55] These examples do not rule out an aesthetic response, but the 'particular lines' and 'general forms of discharge' ensure that the sensation and perception of music is not merely internalized.[56] Rather than passively imbibing sense-data, James implies that only by striving to keep alive the possibility of acting upon the world in new ways can bodily equilibrium and psychic well-being be maintained. However, equilibrium is not to be equated with repetition and bodily stasis or, in Spencer's definition, the 'equilibration' which represents the end or the 'impassable limit' of evolution.[57] Thus, James argues that cultivating good habits is an important aspect of character building, but the extremes of habitual action lead back to the kind of compulsive necessity from which he wishes to escape.

---

[51]   Walter Pater, *The Renaissance: Studies in Art and Poetry*, ed. Adam Phillips (Oxford: Oxford U.P., 1986), p. 152.

[52]   Philip Fisher, 'The Failure of Habit', in ed. Moroe Engel, *Uses of Literature* (Cambridge, MA: Harvard U.P., 1973), p. 3.

[53]   James, *Principles*, vol. 1, op. cit., p. 402.

[54]   Ibid., p. 126.

[55]   Ibid

[56]   Ibid.

[57]   Richard Hofstadter, *Social Darwinism in American Thought* (Boston: Beacon Press, 1992), p. 37.

A Theatre of Simultaneous Possibilities

In the early sections of *Principles* James hovers between two competing notions of selfhood: either the self is conceived as an autonomous and sovereign entity which creatively acts in and upon the world, but is relatively unaffected by it, or it is understood as a flexible structure which fluctuates with changes in environment. In the later pages of 'Habit', James appears to gravitate towards the former version of an active self, but his notion of biological evolutionism rests closer to the other pole. Throughout *Principles* James seeks a middle path between these two extremes by which the self is understood to be both an active entity guided by the formation of good habits at the same time as it fluidly changes and passively adapts to its environment.[58] This notion of the self in flux leads from James' discussion of habit directly into his famous discussion of 'The Stream of Thought'.

The image of the stream follows closely from James' Darwinian description of the plasticity of mind, which moulds itself to adapt to change and mutability in the environment. In 'The Stream of Thought', the mental stream is shown to flow through a landscape of objects, without being entirely distinct from it. Moreover, James suggests an individual consciousness should not be conceived as a seamless whole, because it undergoes temporal breaks and interruptions: 'sleep, fainting, coma, epilepsy, and other 'unconscious' conditions are apt to break in upon and occupy large durations of what we nevertheless consider the mental history of a single man'.[59] Instead of an individual's 'mental history' comprising only conscious life, he proposes that interruptions to this unidirectional history may occur 'where we do not suspect it . . . in an incessant and fine-grained form.'[60] These counter-Enlightenment notions of 'fine-grained' consciousness and flexibility become central ideas in his attempt to conceive of the self midway between the self-determining individual of nineteenth-century bourgeois myth and the predetermined and preconscious bundle of physiological reflexes.

He goes on to claim that the individual does not usually experience these gaps or breaches directly, but becomes aware of them with reference to objective or 'outward time': for example, 'the sight of our wound' after an anaesthetized operation or the awareness of a lapse in chronological time.[61] He remains uncertain how one could decide indubitably whether consciousness is fragmentary or continuous, but, rather than arguing that consciousness sinks to 'a minimal state' or ceases to exist during sleep (as did Descartes and Locke), he

---

58   This sense of fluidity links to Pater's strain of romantic thought: the self is acted upon by the natural environment (or by the work of art) as much as it determines action.

59   James, *Principles*, vol. 1, op. cit., p. 199.

60   Ibid.

61   Ibid., p. 200.

suggests that the self possesses 'a secondary consciousness entirely cut off from the primary or normal one, but susceptible of being tapped and made to testify to its existence in various odd ways'.[62] Acknowledging his debt to the French experimental physicians, Pierre Janet and Alfred Binet, James detects this type of secondary consciousness is most readily apparent in patients with blind-sight or those under hypnosis, who display an awareness of objects even though they appear to be physiologically incapable of doing so.

Binet's research in the 1880s revealed that many hysterics maintain the ability to write 'automatically' while in an unconscious state.[63] Similarly, in his 1896 Lowell Lectures on 'Exceptional Mental States', James explains his own study of automatic writing, in which he postulates 'two simultaneously operating systems of intelligent consciousness, one above the threshold of awareness and one below, with separate characteristics'.[64] He concludes in *Principles* that 'the method of automatic writing proves that their perceptions exist, only cut off from communication with the upper consciousness'.[65] This severance of the two layers of consciousness encourages him to postulate the splitting of identity into the 'upper' and 'under' self, a split which is particularly apparent in hysterics who suffer from 'alterations of the natural sensibility of various parts and organs of the body'.[66] The upper self corresponds to the higher intellectual regions of the brain and usually expresses itself through vocal articulation, while the under or 'subconscious' self reveals itself through somatic symptoms ('pricks, burns, and pinches') and forms of writing which are free from cerebral control.[67] Moreover, James claims that even in non-hysterics these two levels of self to a large extent remain in 'mutual ignorance' of each other and so undermines the Cartesian idea of the self as privileged knower.[68]

If James takes as his model a split and semi-ignorant self then the epistemological certainty of Descartes' knowing subject becomes highly questionable. This position seems to push him towards a position of radical scepticism, but he resists going so far as to reject all grounds for knowledge. As he later argues in *Pragmatism* (1906), humans live in an 'as if' world in which

[62]   Ibid., p. 203.
[63]   Alfred Binet and Charles Féré, 'L'hypnotisme chez les hysteriques', *Revue Philosophique de la France et de l'Étranger*, 19 (1885), 1–25. Binet's paper on 'Visual Hallucinations in Hypnotism' was published in *Mind*, 9(35) (1884), 413–14 and Janet's book *L'Automatisme psychologique* was reviewed in the same journal, 14(56) (1889), 598.
[64]   Taylor, *William James on Exceptional Mental States*, op. cit., p. 6.
[65]   James, *Principles*, vol. 1, op. cit., p. 206.
[66]   Ibid., p. 202.
[67]   Here James parallels the work of Freud and Breuer in *Studies on Hysteria* (1895) and Freud's later metapsychological papers, with his notion of upper self corresponding to Freud's ego and under self with id. Although James came into contact with Freud's work, he rarely wrote in terms of the unconscious, preferring 'sub-conscious' or 'subliminal' self appropriated from the German psychologist F.W.H. Myers.
[68]   Ibid., p. 208.

they must accept certain realities in order to survive and accomplish certain ends. This emerging pragmatic position counters the extreme uncertainty of the sceptics and the absolute autonomy of the Cartesian self; instead of attempting to argue for ultimate foundations James favours 'indirectly or only potentially verifying processes' as more useful for the accomplishment of goals.[69]

In 'The Stream of Thought' he moves away from Spencer's claim that humans have an innate impulse for survival over which they have little control, by arguing that humans actually rely on feelings and sense-perception in order to know about themselves and the environment. But he also suggests that one cannot philosophically rise above the phenomenological world of relations and associations in order to secure an impartial or omniscient view of it. Although James differs from Descartes on this issue, for both of them the first-person 'I' is in a unique position to explore consciousness. As a behavioural psychologist, James relies on third-person observation to study the mind, but he argues only introspective knowledge of one's own particular thoughts can provide a fuller understanding of personal psychology. By opening 'The Stream of Thought' with the proposition that 'we now begin our study of the mind from within', he suggests this is the only tenable position from which to view it.[70]

Whereas Spencer bases his conception of the brain on an economic model of neural discharge by which psychic energy is expended, James develops a more subtle view of the 'waxing' and 'waning' of brain states.[71] This image of the complex interplay and phasing of tonal elements promotes a model of 'multitudinous' brain-states over the atomistic simplicity of Locke's position, which James claims is necessary in order to explain complex thought processes.[72] He also considers closely the temporal plane along which such waxing and waning occurs: 'no state once gone can recur and be identical with what it was before'.[73] Developing the thought of the pre-Socratic philosopher Heraclitus, he stresses that no two experiences can be exactly alike; the relationship between sense and perception inevitably gives rise to the fabric of experience or what James calls 'the river of elementary feeling'.[74]

Experience, then, emerges in the individual's acknowledgement of the subtle differences and variations between perception and sensation. Because it is only possible to speak of 'pure' sensation as an abstraction, complex experience cannot be reduced to the level of simplicity that Locke desires. Because particular experiences are unique, James proposes that it is the responsibility of the individual to attend to his own perceptions: 'experience is remoulding us every moment, and our mental reaction on every given thing is really a resultant of our

---

[69]   James, *Pragmatism*, ed. Bruce Kuklick (Indianapolis: Hackett, 1981), p. 97.
[70]   James, *Principles*, vol. 1, op. cit., p. 224.
[71]   Ibid., p. 235.
[72]   Ibid., pp. 258, 236.
[73]   Ibid., p. 230.
[74]   Ibid., p. 233.

experience of the whole world up to that date', but, because there are different and conflicting levels of consciousness, the individual can only be aware of a minute segment of the whole at any one time.[75] Moreover, as experience moves through a succession of interlocking temporal moments, the individual must increasingly attend to change. James claims that it is only through a refusal, or inability, to acknowledge subtle changes in experience that gives the individual a sense of an unchanging phenomenological world.

James resorts to his painterly vocabulary to suggest that one should be more attentive to the contrasts between different levels of consciousness: 'we feel things differently according as we are sleepy or wake, hungry or full, fresh or tired'.[76] By inferring (rather than directly perceiving) the differences between these stages the individual can begin to understand selfhood and to make the 'first charcoal sketch upon his canvas'.[77] James is not interested in the object as such, but the changes in the whole relational field of consciousness which alter the value or meaning of the object for the individual. Developing the techniques of defamiliarizing the object-world discussed in 'Habit', he recommends that individuals develop the capacity to attend to differences without losing the ability to think 'in a fresh manner'.[78]

The chapter on 'Habit' can be interpreted as a retort to the disinterested pleasure proposed by Pater's aesthetic theory. In 'The Stream of Thought', however, James appears to value the cultivation of good habits only when they are yoked to an awareness of unassimilable sensations which cause contrasts between levels of consciousness. He describes this model explicitly in aesthetic terms: 'when everything is dark a somewhat less dark sensation makes us see an object white' and 'the whole aesthetic effect comes from the manner in which one set of sounds alters our feeling of another'.[79] Here, James directly parallels Pater's need to cultivate reflexive attention in order to focus upon the subtle differences between flickering impressions.

However, in his opening comments on the continuity of thought, James seems to contradict this relational model, defining 'continuous' as 'that which is without breach, crack, or division'.[80] The feeling of the unbroken flow of thought enables the individual to retain a sense of identity with both recent and distant past: 'the natural name for it is myself, I, or me'.[81] The qualities of 'warmth and intimacy' are those with which James associates the function of memory for maintaining a sense of enduring identity within the flow of experience.[82] Thus, the 'stream' of

---

[75]   Ibid., p. 234.
[76]   Ibid., p. 232.
[77]   Ibid., p. 225.
[78]   Ibid.
[79]   Ibid., p. 232; pp. 234–5.
[80]   Ibid., p. 237.
[81]   Ibid., p. 238.
[82]   Ibid., p. 239.

thought describes a continuous and unjointed flow to which the more mechanistic images of 'chain' and 'train' used by the associationists cannot do justice. Some phenomena are 'discrete and discontinuous' (for example, a loud explosion in the midst of calm), but:

> their comings and goings and contrasts no more break the flow of the thought that thinks them than they break the time and space in which they lie. A silence may be broken by a thunder-clap, and we may be so stunned and confused for a moment by the shock as to give no instant account to ourselves of what has happened. But that very confusion is a mental state, and a state that passes us straight over from the silence to the sound.[83]

The sound of thunder might cause a momentary shock which registers as a sense of confusion, but even this confusion is part of the stream of thought. Indeed, based on this model there is no outside with which the stream can be contrasted. Rather, the contrasts and differences between thoughts are contained within the flow of experience: 'what we hear when the thunder crashes is not thunder pure, but thunder-breaking-upon-silence-and-contrasting-with-it'.[84] It is the background of the whole stream which determines the significance and the meaning of the object on which attention is focused. This context refers to all past experiences, whether clearly recalled, dimly recollected or subliminally registered, and an accompanying awareness, no matter how peripheral, of 'bodily position, attitude, condition'.[85] Once more, it is 'our bodily selves' which James claims to be 'the seat of the thinking' and which provide the bedrock or absolute grounds (although clearly these are not explanatory grounds) for these 'phenomena of contrast'.[86] Only by cultivating an awareness of the differences between the formal elements in the compositional field can one begin to understand the structures of thought.

One major problem remains: if one cannot attend to these differences directly, how can one infer them from the plenum of experience? This is where James' ideas of repetition and pattern established in 'Habit' come into play. Like a musical scale, the sense of familiar pattern encourages an anticipatory response, while shifts in the perceptual field represent a modification of a previous pattern, rather than a wholesale change.[87] Instead of positing the repetition of the same, James' is influenced by Søren Kierkegaard's notion of repetition as recollection, or repetition-with-difference. Here, recollection suggests the memory of

---

[83]   Ibid., p. 240.
[84]   Ibid.
[85]   Ibid.
[86]   Ibid., p. 242.
[87]   This is not to devalue the importance of sameness in James' thought: 'sameness in a multiplicity of objective appearances is . . . the basis of our belief in realities outside of thought' (ibid., p. 272) and, as he comments at length in 'The Consciousness of Self', the idea of sameness is crucial to one's sense of enduring identity.

something previously misplaced or forgotten: one can only have a memory of an object or an event if it has been temporarily absent or lost.[88] This notion of repetition-with-difference feeds into James' sense of the continuity and preservation of identity despite flux and mutability. The idea that the organism must adapt itself to changes in the environment implies the impossibility of maintaining a static state of selfhood. Instead, the patterns that constitute mental phenomena (an awareness of a changing environment) are incessantly repeated in different combinations. As an element is rearranged, so must one modify an awareness of 'its relations, near and remote, the dying echo of whence it came to us, the dawning sense of whither it is to lead'.[89]

Rather than trekking through an unexplored phenomenological wilderness, the individual is led by those structures and patterns in which he or she can anticipate the rearrangement of elements. At the same moment, one experiences both repetition and change: the experience of the loss of a particular configuration of elements and the memory of past associations blended with the anticipation of new ones. In order to describe this experience James introduces the word 'fringe' to characterize the 'influence of a faint brain-process upon our thought, as it makes itself aware of relations and objects but dimly perceived'.[90] Such dim perceptions account for the waxing and waning of relational elements as they restructure an individual's perceptual field, as well as those subliminal perceptions only detectable (usually by the scientist or analyst) through somatic signs, dreams or expressed through a medium such as automatic writing.

James' description of 'The Stream of Thought' leads to his poetic claim that the mind is 'a theatre of simultaneous possibilities'.[91] Although consciousness selects the elements by which it has come to recognize and order the world, invading and contaminating these habitual and structuring perceptions are the 'primordial chaos of sensations' in which they are embedded.[92] While habits, customs and cultural codes (such as language systems) encourage individuals to reject the 'swarming atoms' which lie beyond established modes of representation, these constitutive elements remain the raw material of an individual's sense-impressions.[93] James argues pragmatically that by 'rejecting certain portions' of this entropic world it is possible to establish a manageable

---

[88]   In *Repetition: A Venture in Experimenting Psychology* (1843) Kierkegaard's alter-ego Constantin Constantius states 'for what is recollected has been, is repeated backwards, whereas repetition, properly so called is recollected forwards'; Søren Kierkegaard, *Fear and Trembling/Repetition*, ed. and trans. Howard V. Hong and Edna H. Hong (Princeton, NJ: Princeton U.P., 1983), p. 131. This assertion is similar to James' suggestion that memory and anticipation are closely intertwined and prefigures Oliver Sacks' interest in neurological disorders which affect the faculty of memory.

[89]   James, *Principles*, vol. 1, op. cit., p. 255.

[90]   Ibid., p. 258.

[91]   Ibid., p. 288.

[92]   Ibid.

[93]   Ibid., p. 289.

personal environment which serves the individual's requirements.[94] However, the chaotic impressions of which the upper self is ignorant often disrupt an orderly universe in which the individual is the artist-creator. Just as he rejects the determinism of Spencerian psychology, James cannot accept the romantic myth of the artist-creator, because it reduces the hidden complexity of thought to the manifest autocracy of the sovereign individual. Indeed, those critical moments when chaotic impressions invade an orderly sense of reality, imply that one can only maintain provisional control over the self.

The Heave of the Will

In order to outline the ways in which James' redescription of experience in 'The Stream of Thought' bears upon his conception of an active self, this section couples a reading of his discussion of 'will' in the second volume of *Principles* with an analysis of his famous essay 'The Will to Believe' (1896). By addressing the will as a philosophical concept, James stumbles upon one of the most contentious areas of nineteenth-century Anglo-American and German philosophy. However, his notion of 'will-as-activity' is subtly different from either of the philosophical traditions which inform his thinking: that is to say, by taking as his starting point the verb 'to will' he manages to avoid metaphysical discussion of 'the will' as a psychological faculty. This continued desire to bracket off metaphysics represents a positive move towards his emerging theory of radical empiricism, enabling him to describe an embodied and experiential self, rather than the contemplative self of classical philosophy. This theoretical movement away from epistemology towards pragmatism does not suggest that questions of knowledge should be altogether suspended, but implies that no absolute grounds of knowledge will provide individuals with the ultimate foundation for an active or moral life. In short, James' theory of the active willing self (rather than a passive knowing self) enables him to resist Spencer's deterministic position and the logic of his own 'subject-less' stream of thought.

As a representative of the Anglo-American psychological tradition, the British psychologist Alexander Bain, in his influential work *The Emotions and the Will* (1859), considers the will as a faculty which accounts for voluntary activity. Although, like Bain, Mill's argument for volition refutes the immutable model of cause and effect, it does not have a sound philosophical base, constituting a belief about the world. Not surprisingly, the faculty of will was an anathema to the Spencerians whose Darwinian model was founded upon the primacy of involuntary instincts. In many ways, the existence of will is the last line of defence for thinkers like Mill who wished to keep alive the belief in the freedom of reasoned choice. To retain a sense of self-fashioning in a Darwinian

---

94   Ibid.

world, it is important to assert a belief in the ability to make meaningful choices. James adopts and works through these issues, but refuses to be sucked into the futile intellectual task of expounding a theory which verifies the existence of the will. Instead, he advocates 'willing' as a capacity which accounts for otherwise inexplicable aspects of human behaviour.

The other strain of writing which runs through *Principles* and James' later work is his strong connection with romanticism, which furnishes him with a different notion of will as a descriptive term. Early romantic writers often used the term to describe the poet's sense of agency and an inherent design in Nature; for Wordsworth and Emerson the will is viewed as a positive force which, when exercised, can invoke empathic and altruistic feelings for the wider interpersonal, organic and spiritual worlds. However, for both these writers, a guiding force (Wordsworth's 'Nature' and Emerson's 'Over-Soul') is needed in order to temper the capricious drive of the will. Thus, the will lies somewhere between an innate appreciation of, and sympathy for, the natural world and an activity through which the poet can express feeling.

At the other extreme from this expression of optimistic faith in human nature, Schopenhauer's conception of will as an evil force is one indication of the influence of the German tradition on the darker aspects of James' thought. In *The World as Will and Representation* (1819) Schopenhauer argues at length that the will is the primary source of misery and pain in the world: if humans are to liberate themselves from bondage they must cultivate the denial of the will. Thus, life and will are mutually incompatible and the privileging of the one means the negation of the other: the ideal life should be both ascetic and altruistic, repudiating selfhood and self-advancing behaviour. James displayed an ambivalent relation to these ideas: on the one hand, he credits Schopenhauer for addressing 'the concrete truth about the ills of life', but, on the other, rejects his intense pessimism as 'a species of fatalism, in the worst sense . . . an abandonment of the better possibility for sheer inaction'.[95] In this way, Schopenhauer influences the mood of James' religious writings; however, his intensely bleak vision is resisted by the optimistic thrust of the American's writing.

Nietzsche, Schopenhauer's most rebellious disciple, fuses these positive and negative aspects of the will in his notion of will-to-power. This entails the self-conscious overcoming of the limitations of selfhood through exertion of the will, in an ongoing search for spiritual self-fulfilment. In so doing, Nietzsche wrestles the will away from Schopenhauer's tyrannical model towards an affirmatory philosophy in which the will-to-power overtakes the self as the important defining force in human activity. However, although both German thinkers bear upon James' ideas, it is Emerson who provides James' most direct link to a romantic tradition. As the next section outlines, these Anglo-American and

---

[95]   Perry, *The Thought and Character of William James*, vol. 1, op. cit., pp. 721–2.

German intellectual currents merge in James' thought via the transcendental 'New England' Emerson and the 'Germanic' Emerson whom Nietzsche enthusiastically read.

In the chapter on 'The Consciousness of Self' in *Principles*, James shifts away from cataloguing the constituents of self towards a description of the 'feelings and emotions they arouse' and to the 'actions . . . which they prompt'.[96] This movement away from a focus on the individual's passive adaptation to environment to a theoretical position which establishes the centrality of agency follows the wider scope of James' writing and inaugurates the beginnings of romantic science in its twentieth-century guise. As a philosophical manoeuvre which transfers attention from epistemology to pragmatism, the shift marks James' transition towards an explicitly therapeutic form of writing.

At all times James reviles stasis as a condition to avoid both intellectually and practically, because it entails the surrendering to blind forces of instinct or natural laws which lead to a fatalistic view of life. This does not mean that contemplation should be abandoned for an active life devoid of reflection; far from it, the individual should understand that certain aspects of the body 'seem more intimately ours than the rest' and thus constitute a material home.[97] Extending outwards through the circles of family and social recognition (for example, name-giving, shared language and social roles), these zones constitute a habitat in the world. Unlike Spencer, James does not believe that this habitat is wholly predetermined, to which humans must adapt themselves in order to survive. Clearly, an argument for adaptation informs James' thought, but the sense that humans can develop good habits (either through self-development or following educational programmes) implies they have some ability to modify the shape and scope of their habitat.

For James a habitat should not be figured as a fixed zone, or territory, into which individuals are born; rather, the habitat is the shifting ground for the individual's experience of the world. In 'The Stream of Thought' James suggests there is an 'innermost centre' which, although in constant flux, coheres more intimately than the rest of experience.[98] The danger in claiming that there is a centre where 'other elements end by seeming to accrete round it and belong to it' is that it aligns his theoretical position with the associationist model he wishes to resist.[99] However, he claims that 'the active element' actually helps to sustain this 'innermost center' of self and constitutes a sense of identity within a habitat. James calls this centre a 'home of interest', which is at one and the same time a cognitive and a sentient centre: 'that within us to which pleasure and pain, the pleasant and the painful, speak'.[100] On this reading, his description of a 'sanctuary

---

[96]   James, *Principles*, vol. 1, op. cit., p. 292
[97]   Ibid., p. 297.
[98]   Ibid.
[99]   Ibid., p. 298.
[100]   Ibid.

within the citadel' can be redirected away from the notion of an imperial self to an active and flexible self-construction only within the bounds of experience.[101]

James does not dismiss the self as an illusory fiction, because he claims 'this central part of the Self is felt' and 'something with which we . . . have direct sensible acquaintance'.[102] His claim for a sustaining self can be stated in a neo-pragmatic way: even if the self is a fiction, it is one which is constructed in order to make sense of, and give recognizable shape to, the chaotic sense impressions of experience. James counters Hume's denial that there is any one entity, or faculty, which exists above the flow of sense experience by claiming that, because one privileges some aspects of experience over others, it is possible to postulate the existence of an experiencing self. This sense of self is not fixed by the construction of a stable subject-position. In his Pateresque description of the perceptual process James writes: 'I cannot think in visual terms, for example, without feeling a fluctuating play of pressures, convergences, divergences, and accommodations in my eyeballs.'[103] Nevertheless, despite the complexity and mutability of such sensations, the recurring 'portions' of these impressions constitute one's centre of identity.[104]

In order to clarify such an account, James conceptually splits the self into two aspects: the adjusting 'nuclear' self which provides a sense of continuity, and the executing 'shifting' self which enables one to act upon the environment in the pursuit of future goals.[105] In the dynamic interaction between these two, which is both conservative (the cumulative result of following habitual patterns) and projective (the revision of those patterns in the light of new experiences), James locates the primary mark of identity 'I'. This 'I' is both a linguistic structure which enables the individual to express him or herself in language, and a felt centre of activity which exists, despite the mutable fringes of experience, as 'the birthplace of conclusions and the starting point of acts'.[106] In other words, the self is a site where remembering (retaining traces of past activities) and willing (forcing new perceptual stances) meet. Memory does not trace backwards in an unidirectional fashion to connect with a line of identical past selves, nor does the self maintain a stable shape as it pushes into the future. Instead, these fringes of experience which disrupt such a linear sequence cause slight modifications and disruptions in habitual behaviour. This description of repetition-with-difference explains the way in which the self can preserve a dynamic pattern without surrendering to blind habit. For James, knowledge of identity results from retrospective thinking to make connections through the activity of memory. The nuclear self constitutes a site in which events in the present are ordered, but it is

---

[101] Ibid., p. 297.
[102] Ibid., pp. 298–9.
[103] Ibid., p. 300.
[104] Ibid., p. 302.
[105] Ibid.
[106] Ibid., p. 303.

also partly constitutive of, and partly dependent upon, the reconfiguration of past elements and events. This enables James to retain a sense of selfhood (the 'nuclear self') as 'an abstract, hypothetic or conceptual entity' postulated in an act of reflection (either deliberate or involuntary), and a '*Scious*ness' as the active, protean and 'shifting self'.[107]

This model accounts for the manifold phenomenal world and fringes of experience to which the individual must adjust and propels James' account of the self in the direction of a hermeneutic, or interpretative, self which informs the tradition of twentieth-century romantic science. Interpretation occurs most conspicuously at those moments when it is necessary to connect the 'shifting self' with the memory of a formerly postulated 'nuclear self': for example, at times of crisis associated with profound bodily, psychological or environmental changes. The necessary adjustment enables one to locate the bridging dynamic self as the maker of meaning, rather than the epiphenomenon of passive sensation. In this manner, James retains a concept of self compatible with, and not contradictory to, his description of 'The Stream of Thought'.

However, this model is vulnerable to several experiential difficulties. The inability to make sense of an event may result in the repudiation of the present moment for a self located in a safe past. Similarly, loss of memory, or the refusal to remember, may loosen the 'shifting self' from its 'nuclear' moorings into a free-floating world of unconnectable and unassimilable presents. However, these points do not inflict real damage upon James' theory (although they do bear upon Oliver Sacks' extreme medical cases). His experiments with hysterics indicate that these areas of experience which cannot be expressed (either because the patient is neurologically impaired or because the experiencer cannot bear to express his or her ordeal) indicate a subliminal level which contributes to selfhood, but cannot itself be incorporated into a conscious knowledge of identity. These subliminal elements are seen to impinge dimly upon accessible conscious experiences and disrupt the sense of sovereign self. However, these disruptive elements do not preclude or disable most individuals from positing a sense of selfhood which infuses activity with meaning and significance.

James' sustained consideration of will transfers attention from will as faculty (as described by Bain), towards will as the act of attending. He begins from the premise that we can only execute 'voluntary movements' of the body if the movement has already occurred involuntarily prior to mental exertion: thus 'reflex, instinctive, and emotional movements are all primary performances'.[108] Far from reverting to a Spencerian position in which an organism is innately equipped with a set of immutable responses, James suggests involuntary acts are often the result of contingent changes in environment. Similarly, rather than always following a fixed pattern, sometimes the disruption of habits will force the

---

[107]   Ibid., p. 304.
[108]   James, *Principles*, vol. 2, op. cit., p. 487.

individual to take up a different mode of response. Once this reflexive act has occurred, the individual may learn to master the response and cultivate it as a new habit. Memory enables the individual to recollect the consequences of a previous activity in order to make informed choices about the future.

This description seems fairly conventional, but a question raised later in the chapter complicates matters:

> Is the bare idea of a movement's sensible effects its sufficient mental cue.
> . . . or must there be an additional mental antecedent, in the shape of a fiat,
> decision, consent, volitional mandate, or other synonymous phenomenon
> of consciousness, before the movement can follow?[109]

In other words, why does James retain the will if all the self needs is a 'kinaesthetic idea' or the memory of a previous response?[110] He concedes that in certain situations 'the bare idea is sufficient' to stimulate activity; however, when he considers deliberate actions, James suggests some type of 'fiat, mandate, or express consent, has to intervene and precede the movement' in order to resolve internal conflict.[111] Following his earlier description of the mind as a 'theatre of simultaneous possibilities', the will takes the role of a judicial decision-maker in situations in which different options present themselves as equally tenable, or desirable in different ways. However, James departs from the conventional notion of a faculty of will, which mysteriously exists above the flow of experience and is called upon to arbitrate in difficult situations, in his understanding of it as a vehicle by which an act of attention is transferred into bodily kinetic movement: 'the effort to attend is therefore only a part of what the word "will" covers; it covers also the effort to consent to something to which our attention is not quite complete'.[112] The will thus becomes the key term in James' redescription of the Cartesian contemplative theatre of the mind as an embodied and active self.

In explicating this idea, James writes at length about five different types of decision by which individuals settle contradictions: appealing to reason; acquiescing to either internal or external pressures; following a conviction; or resolving to follow a particular course of action. In most situations decision-making provides the individual with few difficulties: after balancing up the options, the gains of one particular course of action usually outweigh the losses. Reasoned choice provides the foundation for this model, but James' understanding of 'resolve' undermines the supremacy of reason as the crucial criterion in the making of decisions; the feeling of exertion encourages him to postulate the existence of will as an inner activity through which conflicting

---

109 Ibid., p. 522.
110 Ibid., p. 493.
111 Ibid.
112 Ibid., p. 568.

impulses are resolved.[113] Unlike other types of decision-making, he suggests that resolve is not a blind act of will, but an activity by which 'in the very act of murdering the vanquished possibility the chooser realizes how much in that instant he is making himself lose'.[114] The creative risk of overcoming internal conflict cannot be explained with reference to the concepts of habit, chance or reason, only, according to James, by activating the 'heave of the will'.[115]

This model is further problematized if willing is viewed as merely another type of habitual response. Instead, he argues that because an individual can simultaneously entertain more than one possibility, it implies that one possesses the capacity to follow a particular course of action, 'whether the act then follows or not is a matter quite immaterial, so far as the willing itself goes'.[116] To characterize this capacity James introduces the idea of an 'ordinary healthiness of will', linking an 'impulsive power' with external criteria like reason, convention or belief, which together may or may not result in kinetic action.[117] Conversely, for an unhealthy will 'the action may follow the stimulus or idea too rapidly, leaving no time for the arousal of restraining associates' (the 'obstructed' will), or 'the ratio which the impulsive and inhibitive forces normally bear to each other may be distorted' (the 'explosive' will).[118] James claims the 'heave of the will' is experienced 'whenever strongly explosive tendencies are checked, or strongly obstructive conditions overcome'.[119]

In summary, he characterizes the activity of will as a type of habit ('essentially a system of arcs and paths, a reflex system') which has entered the realm of conscious life, possessing the capacity to override and legislate for activities which would normally occur beneath consciousness.[120] It is at moments of difficult decision-making when one must recoup a sense of a nuclear self, in order to reflect upon options based on past experiences. This does not imply that James wishes to rescue some transcendent sense of selfhood as a therapeutic prop, only that which can be constituted from, and bears upon, previous experiences. But the capacity to resist, inhibit or override the reflex or the

---

[113] James, *Principles*, vol. 2, op. cit., p. 534.

[114] Ibid., p. 534; James uses 'resolve' in both its senses: firstly, to determine amongst conflicting options, and, secondly, the conviction to persevere with a course of action or belief in oneself. It is this second meaning which maintains James' ethical position *vis-à-vis* the cultivation of a better self (or selves). This notion of resolve pushes James' work in two directions: the resolute thinking of early Heidegger and the 'deep' reading of Stanley Cavell.

[115] Ibid.

[116] Ibid.

[117] Ibid., p. 536.

[118] Obstructed and explosive willing are sometimes useful for extricating the self from particularly exacting situations, but more often they refer to the unhealthy type of will (that which is not good for the self) of which Schopenhauer writes.

[119] James, *Principles*, vol. 2, op. cit., p. 548.

[120] Ibid., p. 575.

habitual 'way' suggests a realm of indeterminacy, which enables the self to switch from one course or path of activity to another.[121] Crucially, this ability to choose between options and consciously adapt to situations is dependent upon bodily constitution (the neurological and physiological base) and mental aptitude (the 'education of the will'), but also upon random movements of 'quasi-accidental reflexes'.[122] Accidental or random discharge are important channels through which 'new paths' are formed, but it is the selective capacity to choose between alternatives which leads James to construct a nuclear self which can act in and upon an 'indeterminate' world.[123] James can be criticized for paradoxically asserting the coexistence of a determined and an undetermined world, but he does discern a philosophical space of indeterminacy within the natural laws of cause and effect in which the willing self can act.

In the preface to *The Will to Believe* (1897), a collection of his short philosophical essays from the 1880s and 1890s, James characterizes 'radical empiricism' as a twofold enterprise which accounts for this hypothesis of a willing self on grounds which are both empirical (the close physiological observation of *Principles*) and philosophical (although not foundational). Firstly, it is empirical because it regards 'its most assured conclusions concerning matters of fact as hypotheses liable to modification in the course of future experience'.[124] In other words, the sense of nuclear self emerges from an engagement with the world, but is revisable in the light of future experiences. Secondly, James claims this is a radical position 'because it does not dogmatically affirm monism as something with which all experience has got to square'.[125] The 'absolute unity' of phenomena is rejected for a pluralistic position in which 'the crudity of experience remains an eternal element thereof'.[126] As such, James sets up a series of incomplete and revisable concepts to account for those fringes of experience that philosophical monists tend to ignore. He realises that the presence of will cannot be established indubitably, because finally it relies on a matter of belief. The hypothesis of the willing self is thus dependent upon whether it is in the individual's interest to entertain a notion of selfhood, by which he or she can order the flux of experience and act meaningfully. For James, the postulate of will does not entail the transgression of the laws of nature, but it does provide the individual with a belief that he or she is capable of aspiring towards goals.

James cites the example of Pascal's wager, in which the individual must weigh up finite loss (the risk of setting up false models) against infinite gain (the benefits which can be accrued through investigating and investing in those same

---

121 Ibid., p. 580.
122 Ibid., p. 579.
123 Ibid., p. 571.
124 James, *The Will to Believe and Human Immortality* (New York: Dover, 1956), p. vii.
125 Ibid. pp. vii-viii.
126 Ibid., pp. viii, ix.

models): 'you . . . may think the risk of being in error is a very small matter when compared with the blessings of real knowledge, and be ready to be duped many times in your investigation rather than postpone indefinitely the chance of guessing true'.[127] Even if the postulated self turns out to be erroneous, James suggests that by avoiding philosophical absolutes individuals can revise their opinions in the light of environmental or bodily change. In any case, he admits errors are inevitable, and should not (in either philosophical or practical terms) be seen as 'such awfully solemn things'.[128] Indeed, James goes on to suggest that the existence of the willing self may be a moral question 'whose solution cannot wait for sensible proof'.[129] That is to say, one needs a sense of identity to continue with life without waiting for an irrefutable argument to confirm its existence. James does not refuse the right for other inquirers to 'wait' for such proof, but for him (as experiencer as well as philosopher) to do so would represent surrendering to the inertia he seeks to avoid. Thus, he claims only by acting, and believing that such activity is meaningful, can 'we' begin to take 'our life in our hands'.[130]

'That Shape Am I'

In one of the most frequently cited passages in *Varieties*, James disguises a lurid description of the mental and spiritual breakdown he had experienced in the autumn of 1872 by attributing it to a (fictional) French correspondent. At the time of delivering his 1901 Gifford Lectures in Edinburgh no evidence existed to suggest this passage was anything but one of the lengthy quotations which characterize *Varieties* as a polyphonic text. Two years after the publication of the lectures in book form, James wrote to Frank Abauzit, who was in the preliminary stages of translating *Varieties* into French, admitting that 'the document on p. 160 is my own case – acute neurasthenic attack with phobia. I naturally disguised the provenance! So you may translate freely.'[131] By disguising the account, James leaves the experience to work dramatically on the reader, rather than resorting to the medical diagnosis of his letter. By concealing the provenance, he throws the experience open to the kind of expansive interpretation he encourages elsewhere in *Varieties* and instead of reducing the description to the level of medical materialism, he encourages a hermeneutic pluralism which mirrors his own

---

[127] Ibid., p. 18.
[128] Ibid., p. 19.
[129] Ibid., p. 22.
[130] Ibid., p. 30.
[131] Gerald E. Myers, *William James: His Life and Thought* (New Haven: Yale U.P., 1986), p. 608. For a recent interpretation of this episode see Louis Menand, 'William James and the Case of the Epileptic Patient', *New York Review of Books*, 45(20) (17 December 1998), 81–93.

growing commitment to an open-ended philosophy, later outlined in *A Pluralistic Universe* (1909).

The central italicized line of the account, '*That shape am I, I felt, potentially*', crystallizes the fear and dread that the correspondent feels in the face of the vision of a mummified epileptic patient whom he had seen in an asylum.[132] The description of the epileptic idiot 'with greenish skin, entirely idiotic' can be seen to draw on the ideas of the German psychologist Carl Carus, whose ground-breaking book *Psyche* (published in 1846 and expanded in 1851) describes epilepsy as an unstable condition which occupies a precarious middle space between insanity (a structural malady) and idiocy (an anatomical affliction): that is, between illnesses of mind and body.[133] The account in *Varieties* depicts a wretched figure sitting motionless in a fugue state, which renders his appearance 'absolutely non-human'.[134] The figure is depicted in traditional nineteenth-century fashion as possessing the outward physiological markers of a profound internal disorder. Sitting prostrate in the asylum like 'a sort of sculptured Egyptian cat or Peruvian mummy', he is portrayed in an analogous fashion to the medical drawings which Sander Gilman suggest characterize the Victorian view of madness as manifesting itself in outward grotesquerie.[135] In his book on *Disease and Representation* (1988), Gilman argues that one of the ways in which medical art in nineteenth-century Europe distanced 'the fear of collapse' and 'the sense of dissolution' from the general public was by exaggerating the features of sufferers so that they appeared to be totally consumed by the disease.[136] In James' account, this type of representational confinement is mirrored in the limited space of the asylum and emphasized by the enclosure of the 'benches, or rather shelves against the wall' and the 'coarse grey undershirt . . . inclosing his entire figure'.[137]

When the correspondent exclaims '*that shape am I*' he associates himself with the figure entombed not only in the cell and his body, but also by the mode of language which stigmatizes him as such.[138] Carus understood the appearance of hallucinatory phantoms to be a projection of the fear of the self: the mental manifestation of an unconscious or barely known double who is nourished on anxiety and psychic trauma (a theme which Otto Rank explores in greater depth). Carus outlines three forms of 'derangement of the conscious by the unconscious' – love, visionary trance and religious ecstasy – all of which characterize the

---

[132] James, *Varieties of Religious Experience*, ed. Martin E. Marty (London: Penguin, 1985), p. 160.

[133] Carus is important for a discussion of James because *Psyche* was one of the first medical accounts to claim the unconscious has a profound influence on conscious mental life.

[134] Ibid., p. 160.

[135] Ibid.

[136] Sander L. Gilman, *Disease and Representation: Images of Illness from Madness to AIDS* (Ithaca: Cornell U.P., 1988), p. 1.

[137] James, *Varieties*, op. cit., p. 160.

[138] Ibid.

experience of 'sinking into a new world' of Otherness.[139] Such psychic 'derangement' is reflected in the acute emotional reaction of James' correspondent:

> There was such a horror of him, and such a perception of my own merely
> momentary discrepancy from him, that it was as if something hitherto solid
> within my breast gave way entirely, and I became a mass of quivering
> fear.[140]

In direct contrast to the enclosed and confined picture of the imagined figure, the emotional reaction reveals solidity disintegrating into 'quivering fear'. The figure is hidden away from the world and constrained by the limited topographical and conceptual space assigned to him, whereas the correspondent's fear is described as a breakdown of solidity into something fragile and unstable. It is possible to interpret this emotional crisis as a threshold state out of which emerges a fresh perspective: a morbid or melancholic view in which the dominant feeling is one of persistent dread and insecurity. When the correspondent explains why he thinks his melancholia 'had a religious bearing', he does not retreat into the security of the scriptures, but reads them as a means of resisting the paralysis personified in his vision.[141]

Later, in his disclosure to Abauzit, when he admits that the vision actually depicts his own youthful experience, James appears to endorse this reading of the passage, in which he juxtaposes his open-ended interpretation with the enclosed and stylized image of the epileptic idiot. This view is supported by the affiliation he displays for the morbid temperament, as opposed to the attractions of the 'healthy-minded' who ignore or dismiss the existence of evil which James perceives as part of the 'very essence' of the morbid-minded life.[142] His philosophical aim to disrupt unity and harmony in order to privilege doubleness and plurality becomes, in the disguised story of the breakdown, a textual manoeuvre which opens up experiential spaces that tend to confine and, in the case of the mummified epileptic idiot, dehumanize the individual. The liberation of these spaces represents James' attempt to inject a more complex philosophy into religious discourse.

The standard interpretation of the passage is that by including a third-person account of his own breakdown James is practising a form of self-therapy. This idea can be extended into the domain of romantic science to suggest the passage dramatizes James' attempt to break down any system of thought which seeks to define and disambiguate rather than liberate. By embedding a notion of open-

---

[139]   James L. Rice, *Dostoevsky and the Healing Art: An Essay in Literary and Medical History* (Ardis: Ann Arbor, 1985), p. 140.

[140]   James, *Varieties*, op. cit., p. 160

[141]   Ibid., p. 161.

[142]   Ibid., p. 113.

endedness within his autobiographical account and interweaving different strands of discourse (psychological, physiological and spiritual) James produces a generically hybrid text. However, unlike the critic Frederick Ruf, who argues that *Varieties* is marked by entropy and dissolution, structures *can* be discerned by which James places temporary limits and boundaries on the unassimilable mass of preconscious flux.[143]

The description of psychic and bodily breakdown also has links with James' resistance to Spencer's deterministic philosophy. In *Varieties*, he continues to stress the possibility of free will and autonomy in order to resist surrendering to the blind forces of biology and neurology. The contemporary medical language of neurasthenia available to James did little to uncover identifiable traits of mental illness. If illness is closely associated with a loss of volition and paralysis, then belief in free will may enable the individual to invest psychic energy in future possibilities, without giving up to the lure of meta-patterns or interpretative absolutes. James' therapy is not confined to the inclusion of the disguised autobiographical passage, but can be detected throughout his work. Instead of the description of his breakdown representing a point of rupture where autobiography leaks into the lectures, the passage is better understood as one of the nodal points of the text where many of ideas played out at length elsewhere are compressed in an explosive dramatic moment.

In order to substantiate this claim, the phrase *'that shape am I'* can be traced laterally through *Varieties* with reference to James' comments on narrative patterns. Narrative is not merely of incidental importance to him: as he goes on to expound in *Pragmatism*, 'things tell a story. Their parts hang together so as to work out a climax.'[144] His account of narrative moves away from a traditional linear structure which neatly resolves itself in a final denouement; instead, he outlines a discontinuous discourse: 'the world is full of partial stories that run parallel to one another, beginning and ending at odd times. They mutually interlace and interfere at points, but we can not unify them completely in our minds.'[145] This form of narrative organizes events, without the story ever reaching a telos or a final moment of stasis. Here, James displays a close connection with modernist writers who resist the closure of dominant nineteenth-century narrative forms in an attempt to do justice to the complexities of experience. But, rather than pushing his thought in the direction of the modernist aesthetics of his ex-student Gertrude Stein, James looks to a romantic template to give voice to his experience of modernity.

In his critical survey of romantic thought, *Natural Supernaturalism* (1971), M.H. Abrams pays special attention to the geometric formations which recur

---

[143] Frederick J. Ruf, *The Creation of Chaos: William James and the Stylistic Making of a Disorderly World* (New York: University of New York Press, 1991).

[144] James, *Pragmatism*, op. cit., p. 67.

[145] Ibid.

throughout romantic poetry and philosophy. He discerns the most recurrent pattern to be the circle, which he detects in the writings of Hegel, Novalis, Hölderlin, Wordsworth and Coleridge, through to Nietzsche's conception of eternal return and to the beginnings of modernist literature. He traces romantic theodicy (which, after Carlyle, he terms 'natural supernaturalism') back as far as St Augustine.[146] In the final book of the *Confessions* (401) Augustine reflects upon the end of time as a return to the beginning and the dissolution of the world. At the end he speaks of the cycle of the days, a cycle which 'has no evening and has no ending', but he completes a formal circle by recalling the beginning of his work and reverting back to his initial praise of God.[147] Because in God is contained both beginning and end and because His 'seeing is not in time', by communing with Him through prayer and rapture, Augustine claims he is able to catch brief glimpses of the eternity which lies beyond the time-bound constraints of perception and sensation.[148]

By transferring this circular pattern from the macrocosm to the private quest of the individual, Augustine paves the way for the kind of circuitous journey found in later religious allegories of Edmund Spenser and Bunyan, through to the romantics and Coleridge's rather more pagan image of 'the snake with it's Tail in it's Mouth' (*sic*).[149] The quotation from Coleridge is particularly significant in his claim that the purpose of narrative (the telling of that journey) is to make 'those events, which in real or imagined History move on in a strait Line, assume to our Understandings a circular motion'.[150] For Coleridge, imaginative art contains within it the 'esemplastic' potential to make whole, or 'to shape into one', a universe which could otherwise only be perceived in sensual fragments.[151] However, for many romantics, the 'circular motion' does not return to its point of origin, but constantly evolves by turning in on itself at the same instant that it moves; Abrams quotes Hegel's claim that 'this circle is a circle of circles . . . which, returning to the beginning, is at the same time the beginning of a new member'.[152] Out of the traditional figure of the circle emerges another more complex pattern, the spiral, helix, or 'ascending circle', which constantly expands at the very moment it turns back on itself.[153]

---

[146] Abrams characterizes the *Confessions* as 'the first sustained history of an inner life' which establishes 'the spiritual vocabulary for all later self-analyses and treatments of self-formation and the discovery of one's identity'; M.H. Abrams, *Natural Supernaturalism* (New York: Norton, 1973), p. 83.

[147] Augustine, *Confessions*, trans. Henry Chadwick (Oxford: Oxford U.P., 1992), p. 342.

[148] Ibid., p. 304.

[149] Abrams, *Natural Supernaturalism*, op. cit., p. 271.

[150] Ibid.

[151] Samuel Taylor Coleridge, *Biographia Literaria*, ed. George Watson (London: Everyman, 1984), p. 81.

[152] Abrams, *Natural Supernaturalism*, op. cit., p. 509.

[153] Ibid., p. 184.

One of Abrams' major achievements in *Natural Supernaturalism* is to chart the historical transition from the neoplatonists who, like Plotinus, follow the mystical circle back to the 'simple, undifferentiated unity of its origin' to the romantics whose art strives to attain 'a unity which is higher, because it incorporates the intervening differentiations'.[154] The major difference between the circle and the spiral is that the latter 'rotates along a third, vertical dimension, to close where it had begun, but on a higher plane of value. It thus fuses the idea of the circular return with the idea of linear progress.'[155] Moreover, whereas the circle is a closed shape which circumscribes a well-defined space, the spiral is open-ended and constantly shifts its position along both its vertical (symbolic) and horizontal (narrative) axes.

Another romantic example of this shift of pattern from the circle to the spiral is found in Emerson's essay 'Circles' (1841). Because, like James, Emerson grounds the circular form in the act of human perception ('the eye is the first circle'), he privileges a pattern arising from personal experience, rather than imposing an abstract system on the world. He sees the circularity to extend in all directions as a spatial pattern which defines his perceptual zone in a world which fades out of view on the peripheries. These momentary perceptions are constantly interconnected and fused to create a capacious sense of the world, but there is no sense of permanence: 'the life of man is a self-evolving circle, which, from a ring imperceptibly small, rushes on all sides outwards to new and larger circles, and that without end'.[156] Emerson claims that the 'force or truth of the individual soul' determines the emergence of a circuitous path, not as a form of will-to-power, but a recognition that certain aspects remain outside the 'mental horizon' of human knowledge.

Here, Emerson suggests that every limit is provisional and will fall through or give way to the next, although the experiencer is not always in a position to see over the horizon. He claims that those who are content with limits feebly reject the virtues of the 'experimenter' and the active and 'energising spirit'.[157] While his philosophy is optimistic and forward-looking, Emerson does not see a simple progression from one state to another as a cumulative process which tends towards 'rest, conservatism, appropriation, inertia'.[158] Movement derives instead from spiritual renewal which may as well go backwards to germination as onwards to the realization of goals. Far from the interconnected circles, or horizons, revolving around a stable axis they are liable to overlap or turn back on each other: 'we now and then detect in nature slight dislocations, which apprise us that this surface on which we now stand is not fixed, but sliding'.[159] These

---

[154] Ibid., pp. 183–4.
[155] Ibid., p. 184.
[156] Ralph Waldo Emerson, *Essays and Poems* (London: Dent, 1986), p. 147
[157] Ibid., pp. 153, 154
[158] Ibid., p. 155.
[159] Ibid., p. 152.

sliding surfaces suggest the circle is never completed on one plane and a return to the original act of perception cannot be accomplished: the circular formation is open and constantly shifts away from the individual's grasp. For Emerson and James, tracing a circle is a heuristic means of ascertaining location, before it is redrawn elsewhere.

Emerson concludes his essay by suggesting that the 'great man' possesses the 'power and courage to make a new road to new and better goals'.[160] Again, this is not accomplished by blind forward movement, but by nourishing the 'enthusiasm' and 'courage' to abandon former goals and resolve not to shy away from changing direction because of the risk involved.[161] The ability 'to draw a new circle' in Emerson's sense does not rest upon the desire to make nature a closed system, but the willingness to give up previously cherished beliefs. The application of volition tempered with a belief in a guiding force (for Emerson, internal impulses directed from without by a benevolent Over-Soul) is a precarious combination by which Nietzschean ruthless autonomy is rejected without surrendering to the blind forces of habit.

In a 1903 address delivered to commemorate the centenary of Emerson's birth, James lionizes Emerson for his unique 'blend' of morality and literary insights.[162] James shares with Emerson a tendency to resist 'consecutive' and systematic thinking in favour of 'gleams' and fragments, but he is critical of Emerson for his belief in 'absolute monism' and 'radical individualism'.[163] However, despite Emerson's idealistic tendencies, James admits he 'never drew a consequence from the Oneness that made him any the less willing to acknowledge the rank diversity of individual facts'.[164] Both Emerson and James root their philosophies in the 'concrete perceptions' of experience, but also turn to a spiritual 'more' beyond the realm of the senses.[165] For James, this does not mean a blind turning to the spiritual authority of orthodox religions, but a laying of emphasis on the primacy of individual experience. James ends his address with a line which echoes Emerson and Blake (who, together with Henry James Snr, were both one-time followers of the eighteenth-century mystic Emmanuel Swedenborg) as well as Thoreau's notion of civil disobedience: 'it follows from all this that there is something in even the lowliest of us that ought not to consent'.[166]

---

[160] Ibid., p. 156. The 'great man' may be interpreted either as one of Emerson's *Representative Men* (1850), as the distinguished public figures of Carlyle's work *On Heroes and Hero-Worship* (1841), or as one who seeks the 'strenuous life' of the saint as outlined in *Varieties*.

[161] Ibid.

[162] James, *Manuscript Essays and Notes*, ed. Frederick H. Burkhardt (Cambridge, MA: Harvard U.P., 1988), p. 316.

[163] Ibid.

[164] Ibid., p. 319.

[165] Ibid.

[166] Ibid.

For these reasons, Ralph Barton Perry suggests James wished to defend experience as the 'real backbone of the world's religious life' and promote the life of religion as 'mankind's most important function'.[167] Religious impulses are understood to derive from the individual's experience of the universe (both the sensual and supersensual world), rather than issuing from systematic theologies and institutionalized religions. For James, 'abstract definitions and systems of concatenated adjectives' should be viewed as 'secondary accretions upon these phenomena of vital conversation with the unseen divine'.[168] If doctrinal thought is rejected for its tendency to restrict the range of possible experiences, then so too should religious myths be rejected as universal templates to follow. Instead of outlining a totalizing narrative, James suggests that for the 'twice-born . . . the world is a double-storied mystery' and the morbid-minded find themselves immersed in 'a universe two stories deep'.[169]

This rejection of established religious narratives pushes James' position toward the romantic desire to renew social language in order to make it personal and authentic. Such a commitment demands that one does not follow ill-considered caprice; this does not imply 'vital conversation' but an instinct which compels the individual to take up a necessary pattern of response. The pattern should not be seen to be totally determined, either culturally or biologically, but neither should it be completely disbanded. Instead, James seems to be searching for a structural 'pattern which avoids the dogmatism of creed as well as the disturbing chaos of a wholly disordered universe:

> When one views the world with no definite theological bias one way or the other, one sees that order and disorder, as we now recognize them, are purely human inventions. We are interested in certain types of arrangement, useful, aesthetic, or moral, – so interested that whenever we find them realized, the fact emphatically rivets our attention. The result is that we work over the contents of the world selectively. It is overflowing with disorderly arrangements from our point of view, but order is the only thing we care for and look at, and by choosing, one can always find some sort of orderly arrangement in the midst of any chaos . . . We count and name whatever lies upon the special lines we trace, whilst the other things and the untraced lines are neither named nor counted.[170]

Here James illustrates the archetypal experience of modernity: the moral and aesthetic compulsion to trace new patterns in the loose sand of what was once the solid stone of traditional religious paths. He recognizes that concepts of 'order' and 'disorder' are actually imposed experientially by the observer, but they are

---

[167]   Perry, *The Thought and Character of William James*, vol. 2, op. cit., p. 327
[168]   Ibid., pp. 446–7.
[169]   James, *Varieties*, op. cit., pp. 166, 187.
[170]   Ibid., p. 438.

necessary if the individual is to arrive at ends-directed plans to follow. While recognition and attention are part of his vocabulary of cognition, their guiding force consists of a mixture of a 'vital conversation' with the unseen world, shaped by an act of will both to foresee goals and to devise plans to obtain those ends. Far from being able to master a fixed repertoire of decisive movements, the demands upon the individual force him or her constantly to revise and reconsider the paths started out upon. The 'morbid-minded' traveller is plagued not only with a dense and contradictory universe, but with the added difficulty that all movement must be tentative and carefully considered, without falling into the other modern trap of endless prevarication. In the tradition of Protestant humility, the traveller must be constantly wary of the vice of pride, but develop the strength to seek the 'strenuous life' which James proposes.

The chapter on 'Mysticism' provides many of the clues which lead towards unravelling James' notion of path-finding by recalling his work on habit. James wishes to rescue mysticism from its general usage as a word of 'mere reproach' by grounding the ineffability and transience of religious experience in a concrete pattern which bridges the inchoate source of inspiration with the demands of the material world.[171] The sense of 'deeper significance' which characterizes such experiences opens the subject to 'vague vistas of a life continuous with our own, beckoning and inviting, yet ever eluding our pursuit'[172] and are often accompanied by 'the feeling of an enlargement of perception':[173]

> We pass into mystical states from out of ordinary consciousness as from a less to a more, as from a smallness into a vastness, and at the same time as from an unrest to a rest. We feel them as reconciling, unifying states. They appeal to the yes-function more than to the no-function.[174]

While such an experience seems to move upwards and outwards, as with Emerson, the perception 'never completes itself': there is no final point of stasis, but only the promise of affirmatory rest. James characterizes mystical states as signifying 'the supremacy of the ideal, of vastness, of union, of safety, and of rest', they open up a 'more extensive and inclusive world' which cannot be known and bounded by systematic thought.[175] While some may choose, or be

---

[171]   Ibid., p. 379.

[172]   Ibid., p. 383.

[173]   Ibid., p. 384. James quotes the phrase 'dreamy states' from an article by Sir James Crichton-Brown published in *The Lancet* (6 and 13 July 1895), but he takes issue with Crichton-Brown's dualistic position which reduces personal significance to medical formulae. James claims that his 'path pursues the upward ladder chiefly' (a romantic movement upwards and outwards) in an attempt to find significance in all 'of a phenomenon's connections'; James, *Varieties*, op. cit., p. 384.

[174]   Ibid., p. 416.

[175]   Ibid., p. 429. This structure closely echoes the central aspects of Buddhism: at every limit there is a higher stage, or higher state of contemplation (*dhyana*), the pursuit

compelled to follow, a fixed or repetitive pattern, only by loosening these strictures can the idea of a healthy will be redescribed. Similarly, only the morbid-minded traveller who seeks a fuller conception of the universe by pushing back the boundaries of the old, can forego the material comforts of a fixed home in the pursuit of this final, but always elusive, homecoming.

However, this reading begs several questions. If one cannot invest importance in the old and outworn religious paths because they only lead to the strictures of earthly creeds, then in what direction do transient mystical experiences lead? In other words, where can conceptual redescription take the individual in pragmatic terms? James' answer is complicated, but he suggests the marks of mystical experience bring about both a new orientation to life and an inner conviction that the individual should eschew the worldly goals of comfort and security, in order to pursue an Emersonian path which is constantly merging with higher routes too difficult to perceive directly. This path is not wholly other-worldly: the actions of those who follow the saintly path can have direct effects on the worldly environment around them. As 'vivifiers and animators of potentialities', saints have the strength of will to transfer their vision of the ideal into the realm of the real and thereby 'energize' those around them to produce 'practical fruits'.[176] Whilst the initial orientation is melancholic, from this emerges a 'denial of the finite self' and a spiritual need to identify with an 'always enlarging Self'.[177] This idea may seem to lead back to the undifferentiated unity of the neoplatonists and forward to the universal consciousness of Carl Jung, but James (like other romantic scientists who follow him) is distinct in insisting that the journey to this state is always under way and incomplete. As a result, the narrative of the journey can never be finally told and must be continually revised *en route*.

James concludes his lectures by commenting on the irreducibility of personal experience: 'I turn back and close the circle which I opened in my first lecture.'[178] Rather than actually closing the circle, he goes on to add that 'I might easily, if time allowed, multiply both my documents and my discriminations.'[179] This suggests that the circle is perpetually open and its dimensions are always provisional and, therefore, revisable. In a footnote to his concluding lecture, he discusses the possibilities of romantic science in bridging the chasm between 'scientist facts and religious facts'.[180] This should be done, he says, by giving up the notion of 'final human opinion' and by reverting 'to the more personal style,

---

of which takes the experiencer past the semblance of rest. As discussed in Chapter 5, this is analogous to Oliver Sacks' conception of health, which contains within it the promise of homecoming as a spur to encourage the patient to move in a direction which liberates rather than limits and restricts.

[176] Ibid., pp. 358, 259.
[177] Ibid., p. 418.
[178] Ibid., p. 484.
[179] Ibid.
[180] Ibid., p. 501.

just as any path of progress may follow a spiral rather than a straight line'.[181] Not only is the circle forever open, but the figure of the spiral gestures back to the patterns Abrams detects throughout European romantic writing. The spiral prohibits either a simple return to origins or a straightforward progressive movement; it may seem to move upwards and outwards away from a material world, but, as it circles round, it cannot leave the body behind for more than a fleeting moment of transcendence, suggesting an unbreakable link between the idealistic aspirations of individuals and an intractable organic embeddedness in the world.

Throughout *Varieties* James hints at a therapeutic path upon which the convert can move away from the total paralysis dramatized in the hunched and lifeless figure bereft of will and self-definition. When the correspondent retches forth the apocalyptic words *'that shape am I, I felt, potentially'* the reader is made aware both of an identification with the figure and a dissociation from the state of utter wretchedness. *Varieties* can be read as James' attempt to find an alternative shape, or pattern, which can lead away from determinism without leaving behind the organic body for abstract metaphysical musings. For Emerson, the 'act' of perception is central to the individual's sense of his or her place in the world and, without that initial act, a narrative of self-overcoming cannot unfold. Similarly, for James, without attention to and acknowledgement of narrative possibilities the individual cannot hope to locate him/herself in an open-ended universe.

Releasing Hidden Energies

In *Varieties* James sketches out an emergent spatial pattern which projects self-knowledge gained through personal experience along a temporal plane, but he cautions against abandoning the moment of insight for linear progression away from it. Like Emerson, James wishes to abort the doctrinaire aspects of Puritanism and derive a new spirituality from a transcendent moment of personal experience which has no 'reference to external authority'.[182] However, James parts company with Emerson's romanticism in his preoccupation with the corporeal body which encroaches on the transcendent 'bliss of spiritual exchange' and 'apprehension of the divine'.[183] This section proposes that James' conception of the spiritual path is intrinsically therapeutic in its capacity to motivate the body into action, as well as energizing the mind. However, his insistence on the corporeal body permits only glimpses of a spiritual realm by which material limitations can be transcended. Bodily energy is seen as a motivating force which he derives from discussions of physiology and

---

[181]   Ibid., p. 501.

[182]   Irving Howe, *The American Newness: Culture and Politics in the Age of Emerson* (Cambridge, MA: Harvard U.P., 1986), p. 9.

[183]   Ibid., p. 11.

consciousness in *Principles*, links to his discussion of the will and is granted a central place in his understanding of religious experience.

The conceptual differences between Emerson and James help to clarify the different moods of the two thinkers and push James closer to the European romantics. If, as Nietzsche claims, Emerson is an 'enlightened', 'contented' and 'cheerful' thinker, then James shares with Kierkegaard and Schopenhauer a darker and more troubled view of the self's relation to the universe.[184] In his two lectures on 'The Sick-Soul' in *Varieties*, James starkly characterizes the difference between the two tempers: 'the sanguine and healthy-minded live habitually on the sunny side of their misery-line, the depressed and melancholy live beyond it, in darkness and apprehension'.[185] James reverses the Enlightenment optical dichotomy of light-dark, to suggest a mysterious and fine-grained universe which cannot be clearly illuminated, while his image of the 'misery-line' stresses the primacy of an emotional engagement over rational cognition. For James, the 'ennobling sadness' of the European writers coincides with his 'solemn, serious, and tender' attitude to religion.[186] But, he admits that it 'is almost as often only peevishness running away with the bit between its teeth'; the 'sallies' of Schopenhauer and Nietzsche 'remind one, half the time, of the sick shriekings of two dying rats'.[187] James resists such primal extremes by refining an in-between, or purgatorial, tone which imbues his writing with the solemnity of 'religious sadness'.[188]

Although Germanic romantic gloom asserts a strong influence on his writing, it is tempered by the influence of Emersonian optimism which, rather than descending into the abyss of nihilism, results in motivating movement and activity. This is not the 'sky-blue' optimism of the New England mind-curists, but an inherent piety which can protect 'all ideal interests and keep the world's balance straight'.[189] James' notion of a pluralistic universe is more heterogeneous and complex than Emerson's 'divine soul of order', but both thinkers share the motivation to create anew. Thus, *Varieties* can be understood as James' contribution to the American romantic project which seeks to relocate spiritual life outside the support of religious institutions and attempts to redefine spirituality within the terms of modernity.

The 'new life' which emerges out of the French correspondent's abject vision of the epileptic patient represents a departure from, and a supplement to, the old narrative pattern. James calls such moments of insight 'temporary "melting moods", into which either the trials of real life, or the theatre, or a novel

---

[184] Nietzsche, *Twilight of the Idols/The Anti-Christ*, trans. R.J. Hollingdale (London: Penguin, 1990), p. 85.

[185] James, *Varieties*, op. cit., p. 135.

[186] Ibid.

[187] Ibid., p. 38.

[188] Ibid.

[189] Ibid., p. 33.

sometimes throw us'.[190] At these moments, the self stands at the threshold between two possible patterns, with the outcome resulting from individual choice. By choosing the easy option, whether it is the comfort of simple ecclesiastical piety or a backsliding into drunkenness and vice, the liberation of transformation is eschewed for the safe and the habitual. Conversely, by choosing the more complex pattern, the experience contributes towards the pursuit of the strenuous life as ego-boundaries dissolve and reform in a different configuration.

James does not reserve this rejection of openness and plenitude for those who safeguard themselves against religious conversion, for the 'narrow' excesses of sainthood (characterized by absorption, withdrawal and fanaticism) run tangential to his embracing vision of mysticism. He argues that only the precarious blend of energetic action and surrender to the subliminal self will propel the individual along the 'extensive' and 'inclusive' path: these two forces of action and yielding converge by affirming 'the new centre of personal energy' and allowing it to 'burst forth'.[191] This active and creative element represents a will to believe in an inner voice, to break through the constraining force of inhibitions and conventions and to cultivate the ability to say 'Yes! yes!'.[192] Like Carlyle's Teufelsdrökh in *Sartor Resartus*, James suggests this affirmation passes 'from the everlasting No to the everlasting Yes through a "Centre of Indifference"'.[193] While James detects links between his and Carlyle's affirming voices, the temper of the American's writing positions him nearer to the Dionysian exuberance of the 'Yes-saying' Nietzsche than the strong conservative politics which inform Carlyle's thought. But, while James' notion of will to believe overcomes the passivity of mystical experience, he does not share Nietzsche's ruthless drive to overcome the self at the expense of human fellowship. His sharp criticism of those activities which tend towards self-absorption and withdrawal from the world lend his writing a note of altruism and compassion lacking in Nietzsche.

For James the cries of 'Yes! yes!' accompany the release of hidden energies, but energies guided by an almost pantheistic feeling. James presents a negative example to characterize his unfolding moral framework. He argues that the impulse of the alcoholic which encourages him to imbibe more alcohol may seem to affirm the self, but this would only be a surrender to habit: a yielding without action and a refusal to pursue an ideal. Only by affirming the nuclear self (in its shifting relation to the subliminal self) and giving up the restraints of fixed identity can the false voices be overcome: 'the psychological basis of the twice-born character [is thought] to be a certain discordancy or heterogeneity in the native temperament of the subject, an incompletely unified moral and intellectual

---

[190] Ibid., p. 267.
[191] Ibid., p. 210.
[192] Ibid., p. 261.
[193] Ibid., p. 212.

constitution'.[194] Inspiration derives from a sudden and often violent revelation or manifestation of the subliminal self: 'the higher condition, having reached the due degree of energy, bursts through all barriers and sweeps in like a sudden flood'.[195] Such a revelation may force the individual to re-evaluate and redirect his or her former life towards new goals.

James expands upon his theories of latent motivating forces in his 1906 Presidential Address 'The Energies of Men'. Here he detects a parallel between Carlyle's 'Centre of Indifference' and the colloquial Adironrackian term '*oold*': a feeling of 'intellectual or muscular' staleness or fatigue.[196] This notion is analogous to the state of '*adhedonia*' or spiritual apathy discussed in *Varieties*: a 'mere passive joylessness and dreariness, discouragement, dejection, lack of taste and zest and spring'.[197] Such spiritual vacuity represents a state of absolute depletion and inertia, symptoms which were thought to comprise neurasthenia. Popularized by the New York neurologist George Beard in the 1880s (and recommended to James' younger sister Alice) the fashionable cure for the nervous exhaustion brought on by modern life was considered to be rest and the conservation of energy. James' argument that individuals should not surrender to general feelings of fatigue directly counters Beard's rest-cure:

> [fatigue] gets worse up to a certain critical point, when gradually or suddenly it passes away, and we are fresher than before. We have evidently tapped a level of new energy, masked until then by the fatigue-obstacle usually obeyed. There may be layer after layer of this experience.[198]

By tapping into the hidden resources of the subliminal self the individual is able to draw upon latent motivating forces which can energize the body into renewed activity. This is a very different kind of abandonment from the rest-cure. In a distinction suggested by the political theorist William Corlett, this would represent a loss of self-control or a 'giving in to' impulses which exist beneath the level of conscious control, without 'giving up' to the passivity of Beard's recommendation.[199] This distinction does not necessarily imply a model of the unconscious, but of an inner bodily energy which dwells latently in the deeper

---

[194] Ibid., p. 167. James' phrase 'native temperament' seems to work in tension with his argument that former paths can be diverted or one can alter the ideals which one pursues. However, rather than personality being entirely determined or genetically prescribed, James' view of 'temperament organically weighted' does not preclude the affect of circumstance and experience on disposition; ibid., p. 135.

[195] Ibid., p. 216.

[196] James, *The Moral Equivalent of War and Other Essays*, ed. John K. Roth (New York: Harper & Row, 1971), p. 34.

[197] James, *Varieties*, op. cit., p. 145.

[198] James, *The Moral Equivalent of War and Other Essays*, op. cit., p. 34.

[199] William Corlett, *Community Without Unity* (Durham, NC: Duke U.P. , 1993), p. 3.

recesses of the self. By making contact with the subliminal self the release of spiritual inspiration has the capacity to galvanize the body into a more strenuous life. Echoing Nietzsche, by tapping into a Dionysian energy source James envisages a self liberated from the rigid structures of egoism.

Once again, James' ideas touch upon Buddhistic thought (as well as Stanley Cavell's version of moral perfectionism discussed at length in the next chapter) which encourages the overcoming of mental barriers in the pursuit of spiritual enlightenment. Moreover, the spiritual pursuit of Buddhism, like James' strenuous life, is never completed: the pursuit is always under way with the ultimate achievement (Nirvana or sainthood) always deferred. Again, despite the similarities, one crucial difference between the two positions is that, whereas Buddhists internalize energy and deny the body in the search of higher levels of consciousness, James suggests the release of hidden energies should sustain our 'inner as well as our outer work'.[200] He develops his ideas of hidden energies towards a recommendation for living at maximum energy and performing to the optimum, when he writes of 'excitements, ideas, and efforts' which can 'carry us over the dam' of apathy or of 'chronic invalidism'.[201] James argues that the source of this energy (biochemical, psychic and spiritual) must stem from the individual's will to confront inertia. However, the problem which remains unanswered in 'The Energies of Man' and *Varieties* is in what kind of discourse is it possible describe these energizing forces?

There are at least three available discourses in which James can describe these hidden energies. The first is the hydraulic system outlined by Spencer in his essay on 'The Physiology of Laughter' (1860) and Freud's abandoned 'Project for a Scientific Psychology' (1895). Both thinkers describe the build-up of nervous energy until it reaches a critical point at which it seeks release through a bodily outlet. Freud outlines his model of the mind 'in terms of increase, diminution, displacement and discharge of energy or "quantity" conceived as flowing through and accumulating within a differentiated network of neurones'.[202] Although Freud later moved away from a natural-scientific standpoint and developed a more explicitly metaphorical understanding of psychic apparatus, the language of discharge continues to inform works such as *Jokes and Their Relation to the Unconscious* (1905) in the shape of laughter conceived as a cathartic channel of psychic relief.

The second option available to James links directly with the mood and experimentation of early nineteenth-century German romantic science. In his 'Condition of England' essay, 'Signs of the Times' (1829), Carlyle makes an impassioned plea for the emergence of a dynamic science which would counter

---

[200]   Ibid., p. 37.

[201]   Ibid., p. 38.

[202]   Peter Dews, *Logics of Disintegration: Post-Structuralist Thought and the Claims of Critical Theory* (London: Verso, 1987), p. 45.

what he deems to be the dehumanizing philosophy of utilitarianism and rationalism. Carlyle proposes to revitalize science based on rigid Newtonian laws: a dynamic science 'which treats of, and practically addresses, the primary, unmodified forces and energies of man, the mysterious springs of Love, and Fear, and Wonder, of Enthusiasm, Poetry, Religion, all which have a truly vital and infinite character'.[203] If Carlyle's science could be developed it would enable James to speak of ineffable phenomena in terms of emotions, without reducing them to the level of undifferentiated psychic energy.

The third option is a discourse of relational structure in which energies are described as part of a transitive system, homologous to the perpetually moving stream of consciousness outlined in 'The Stream of Thought'. For James, energy is essentially biochemical, but he also speaks figuratively of poetic, spiritual or religious energy as the particular end toward which biochemical energy is channelled, without reverting back to faculty-based philosophy. By exercising the will it is possible to redirect energy along one particular channel, which represents both a rupturing of the old pattern and the facilitation of new pathways, as either a willed or an accidental deviation from the rigid pattern of habit.[204] However, in order to speak in terms of religious experience James must first articulate a conception of energy, as a real, but latent, force (like the will, energy can only be detected empirically in its kinetic form) and as a metaphor for action.[205] When he writes of 'the threshold of a man's consciousness in general, to indicate the amount of noise, pressure, or other outer stimulus which it takes to arouse his attention at all', he implies that enough excitation will cause one mental state to pass into another.[206] Mystical experience often occurs at such liminal moments, but rather than merely representing the epiphenomena of bodily excitation, James claims these revelations offer visions of alternative paths to follow.

Whereas Carl Carus links peculiar mental states with a visionary revelation of a double self, Freud understands hallucinations to be merely a manifestation of psychopathology.[207] However, James, like Carus (and later Binswanger), writes in

---

[203] Thomas Carlyle, *Selected Writings*, ed. Alan Shelston (London: Penguin, 1987), p. 72

[204] This parallels the post-structuralist reading of Freud's theory of 'breaching' or 'path-breaking' described in 'A Note Upon the "Mystic Writing Pad"' (1924). Derrida interprets 'path-breaking' as 'a metaphorical model and not as a neurological description': the 'breaching, the tracing of a trail, opens up a conducting path' which rechannels the old path towards a new end; Jacques Derrida, *Writing and Difference*, trans. Alan Bass (London: Athlone, 1978), p. 200.

[205] Tim Armstrong argues that James' interest in energy flows conforms to the turn-of-the-century preoccupation with electrification in America; Armstrong, *Modernism, Technology and the Body*, op. cit., pp. 13–41.

[206] James, *Varieties*, op. cit., pp. 135, 134.

[207] See, for example, 'Dostoevsky and Parricide' (1927) in Freud, *Art and Literature*, ed. Albert Dickson, trans. James Strachey (London: Penguin, 1990), pp. 441–60.

romantic terms about the exceptional and extraordinary nature of mystical experiences, rather than attempting to characterize these experiences in terms of pathology. For James, the threshold between one mental state and another, where the new state leaves a permanent mark on the psyche, would be the moment of rupture of an old pattern and the emergence of a different configuration. This 'giving in to' the revelations as they occur is not a passive process; the search for a strenuous life sets the tone for the energetic excess of these experiences. Indeed, the choice to heed these insights takes a resolute strength of will, because invariably they point toward the more arduous of the two directions. Rather than stimulating a backsliding to an old pattern, the individual should realize that self-perception is always temporary and incomplete and the narrative of the self always under way. Such a recognition may initiate a striving for a fuller appreciation of the human condition, which, for James, should link an awareness of the subliminal self with the urge to overcome 'the habit of inferiority to our full self'.[208]

James raises two important questions which emerge directly from this desire to be other than, or more than, the existing self. Firstly, he asks 'to what do the better men owe their escape? and . . . to what are the improvements due, when they occur?'[209] In answer he suggests that 'either some unusual stimulus fills them with emotional excitement, or some unusual idea of necessity induces them to make an extra effort of will. Excitements, ideas, and efforts, in a word, are what carry us over the dam.'[210] Whether it is 'stimulus' or 'idea', only by cultivating a sensitivity to the qualities of the 'unusual' and the peculiar is it possible to shake off 'an inveterate habit' and the *adhedonia* of daily life. According to James, daily life tends to impress upon individuals a routine pattern which encourages 'habit-neurosis'.[211] One can make an exerted effort to avoid such repetitions by attending to the ordinary and the habitual, but the abrupt and unpremeditated nature of the revelations indicate 'a deeper kind of conscious being than could [be enjoyed] before'.[212]

Secondly, James suggests that conversions can be empowering and can 'unify us', even if only temporarily, resulting in 'freedom, and often a great enlargement of power'.[213] By following the arduous path, the individual may encounter a degree of freedom unknown from the perspective of daily routine. Moreover, the tracing of a visionary path may lead to an understanding which enables the

---

[208]   James, *The Moral Equivalent of War and Other Essays*, op. cit., p. 38.

[209]   For James these 'better men' are those 'twice-born' religious believers who have written of their conversion experiences – Ignatius Loyola, St Theresa, St John of the Cross, Swedenborg, Tolstoy, Bunyan and Emerson – a tradition of religious biography developed later in the twentieth century in Erikson's psychobiographical work; ibid., p. 38.

[210]   Ibid.

[211]   Ibid., p. 39.

[212]   James, *Varieties*, op. cit., p. 157.

[213]   James, *The Moral Equivalent of War and Other Essays*, op. cit., p. 47.

individual to confront, or to escape from, imprisoning or habitual conditions, even if it is at the risk of socially defined integrity: 'whatever it is, it may be a high-water mark of energy, in which "noes," once impossible, are easy, and in which a new range of "yeses" gains the right of way'.[214] This notion of rebellion against social conformity and clinical categories of illness seems to indicate a higher realm of morality on an existential plane which eschews class-bound morality, 'intellectual respectability' and 'decorum' for their tendency to restrict the ends to which humans can direct energy: 'locking up the rest of his organism and leaving it unused'.[215] The metaphors of 'locking' and 'unlocking' the self suggest an activity more refined than the simple unblocking and release of energy and is procured only through constructing an experiential narrative which incorporates vivid and intense perceptual moments.

For James, a life of moral virtue is not necessarily a stoical life. Anything which excites the 'Yes function in man' and which 'makes him for the moment one with truth' should be encouraged.[216] Anticipating Aldous Huxley's experimentation with mescalin, James suggests that to stimulate mystical experiences one should make use of alcohol, nitrous oxide and the search for the extraordinary in order to unlock this 'higher' life.[217] Where James and Nietzsche part company however is in James' stress upon ethical boundaries which the pursuit of intense experiences should not break, the individual being directed by a strong sense of morality and human compassion.

James' writing on sublime experiences proves richer for a consideration of romantic science than the early romantics and the mind-curists, primarily because of his emphasis on the body. He begins both *Principles* and *Varieties* with reference to physiology and neurology: 'there is not a single one of our states of mind, high or low, healthy or morbid, that has not some organic process as its condition'.[218] Although he wishes to shift his consideration to the subliminal processes of mind, he constantly returns to the body as the material vehicle for mental experiences: 'every one of them without exception flows from the state of their possessor's body'.[219] While one may transcend the corporeal to maintain an unmediated union with God (or a similar transcendental signifier), James is aware of the limits and boundaries which infringe upon and constrain these

---

[214] Ibid., p. 47.

[215] Ibid.

[216] James, *Varieties*, op. cit., p. 387.

[217] In *The Doors of Perception* (1954) and *Heaven and Hell* (1956) Huxley reflects upon his attempts to extend the mystical tradition with 'controlled' use of hallucinogenic drugs. In an appendix to the 1990 edition of *Awakenings*, Sacks pays tribute to three early medical experimenters, Freud, James and Havelock Ellis, who toyed with the 'notion of a drug which will banish sadness and fatigue, increase energy, expand consciousness, imbue or reimbue the world with wonder'; Oliver Sacks, *Awakenings* (London: Picador, 1991), p. 323.

[218] James, *Varieties*, op. cit., p. 14.

[219] Ibid.

metaphysical flights of the mind. As such, he criticizes the limited healthy-minded perspective which, in the search of a sanitized 'perfect health', tends to 'divert our attention from disease and death' to create a hygienic 'poetic fiction far handsomer and cleaner and better than the world really is'.[220] In a 1897 tract entitled *In Tune With the Infinite*, written by one of New England's chief mind-curists, Ralph Waldo Trine, this would be achieved by a simple 'exchange [of] dis-ease for ease, inharmony for harmony, suffering and pain for abounding health and strength'.[221] For James, such a denial of evil or strife in the universe overlooks the complex nature of man's finite existence in the world.

James quotes another mind-curist, Henry Wood who suggests in *Ideal Suggestion Through Mental Photography* (1899):

> If we will we can turn our backs upon the lower and sensuous plane, and lift ourselves into the realm of the spiritual and the Real, and there gain a residence . . . the spiritual hearing becomes delicately sensitive, so that the 'still, small voice' is audible, the tumultuous waves of external sense are hushed and there is a great calm.[219]

These words echo James' recommendation of exercising the will, together with his attention to qualitative detail (the 'still, small voice') amongst the noise of sensual tumult. However, Wood's rhetoric suggests a strain of divine healing which encourages a shedding of corporeality for a purely spiritual realm. Thus, James states 'in mind-cure circles the fundamental article of faith is that disease should never be accepted. It is wholly of the pit.'[223] While he would encourage the pursuit of goals to lift humans out of repetitive and habitual patterns, he realizes that only by accepting the 'dirt' of the 'pit' can one begin to understand mortality. James evokes the supernatural elements of *Macbeth* as well as the universal morality dramatized in medieval *Everyman* plays when he tropes on a fine phrase: 'still the evil background is really there to be thought of, and the skull will grin in at the banquet'.[224] It is as if the body acts as a corporeal regulator for mental life which anchors spiritual flights to the material world. Indeed, James claims that biochemical energy provides the initial stimulant of mystical experiences. Thus, spiritual flight is inscribed in the space of the body. James does not suggest that all mystical experiences occur in a state of bodily excitation, but only that the body represents the ultimate site of the individual's experience. Conversely, although psychic experience is seen as a function of the brain, it is irreducible to the bodily states which determine it. James does not recommend an escape from the embodied self, because physical energy is a source of life and

---

220   Ibid., pp. 183, 90.
221   Ibid., p. 101.
222   Ibid., p. 117.
223   Ibid., p. 113.
224   Ibid., p. 140.

vitality. Instead, he encourages a conceptual shift away from stasis to an open and incomplete universe and a pragmatic movement towards a narrative of spiritual plenitude.

Chapter 2

Otto Rank: The Creative Romantic

Had William James published *The Principles of Psychology* a decade later than 1890 he would certainly have been influenced by European psychoanalysis, which as a coherent body of thought did not reach America until the early years of the twentieth century.[1] However, by James' death in 1909 schisms had already started to emerge in the ranks of European analysts, stimulated mainly by Freud's determination to keep control over the direction of the psychoanalytic 'cause' after the retirement of his one-time mentor and collaborator on *Studies in Hysteria* (1895), Josef Breuer.[2] The American critic Lionel Trilling claims in his introduction to Ernest Jones' standard biography of Freud that, apart from Breuer, none of the other European analysts 'contributed anything essential to the theory of psycho-analysis'; thus, for Trilling, 'the basic history of psycho-analysis is the account of how it grew in Freud's own mind, for Freud developed its concepts all by himself'.[3] This received opinion has been challenged consistently since the 1960s by intellectual historians, post-structuralist theorists and second-wave feminists, who have questioned the legitimacy of Freudian vocabulary and the institutional status of psychoanalysis. However, the early fractures in the psychoanalytic circle are worthy of consideration because they prefigure some of the later criticisms of psychoanalysis and open the door for discussing the forms romantic science underwent in the early twentieth century.

The Austrian psychiatrist Alfred Adler and Swiss analyst Carl Jung usually receive most attention as rivals and critics of Freud, but Otto Rank (1884–1939) and, in the next chapter, Ludwig Binswanger provide alternative perspectives on the early developments of the movement. Born into an Austrian-Jewish family in Vienna in 1884 and named Otto Rosenfeld at birth, Rank was determined to mould an identity for himself outside the traditional expectations of nationality and ethnicity. Intensely interested in the artist-figure from a young age, he explored the world of legend, myth and imagination more extensively than any other psychoanalyst. He first met Freud in 1905 and became the Secretary of the

---

[1]   For the reception of psychoanalysis in America see Nathan Hale's two-study *Freud and the Americans* (New York: Oxford U.P., 1971 and 1995) and for a detailed account of Freud's only visit to America see Saul Rosenzweig, *The Historic Expedition to America (1909): Freud, Jung and Hall the King-maker* (St Louis: Rana House, 1992).

[2]   In 1909 Freud claimed 'I do everything only for the cause, which, again, is basically my own'; E. Brabant et al. (eds), *The Correspondence of Sigmund Freud and Sandor Ferenczi. Volume 1, 1908–1914* (Cambridge: Harvard U.P., 1993), p. 33.

[3]   Lionel Trilling, 'Introduction' to Ernest Jones, *The Life and Work of Sigmund Freud*, eds Lionel Trilling and Steven Marcus (Harmondsworth: Penguin, 1964), p. 12.

Vienna Psychoanalytic Society in 1906, but had already begun work on his first two books, published as *The Artist: Toward a Sexual Psychology* (1907) and *The Myth of the Birth of the Hero* (1909), before his first meeting with Freud.[4] Freud admired Rank's work and, after the defection from the International Psychoanalytic Association (IPA) of Adler in 1911 and Jung in 1913 he became Freud's favourite for carrying the beacon of psychoanalysis. However, with the publication of his key work *The Trauma of Birth* (1923) and, in collaboration with Sandor Ferenczi, *The Development of Psychoanalysis* (1924), Rank broke from Freud's emphasis on the father as primary influence on the child's psychic and sexual identity by concentrating on the infant's primal relationship with the mother. Soon afterwards he left the psychoanalytic circle and began to develop an alternative therapeutic practice, moving to Paris and New York and becoming a major influence on the creative development of the internationalist writers Henry Miller and Anaïs Nin.

Rank's reputation suffered after the publication of Jones' biography of Freud; he is described as having 'unmistakable neurotic tendencies', which emerged during the war and 'wrecked' him as a promising analyst.[5] Although Freud took Rank's ideas more seriously than Jones or Karl Abraham (the secretary of the IPA at that time), he later claimed 'I am getting further and further away from the birth trauma' as the tensions between members of the Association intensified.[6] However, both Freud's ambivalent feelings to, and Jones' bitter rivalry with, Rank do not disguise a thinker of real merit whose fusion of natural-scientific understanding and aesthetic creativity moved romantic science and James' insights in a fresh direction. Next to Schopenhauer, Nietzsche and Ibsen (after whose character in *The Doll's House* (1879) Rank named himself), James was one of the most profound, but also least acknowledged, influences on Rank's development of will therapy as a reaction 'to a completely atomistic, mechanistic, and deterministic view of personality'.[7] This chapter focuses on aspects of Rank's thought which differentiate his work from Freudian orthodoxy and introduces a set of criticisms explored further in relation to Binswanger's existential analysis.

---

[4]   Even though Rank later admits that his 'terminology was at that time essentially Freudian' his adolescent preoccupation with the artist-figure pervades his early published work; Rank, *Art and Artist*, trans. Charles Francis Atkinson (New York: Norton, 1989), p. xxi.

[5]   Jones, *The Life and Work of Sigmund Freud*, op. cit., p. 420.

[6]   Ibid., p. 531.

[7]   Fay B. Karpf, *The Psychology and Psychotherapy of Otto Rank* (New York: Philosophical Library, 1953), p. 55. One of Rank's lecturers at the University of Vienna was a dedicated Jamesian and both Rank and James place a strong emphasis on the therapeutic possibilities of the will; E. James Lieberman, *Acts of Will: The Life and Work of Otto Rank* (New York: The Free Press, 1985), p. xxxviii.

'A Human Soul under the Microscope'

At the turn of the century Rank was deeply enmeshed in a period of turbulent adolescence, marked by withdrawal from family life, vivid nightmares and profound depression. He documented these experiences in a series of daybooks (*Tagebücher*), in which he questioned his self-identity and charted a series of suicidal fantasies up to the age of twenty. The daybooks portend Rank's interest in romantic science by recording, in his words from 1903, 'an image of a quivering human soul under the microscope'.[8] Born into a poor Jewish family in Vienna as the youngest child, Rank felt uncomfortable about his parents' expectations for him to become an apprentice mechanic. He read widely in his spare time and, although he had no single conversion experience like James, his despair inspired him to write romantic poems, dramatic acts and novel fragments, which complemented his growing passion for theatre and opera.[9] Despite displaying feelings of self-loathing, his diary reveals a thinker with belief in his own heightened consciousness which distanced him from the mercantile and practical types he observed in cosmopolitan Vienna. One entry in his daybook that casts light on his conception of the artist as a generic or collective type also expresses his psychological condition and aesthetic aspirations: 'I am an artist, even if I fail to bring forth a single work of art.'[10] Although Vienna had become one of the great artistic centres by 1900, Rank detected a slavish pursuit of material success among the Viennese bourgeoisie which detracted from the committed life of the artist. Nor did he have much respect for conventional religion; he reserved his energies for the Nietzschean cultivation and celebration of the inner self unshackled from social and religious expectations. His daybook allowed him to release the energies of his 'heart's blood' in written form and to analyse his own psychic development: 'self-observation is a prime essential and to that end I am making these notes. I am attempting to fix passing moods, impressions, and feelings, to preserve stripped off layers that I have outgrown.'[11]

In a 1903 entry in his daybook Rank outlines an eight-rung ladder of human development which is remarkably similar to James' moral hierarchy of social types. In *Principles*, James describes an ascending scale, moving from drunkards and tramps, through bachelor, father, patriot and upwards to the lofty heights of

---

[8]   Jessie Taft, *Otto Rank* (New York: Julian Press, 1958), p. 28.

[9]   Two of Rank's early poems translated by Miriam Waddington as 'Weltscmerz: Lines Before Breakfast' (1903) and 'School for Preparing' (1904) have been published in *Journal of the Otto Rank Association*, 1(1) (Fall 1966), 26–9. He also began work on a four act 'anti-civilization' play in 1903, 'Götzen: Vier Acte aus dem Schauspiel des Lebens' ('Idols: Four Acts Taken From the Spectacle of Life'), of which only the first act is extant in the Rare Books and Manuscript Collection, Columbia University, New York.

[10]   Lieberman, *Acts of Will*, op. cit., p. 18.

[11]   Quoted in an epigraph to a special issue on the novel, *Journal of the Otto Rank Association*, 5(1) (June 1970); Taft, *Otto Rank*, op. cit., p. 4.

philosopher and saint. At the base of his ladder Rank places religion, the bourgeois 'respect for art . . . which soon cloys and disgusts' and worship of women.[12] Further up the ladder, Rank positions inert 'practical people' and the 'grumblers, sick and morose' and then ascends through the ranks of scholars, sceptics and philosophers, with the latter representing the ideal of human achievement.[13] The philosopher (a figure he later conflates with the poet) is seen as the ideal artistic type, but he admits that 'seldom does one climb so high. That air is too rare and thin the thoughts also. His powers fail before then.'[14] Like Nietzsche and the architect Solness in Ibsen's play *The Master Builder* (1892), those who attempt to climb to this last rung usually turn their back on bourgeois art, but are always in danger of falling into psychosis or overreaching their human capabilities.

As with James, such a developmental model of human creativity positions the individual as being firmly in charge of his or her destiny, rather than being controlled by a set of deterministic responses. However, Rank was not such a liberal as to deny the powerful influence of unconscious forces on psychic life: he recommends that psychic energies should be channelled by an artistic vision of a qualitatively different life. This is evident in a quasi-apocalyptic (and almost Emersonian) 1905 entry in his diary when he claims that 'instead of "Know Thyself", I would write over the gates of the temples of modern thinkers: "Seek to Know Thyself." The search – that's the most important thing.'[15] In other words, the path to self-realization is always under way and threatened by impulses which seek to distract or unsettle the individual from the pursuit. On this model, the committed thinker is plagued by self-doubt and uncertainty, but strives to open him or herself up to experiences which individuals further down the ladder disdain or cannot comprehend. This may seem like an elitist romantic view, but there is nothing in either James' or Rank's work to prohibit *any* individual from ascending the ladder. Indeed, the amateurism which both thinkers pioneered – James in his private therapy sessions and Rank's status as the first lay psychoanalyst – suggests that the world of the troubled soul is accessible to anyone who considers their existence seriously.[16] To recall James (and prefiguring

---

[12] Rank was heavily influenced by the Viennese philosopher Otto Weininger, whose shocking work *Sex and Character*, published in 1903 just before Weininger shot himself, is heavily misogynistic and draws from Freud's and Breuer's work on hysteria from the 1890s. However, Rank's later emphasis on the child's primary relationship with the mother and his work with Anaïs Nin provides evidence to suggest he developed into a proto-feminist analyst.

[13] Lieberman, *Acts of Will*, op. cit., p. 12.

[14] See 'The Psychology of the Poet', *The Don Juan Legend*, trans. David G. Winter (Princeton: Princeton U.P., 1975), pp. 120–6; Lieberman, *Acts of Will*, op. cit., p. 13.

[15] Ibid., p. 43.

[16] Freud encouraged Rank to turn away from the practical life his parents laid out for him and to complete his education at the University of Vienna, but advised him against becoming a qualified doctor.

Rank's work on the *Doppelgänger*), this is a world 'two-stories deep' in which the self is haunted by alternative life-styles and shadowed by glimpses of self-becoming.[17] The risk is that the individual will be left without support from others who cannot, or do not have the courage to, accompany the individual on this journey. From this romantic view, the artist is often a solitary or self-absorbed figure, prevented from intimate communication and pursued by neuroses to which he or she constantly seeks to give material form.

Rank began to clarify these ideas in his first complete essay *The Artist* (which he later revised with modifications suggested by Freud) and developed them more fully in his psychological study *The Double* (1914) and in his late book *Art and Artist* (1932). When *The Artist* was first published in 1907 in the psychoanalytic journal *Imago* it was the first essay on psychoanalysis not published by Freud and foreshadowed some of Jung's ideas on the transformation of the libido. In this long and complicated essay, Rank based his theory of the artistic temperament on the development of sexual drives, heavily influenced by Freud's *Three Essays on the Theory of Sexuality* (1905) which he had energetically read during the composition of *The Artist*. Rank aligns himself with Freud when he quotes from *The Psychopathology of Everyday Life* (1901): 'the dim recognition of psychic factors and relations is mirrored in the construction of a supernatural reality which shall be re-transformed by science into *a psychology of the unconscious*' which is to be achieved by translating '*metaphysics* into *metapsychology*'.[18] Freud's desire to ground ineffable phenomena in a knowledge of inner motivations made him sceptical of all superstitious beliefs: 'a large part of the mythological view of the world, which extends into the most modern religions, *is nothing but psychology projected into the external world*'.[19] Whereas James worried that his psychological work would dissolve into metaphysical musing, by developing a theory of the psyche based on a biological model of drives and energies, Freud did much to ward off the spectre of metaphysics. Rank shares with Freud a desire to ground experience in a theory of psychic drives, but he argues that aesthetic and supernatural impulses are not epiphenomenal discharges but actually possess a vital life of their own.

Rank's early contribution to the field of psychoanalysis lies in his understanding of the relation between psychic impulses and an artistic temperament which has the capacity to transform inner energies into the outward form of art-work. Although he adopted much of Freud's sexual-drive theory, in a 1925 preface to *The Artist* Rank describes an 'originally uniform energy' possessed by all humans, which normally develops into libido, but which may

---

[17]   James, *Varieties*, op. cit., p. 187.

[18]   Quoted in Rank, *The Artist*, trans. Eva Salomon, *Journal of the Otto Rank Association*, 15(1) (1980), 16; slightly modified translation in *The Psychopathology of Everyday Life*, ed. Angela Richards, trans. Alan Tyson (London: Penguin, 1991), pp. 321–2.

[19]   Ibid., p. 321.

also be rechannelled by force of effort into artistic creation.[20] Rank's main interest is not artistic production as such, but 'the artist in the process of becoming' him or herself.[21] He argues that the artist figure leads a precarious existence on the borders between inner (self-immersed) and outer (social) worlds and must mediate between the two: 'externally he provides people with a new individual ideal' which 'he is urged to form . . . by his inner conflict'.[22] Rank claims that in many ways the artist cannot help him or herself: a compulsion from within forces the artist to produce, but the effort to rechannel psychic conflict in a fresh direction must be consciously furnished by the artist. Not only does the idea of rechannelling psychic energy to produce fresh patterns echo James' recommendations for individualistic path-finding, but Rank elides the artist 'as the ideally creative type' with the hero as 'the pragmatically creative type'.[23] Both figures are doers or makers, whose actions serve to change the determined course of events: in other words, both artist and hero are productive types. Writing during a period of expanding industrialism when production was seen in utilitarian terms, the kind of productivity which Rank describes is antimodernist in his recommendation of active self-fashioning. But, this idea of producing or 'giving forth' to create a change in the psychic and social environment actually enables the self-asserting individual to avoid necessity and provides a remedy for neurosis (*Heilkünstler*).

The artist is also a liminal figure in another sense. Rank positions the artistic individual between two other psychological types: 'the dreamer', who leaves fantasy behind in sleep in order to accept a socially defined reality, and 'the neurotic', who cannot distinguish between fantasy and reality enough to deal with either.[24] The artist not only steers a course between these two, but has the potential to become a prophetic figure (with shades of Nietzsche's *Übermensch*) who can speak to and positively influence others. The dreamer is too heavily dependent on social reality to engage in artistic pursuits, while the neurotic is a failed artist (an '*artiste manqué*') who represses socially unacceptable impulses in order to preserve the self.[25] As a consequence, the neurotic's life is pulled in two directions, one which is self-determining and the other dictated by social and psychic pressures. According to Rank, only by separating oneself from the pressures of normalization can the neurotic transform him or herself into an artist, but the risk is that he or she will be labelled as being sick. This premise is very different from the Freudian cure which attempts to reintegrate the individual in society by curbing wayward instincts and working through a set of impulses which disable the self from functioning normally. Whereas Freud recommends

---

20   Rank, *The Artist*, op. cit., p. 7.
21   Ibid., p. 8.
22   Ibid., p. 9.
23   Ibid., p. 7.
24   Ibid., p. 40.
25   Rank, *Art and Artist*, op. cit., p. 25.

that the psychoanalyst should be a rigorous scientific investigator, Rank envisions the analyst as coextensive with the artist, emphasizing the art of the healer and a belief in self-transformation.

Much of Rank's early work focuses on psychic barriers which block the channels of creative expression: the neurotic often suffers from 'an excessive check on his impulsive life' or 'fear-neurosis' which thwarts the self from developing the capacities which would lead to a creative life.[26] But Rank maintains the life of the artist actually emerges out of neurotic life, rather than being categorically distinct from it: artistic production represents the 'creative development of a neurosis in objective form' and 'neurotic collapse may follow as a reaction after production, owing either to a sort of exhaustion or to a sense of guilt arising from the power of creative masterfulness as something arrogant'.[27] As Rank elaborates upon in *The Double*, often writers project their own anxieties onto characters whose triumph or tragedy suggest a staging, rather than the resolution, of psychic conflict. Artistic creation does not resolve neurosis, but provides a channel for the 'fear of life' from which it emerges. Although Rank's model of the self is often divided and he focuses on the split-consciousness of the neurotic and artistic types, his philosophy is not inherently dualistic. In his work on will therapy in the early 1930s, Rank developed a holistic model of the self in which the will acts as a unifying force to facilitate an individual's creativity. He maintains that the artist has the potential to unify mind and body (although this potential is rarely fully realized) and check the bifurcated life of the neurotic: 'my feeling is insistent that artistic creativity, and indeed the human creative impulse generally, originate solely in the constructive harmonising of this fundamental dualism of all life'.[28]

Rank's discussion really only applies to the romantic artist whose exceptional creative powers are brought about by accidents of birth and circumstance as much as through wilful effort.[29] In *Art and Artist*, Rank positions the romantic artist closer to the idiosyncratic world of the neurotic than to the craft and moral purpose of the classical artist because 'his personal ego and his experience are more important than, or as important as, his work'.[30] The romantic artist suffers for, and is burdened by, his or her art but 'does not understand much more about it than the dreamer'.[31] However, although a rebirth experience seems to occur passively to such individuals, they must desire to explore their unconscious if

---

[26]   Ibid., pp. 41–2.

[27]   Ibid., p. 43.

[28]   Ibid., p. xxii. This idea of a regenerative rebirth experience is directly comparable to the psychodrama experienced by James' fictional French correspondent.

[29]   Rank acknowledges this in his review essay 'Literary Autobiography', *Journal of Otto Rank Association*, 16(30) (1981), 4.

[30]   Rank, *Art and Artist*, op. cit., p. 50.

[31]   Rank, *The Artist*, op cit., p. 59.

they wish to transform the insights into what Rank calls 'soul-products'.[32] He argues that most people deal with the pressures of the unconscious during sleep, whereas romantic artists experience the conflict between conscious and unconscious 'projected on their "ego" in its highest individual potency, *when it has already become over-ripe for the dream, but not yet pathogenic*, and they strive to free themselves from it through phantasy-creations'.[33] The created art-work also serves a double 'prophylactic' function as it prevents the artist fostering uncontrollable neuroses and provides 'the psychotherapeutic effect of art enjoyment' for unproductive people.[34] According to Rank, the successful artist has learnt to 'free himself from painful feelings when they press him' (although this is rarely sustained freedom), which is 'different from the neurotic who *cannot, but wants to*, and the dreamer who *lets it happen*': in other words, 'the neurotic wants, if one may say, to digest the painful, the artist spits it out, the dreamer sweats it out'.[35] This notion of 'spitting out' through wilful effort is reminiscent of James' 'heave of the will' as a retching forth of psychic material into material shape. The period of suffering which often coincides with the process of artistic production is seen to make the artist 'more whole' and provides insights into the barriers which prevent neurotic individuals from being productive.[36] Crucially, knowledge of suffering is not just scientific or technical knowledge of illness, but also a 'knowing' acknowledgement or solidarity with the neurotic.

Although Rank maintained a belief in the exceptional qualities of imaginative individuals, by studying larger mythic patterns he argued, like Jung, that there are particular archetypes which define the role of the modern artist. Even at this early stage the difference between Freud and Rank can be discerned. For example, in *The Artist* Rank begins his exploration of the links between the artist-figure and the dramatic hero: he argues that the romantic artist (although here he seems to be discussing the novelist) 'idealizes himself in his hero' and 'whatever failings and weaknesses there are in the artist he attributes to the personified resistances (villain)'.[37] This idea of displacing or jettisoning psychic material suggests that art-work, like dreams, represents a dramatic staging, or re-staging, of unresolved psychic conflict, rather than an essentially unproductive sublimation, as Freud argued in his writings on literature and art. Moreover, although there are aspects of classical myths which substantiate Freud's hypothesis about the desire for vengeance which the infant imaginatively recreates during the Oedipal crisis,

---

32   Ibid., p. 38.

33   Ibid.

34   Ibid., p. 40. This notion of an exemplary life defined in terms of romantic distinctiveness, rather than conformity to a socially-defined moral virtue, is developed by Erikson in his psychohistorical studies.

35   Ibid., pp. 51, 48.

36   Ibid., p. 60.

37   Ibid., p. 56.

Rank claims that most myths actually restage the birth experiences of individuals on the threshold of a new life. Although they usually possess more nobility than their modern literary counterparts, classical heroes are vulnerable characters constantly threatened by temptations which challenge their heroic status.

In *The Myth of the Birth of the Hero* Rank attacks the view that myths have an 'astral' or metaphysical origin, in favour of the Jungian view that mythic stories chart the 'general movements of the human mind'.[38] He argues that myths derive from collective human origins and dramatize 'terrestrial' conditions of existence across diverse cultures.[39] To substantiate his claim that most stories draw off similar mythical patterns, Rank outlines fifteen world myths which deal with the early development of the hero, ranging from the founding Babylonian myth of Sargon the First, the Biblical myths of Moses and Jesus, the birth of Karna in the Hindu epic Mahabahata, the Greek myths of Oedipus and Perseus and the European myths of Tristan and Siegfried. Rank argues that these myths all involve a male child whose 'origin is preceded by difficulties' and whose conception is often mysterious: lost or abandoned in infancy and saved by humble creatures he eventually rediscovers his 'distinguished parents, in a highly versatile fashion'.[40] He claims that the mythic hero is comparable to the artist in the respect that he or she 'should always be interpreted merely as a collective ego', whose exaggerated traits can be perceived in all individuals.[41] As such, the individual's life-story is structured by an inherited skeletal pattern and cannot simply be created by an individual act of will. He uses this mythic template to chart the growth of the child away from the parent and to distinguish the differences between neurotics and artists: the preoccupations of the neurotic are 'the uniformly exaggerated reproductions of the childish imaginings', whereas the artist emerges from the distressing circumstances of his childhood into a self-determining maturity.[42] In this way, myths (like dreams) can be 'progressive rather than regressive' by helping to resolve problems rather than 'perpetuate' them.[43]

Rank develops Freud's position in *Three Essays on the Theory of Sexuality* in relation to the child's alienation from the parents and the ensuing 'family romance' (a term supplied by Freud), in which the child's play and dreams act as

---

[38]   Rank, *The Myth of the Birth of the Hero and Other Writings*, ed. Philip Freund (New York: Vintage, 1959), pp. 6, 8.

[39]   Ibid., p. 10.

[40]   Ibid., p. 65.

[41]   This claim is slightly exaggerated: not only is there a gender prejudice in Rank's early work, but the myths Rank analyses are selective examples of 'young hero' myths. Robert Segal suggests that 'mature hero' myths like Odysseus and Aeneas do not conform to Rank's pattern; ed. Robert Segal, *In Quest of the Hero* (Princeton: Princeton U.P., 1990), p. xv.

[42]   Ibid., p. 67.

[43]   Rank, *In Quest of the Hero*, op. cit., p. ix.

outlets for the fulfilling his 'erotic' or 'ambitious' desires.[44] Rank stresses that the hostility towards the parents results from the loss of an ideal:

> the child's longing for the vanished happy time, when his father still appeared to be the strongest and greatest man, and the mother seemed the dearest and most beautiful woman . . . The overvaluation of the earliest years of childhood again claims its own in these fancies.[45]

The major difference between the mythic story and the family romance is that in myths the father often seeks to rid himself of the child (or, at least, loses him), whereas in the family romance the child directs hostilities toward the parent. Rank suggests that myths often work by equating the father with the tyrant and projecting the 'real' child's own distress onto the hostility which the father/tyrant displays towards his 'mythical' child, thus serving to justify the child's own resentment. On this account, myths are 'paranoid' structures which provide an insight into the psychic world of the child through a series of projections, distortions and displacements.[46] Moreover, he suggests that myths are the result of the adult storyteller's own 'retrograde childhood fantasies', in which the hero is invested 'with their own infantile history'; but because they are (potentially) parents themselves, their hostility toward the original father redirects internal anger toward an oblique image of themselves. In this way, Rank's theory of child development is similar to Freud's, deriving from the child's (and later the adult's) unresolved feelings about his or her individual autonomy. These 'young hero' myths prefigure Erikson's psychohistory *Young Man Luther* (1958) by exploring the child's (or adolescent's) ambivalent feelings of independence from, yet attachment to, the parent.

Where Rank differs from Freud, is his emphasis upon the productive use of childhood fears in the creation of art-work: if 'play' is a mechanism which enables the child to cope with fear and ward off 'unpleasure', then 'normal adults' can be seen to indulge in the same kind of play in 'day-dreaming' or 'phantasizing'.[47] In his Rankian book *The Denial of Death* (1973), Ernest Becker argues that 'heroism is first and foremost a reflex of the terror of death', in which the archetypical hero descends into the spirit world and returns alive.[48] Similarly, for Rank, the artist can develop the raw material of myth to purge him or herself of masochistic feelings or resolve complicated emotions for another person.

---

[44]   Ibid., p. 69.
[45]   Ibid., p. 71.
[46]   Ibid., p. 78.
[47]   Rank, *The Artist*, op. cit., p. 49. Rank argues that psychoanalysis cannot explain feelings of fear and guilt as pathologies or 'mere exaggerations or intensifications of the normal'; rather, fear is a self-perpetuating anxiety about failure and 'incapability'; Rank, 'Self-Inflicted Illness', *The Proceedings of the California Academy of Medicine* (February 1935), 12–13.
[48]   Ernest Becker, *The Denial of Death* (New York: Free Press, 1973), p. 11.

Because the artist is the 'purest type' of creative individual, he or she has the capacity to transform 'infantile play' into 'theatre play' or 'egocentric daydream into thrilling novel'.[49] Rank reverts back to his adolescent love of theatre when he argues that dramatic 'play' is 'the most direct kind of presentation' (in which there is no mediating narrator) because 'it comes nearest the dream form and even borders closely on the action of the hysterical attack'.[50] The dramatic staging of unresolved conflict enables the artist to steer a course between internal pressures (the release of psychic energy) and external social and cultural forces which impinge upon his or her world. The translation of this kind of psychic material is of great interest to Rank and enables him in his later work to develop a series of links between artist, hero, dramatist, actor and physician whose collective task, he argues, is to play the role of the 'good interpreter'.[51]

The Hovering Double

Rank develops his exploration of myths in *The Myth of the Birth of the Hero* in his essay on *The Double*, by directing his attention towards more recent European literature. He argues that many late nineteenth-century writers sought to twist or transform well-known myths and their 'poetical offshoots, the fairy tale, the saga, and the epic', rather than offering a clean break from the past.[52] In order to substantiate this hypothesis, he outlines a genre of post-romantic stories which explore themes of creativity, narcissism and mortality and in which characters are 'more or less pathologically divided between reality and fantasy'.[53] The emergence of the split-self or *Doppelgänger* is particularly relevant to Germanic myth, but arose as a major theme throughout European literature as writers reacted against the presuppositions of literary realism and social progressivism. Fyodor Dostoevsky's *The Double* (1846), Robert Louis Stevenson's *The Strange Case of Dr. Jekyll and Mr. Hyde* (1886) and Oscar Wilde's *The Picture of Dorian Gray* (1890) were among the most popular publications imaginatively probing the hitherto unexplored shadowy depths of the respectable bourgeois psyche in an effort to expose psychic contradictions and growing philosophical uncertainties.

First published in the psychoanalytic journal *Imago* in 1914, Rank's extended essay *The Double* analyses these narratives (among others) in order to assess why, and to what psychic end, the double as a literary figure became prevalent at this time. Through these examples Rank began to outline the aesthetic parameters and the ethical implications of the creative self as a development of his earlier

---

[49]   Rank, *The Artist*, op. cit., p. 49.
[50]   Ibid., p. 55.
[51]   Ibid., p. 57.
[52]   Rank, *Art and Artist*, op. cit., p. xxi.
[53]   Andrew Webber, *The Doppelgänger: Double Visions in German Literature* (Oxford: Clarendon, 1996), p. 1.

romantic conception of the imaginative artist. Although Rank did not discuss romanticism as an aesthetic movement explicitly until the early 1930s, his engagement with post-romantic literature can be discerned throughout *The Double* in his discussion of writers whose art-work is 'a forcible liberation from inward pressure'.[54] *The Double* reveals marked similarities with James' work, but Rank's romantic-scientific premise is diametrically opposed to James: he begins from an aesthetic position which feeds into a science of the mind, whereas James begins from a natural-scientific position from which emerged his theory of radical empiricism.

Written two years after he graduated from the University of Vienna, *The Double* was stimulated by the hugely popular German film *The Student of Prague* (*Der Student von Prag*, 1913), written by best-selling German horror writer Hanns Heinz Ewers, directed by the Dane Stellan Rye and starring Paul Wegener. Ewers' work explores the life of a double-self, appropriated from stories by Poe and Hoffman and fused with a reworking of the Faust myth. Produced in the style of German Expressionism, the film portrays a gothic psychodrama of a young student, Balduin, who sells his mirror-image to a sorcerer, Scapinelli, for endless wealth. After the Faustian pact, Balduin's reflection takes on an independent double life with the malevolent intent to destroy him, a theme which Siegfried Kracauer reads as a manifestation of 'the deep and fearful concern with the foundations of the self' which was becoming increasingly prevalent in Germanic culture at the beginning of the century.[55] Reacting against the nineteenth-century *Bildungsroman*, German Expressionist film (together with the pre-war literature of Thomas Mann, Arthur Schnitzler and Franz Kafka) explored the underside of the Germanic culture of self-development: psychic exploration was no longer equated with progressive social mobility, but with disintegration and dissolution of identity.[56]

Rank argues that the experience of the double is born as a phantom of the disturbed psyche: an unresolved conflict which cannot be dealt with internally, nor contained within the world of dreams. In this way, the *Doppelgänger*-experience is a manifestation of James' troubled or sick soul, as opposed to the clarity and simplicity of the healthy-minded. *The Student of Prague* not only reworked the Faust myth for the new century, but suggested to Rank something distinct about the dominant psychological trends in modern Germanic culture.[57] The preoccupation with the artist-figure in modernist literature (in Gide, Mann,

---

[54] Rank, *Art and Artist*, op. cit., p. 51.

[55] Siegfried Kracauer, *From Caligari to Hitler: A Psychological History of the German Film* (Princeton: Princeton U.P., 1947), p. 30.

[56] On meeting Rank in 1933, Anaïs Nin commented on his 'Dr. Caligari body', referring to Robert Wiene's Expressionist film *The Cabinet of Dr. Caligari* (1919); Noël Riley Fitch, *The Erotic Life of Anaïs Nin* (Boston: Little, Brown, 1993), p. 163.

[57] Lotte Eisner, *The Haunted Screen*, trans. Roger Greaves (Berkeley: University of California Press, 1969), p. 43.

Proust, Joyce, Woolf and Kafka) is equalled only by the emergence of the neurotic figure who, as Rank claims, 'represents the artist's miscarried counterpart'.[58] The critic Andrew Webber convincingly argues that the *Doppelgänger* is 'a ghostly presence, but also . . . a radical absence, which terrorizes the post-Romantic subject with a mixture of Romantic agony, ecstasy, and irony'.[59] On this argument, rather than representing an absolute break from history, modernist culture recycles and reuses elements from the past, presenting 'the primitive complexes of mankind, in their actual state of repression' within the context of a post-Darwinian world.[60] Indeed, the modernist preoccupation with temporality suggests that the past is a spectre which challenges and erodes the solidity of the present and the promise of a desirable future: 'a person's past inescapably clings to him and . . . becomes his fate as soon as he tries to get rid of it'.[61] For Rank, modern psychology should move away from the traditional view of primitivity as the remnants of 'fossilized' cultures, to examine the coexistence of the archaic and the modern in each individual. As such, *The Student of Prague* visually dramatizes the motif of the double as a way of 'reapproaching, intuitively, the real meaning of an ancient theme which has become either unintelligible or misunderstood through the course of tradition'.[62]

Like *The Artist*, Rank begins his study from a traditionally Freudian perspective. As Freud elaborates in *The Psychopathology of Everyday Life* (1901), 'deeply buried and significant psychic material' can be uncovered in what Rank calls the most 'random' and 'banal' subjects to reveal the 'meaningful problems of man's relation to himself'.[63] Well before Freud developed a broader psychoanalytic view of culture in *Totem and Taboo* (1918) and *Civilization and its Discontents* (1929), Rank argues that there is great value in recounting the 'history of an old, traditional folk-concept' because this may provide insight into a bewildering present. Even though many modernist artists expressed their opposition to tradition, writers like Joyce, Eliot and Broch structured their expressions of modernity with reference to the fragments of myths and well-worn stories. As such, Rank takes *The Student of Prague* as his starting point because its 'prominent patterns' visually recycle the late nineteenth-century *Doppelgänger* motif which he traces back to 'superstitious notions associated with the shadow which even today are encountered among us'.[64] Significantly, in *The Student of Prague*, Balduin's double haunts him only during his trysts with his beloved 'as a silent figure of warning', echoing the type of love-sickness Carl Carus outlines in *Psyche* as one of the 'psychopathological phenomena . . .

---

58   Rank, *Truth and Reality*, trans. Jessie Taft (New York: Norton, 1978), p. 4.
59   Webber, *The Doppelgänger*, op. cit., p. 11.
60   Rank, *The Don Juan Legend*, op. cit., p. 120.
61   Rank, *The Double*, trans. Harry Tucker Jr (London: Karnac, 1989), p. 6.
62   Ibid., p. 4.
63   Ibid. pp. 3, 7.
64   Ibid., p. 49.

precariously governed by occult forces lodged in the self, in others, and in the cosmos'.[65] This is corroborated as Balduin's beloved is also shadowed by her double, in the guise of the gypsy Lyduschka who is intent at coming between the two. When Balduin finally comes to shoot his alter-ego he actually murders himself and his double survives as the diabolic companion of Scapinelli. The plot suggests that Balduin's excessive narcissism, or elevated sense of his own self, actually prevents him from realizing or consummating his love for another. His alter-ego is both his adversary and an unshakeable manifestation of his conscience: in this way, his final demise is shown to be a direct result of his 'narcissistic fixation'.[66]

Freud quotes Rank's essay 'A Contribution to the Study of Narcissism' (1911) at the beginning of his own influential essay 'On Narcissism' (1914), published the same year as *The Double*. Whereas Freud treats narcissism as a manifestation of perversion and 'the attitude of a person who treats his own body in the same way in which the body of a sexual object is ordinarily treated', Rank suggests 'it might claim a place in the regular course of human sexual development'.[67] Indeed, in *The Artist* Rank argues that a degree of narcissism is one of the chief 'character traits' which differentiates the productive artist from the helpless neurotic.[68] Freud admits that, in this sense, narcissism is connected to 'self-preservation, a measure of which may justifiably be attributed to every living creature', but then concentrates on the manner in which some individuals become pathologically narcissistic in their inability to detach themselves from infantile satisfaction.[69] In contrast, throughout his work Rank is more interested in the 'normal' role of narcissism. Nevertheless, *The Student of Prague* provides an example of a self-absorbed individual who is driven by material greed and inhibited by a narcissistic inability to develop the frustrated relationship with his beloved: his shadow self acts as a consciously undesirable safety check which blocks him from any productive life with another. Rank's study of the double is driven by his desire to understand the cultural phenomenon of narcissism without resorting to the libido theory underpinning Freud's psychoanalytic model: if the 'life of the double is linked quite closely to that of the individual himself', then

---

[65] James L. Rice, *Dostoevsky and the Healing Art*, op. cit., p. 140.

[66] The film critic Heide Schlüpmann links the theme of 'narcissistic fixation' to the visual medium of film, which Rank overlooks in his narrative analysis of the film; Heide Schlüpmann, 'The First German Art Film: Rye's *The Student of Prague* (1913)', in ed. Eric Rentschler, *German Film and Literature: Adaptations and Transformations* (New York: Methuen, 1986), p. 17.

[67] Freud, 'On Narcissism: An Introduction' (1914), in ed. Angela Richards, trans. James Strachey, *On Metapsychology: The Theory of Psychoanalysis* (London: Penguin, 1991), p. 65.

[68] Rank, *The Artist*, op. cit., p. 52.

[69] Freud, 'On Narcissism', op. cit., p. 66.

it is fruitless to attempt to find the root cause of narcissism in order to cure the individual from it.[70]

Rank outlines two major manifestations of the double in literature: firstly, the physical double embodied by the twin or the replica, 'a likeness which has become detached from the ego and become an individual being (shadow, reflection, portrait)'; and, secondly, 'double consciousness' which manifests itself in individuals whose lives have been punctuated by amnesia or 'the spontaneous subjective creation of a morbidly active imagination'.[71] He pays more attention to this second category, suggesting that the double most often 'works at cross-purposes with its prototype', accompanied by 'persecutory delusion' and preventing the fruition of sexual relationships.[72] Thus, the double is invested either with the individual's fear of being pursued, or is associated with self-doubt and the threat of finitude.

While Rank adds much to the understanding of the double motif in literature, he often makes the error of associating these delusions and examples of narcissism with the authors of the work, instead of viewing them as fictional explorations of psychic disturbances. However, although he characterizes a pantheon of male writers (Goethe, Poe, Dostoevsky, Hoffman, Maupassant, Wilde) as having 'pathological personalities' and eccentric lifestyles ('whether in the use of alcohol, narcotics, or in sexual relations'), he does not reduce textual complexity to a determinate principle in the life of the writers as does Freud in his essay on 'Dostoevsky and Parricide' (1927), in which he argues that the Russian writer's hostility toward his father finds a sublimated outlet in his fiction.[73] Rank detects that an 'overexaggerated attitude towards one's own ego' may well be reflected in the literature written by morbidly predisposed writers. This principle cannot serve to explain the work, but may aid an understanding of it:

> [the] recurring ways in which these forms appear do not become intelligible from the writer's individual personality. Indeed, to a certain degree they seem to be alien to it, inappropriate, and contrary to his way of otherwise viewing the world.[74]

However, although connections are clearly evident between writer and art-work, Rank admits that 'we do not quite understand it'.[75]

---

[70]   Rank, *The Double*, op. cit., p. 11.

[71]   Ibid., p. 20.

[72]   Ibid, p. 33. Although Rank's literary examples focus almost exclusively on men, two of Binswanger's cases discussed in the next chapter provide examples of persecutory delusion in women: Lola Voss and Ellen West.

[73]   Freud, 'Dostoevsky and Parricide' in Albert Dickson, ed., *Art and Literature*, op. cit., pp. 437–60.

[74]   Ibid., p. 48

[75]   Ibid.

For this reason, taking his lead from the Scottish anthropologist James Frazer, Rank moves on to examine the ways in which the double has arisen repeatedly in folklore and myth.[76] Rank detects in myth a series of ancient fears, beliefs and superstitions which often lead the artist to project the fear of the Other (either an external or internal threat) as a double image or a shadow figure. Narcissism does not always result from vanity and self-importance but also from anxiety or the threat of loss. For example, Frazer argues that the shadow is traditionally seen as a sign of security which, when it shortens or becomes detached from the individual, suggests the bodily anchor in the world has faded or vanished: 'if it is trampled upon, struck, or stabbed, he will feel the injury as if it were done to his person; and if it is detached from him entirely (as he believes that it may be) he will die'.[77] As such, the diminishment of the shadow is associated with a fear of bodily death and 'a corresponding decrease in the vital energy of its owner'.[78] In the proto-existentialist literature which Rank analyses, the figure of the double provides a way of dramatizing a character's fear of mortality or the anxiety of being contaminated by a desirable, and yet potentially, dangerous Other.

Rank concludes *The Double* by linking the individual's fear of death to his theory of narcissism. Stories in which 'the hero seeks to protect himself permanently from the pursuits of the self' by killing his double are really tales of philosophical suicide: 'the suicidal person is unable to eliminate by direct self-destruction the fear of death resulting from the threat to his narcissism'.[79] In this way, 'the double is part of the ineradicable past', which the individual attempts to jettison because he 'esteems his ego too highly to give it pain or to transform the idea of his destruction into the deed'.[80] Similarly, Rank argues that belief in immortality, or a second life, reflects the individual's narcissistic belief in his or her own self-importance and a repudiation of the finitude of existence. Rank does not believe the individual can actually separate him or herself from these fears or superstitions, but argues that he or she has the capacity to acknowledge and potentially unify the doubleness and uncanny nature of existence, not by shoring up ego-boundaries but living a precarious multiple existence. Such a multiple existence does not equate with clinical schizophrenia (a condition which Binswanger examines in some detail), but an ability to respond appropriately to a variety of experiences.

---

[76]   The anthropological chapter of *The Double* owes more to Frazer's *The Golden Bough* (1890–1915) than to direct empirical evidence from ethnographic study. However, Frazer was more interested in the literal interpretation of myths, whereas Rank was motivated by symbolic meaning.

[77]   James Frazer, *The Golden Bough: A Study in Magic and Religion* (London: Wordsworth, 1993), p. 189.

[78]   Ibid., p. 191.

[79]   Rank, *The Double*, p. 79.

[80]   Ibid., p. 80. Rank had entertained the idea of suicide in his adolescence but used this experience as a launching pad for reconciling himself with his own finitude; Lieberman, *Acts of Will*, op. cit., p. 27.

As the critic Paul Coates discerns, the typical fictional world of post-romantic fiction is composed of both substance and shadow, revealing an uncanny existence which cannot be explained by a single frame of reference.[81] Rank's interest in the double is a way of rejecting a uniform model of psychic life which explains certain drives with reference to a single determining principle. Although it can be argued that Freud's libido theory is such a totalizing model of mind, there are important similarities between Rank's *The Double* (revised in 1925) and Freud's essay 'The Uncanny' (1919) which suggests Freud's metapsychology is more subtle than Rank acknowledges in his later work. However, in a series of manuscript notes from 1930, collected under the title 'Literary Autobiography', Rank claims that 'The Uncanny' was 'an (unconscious) attempt on Freud's part to substitute my explanation (fear of death) by his old castration theory'.[82] Whether the accusation against Freud is fair or not, Rank's comment reveals an important connection between his notion of fear of death and the psychic uncertainty which characterizes Freud's description of the uncanny.

In one of the exemplary studies of structuralist poetics *The Fantastic* (1973), Tzvetan Todorov outlines the generic parameters of fantastic literature, which he positions between the categories of 'the marvellous' and 'the uncanny'. Just as Rank emphasizes the reader's confusion about the precise borderline between the natural and the artificial, Todorov suggests that the chief feature of the fantastic genre is that the reader hesitates 'between a natural and a supernatural explanation of the events described': the *'reader's hesitation* is therefore the first condition of the fantastic'.[83] Todorov indicates central differences between the three terms – marvellous, fantastic and uncanny – but he allows for hybrid forms of writing evident throughout the post-romantic literature with which Rank deals. Responding to Todorov's distinctions, the literary critic Rosemary Jackson characterizes fantastic narratives as transporting 'the reader of the known and everyday world into something more strange'.[84] Unlike the marvellous which tends towards a non-referential mode of writing, the fantastic insists on its 'interstitial placing' in a manner which can be used to question the categories of 'real' and 'imaginary'.[85] Jackson's central argument is that, rather than circumscribing an escapist fiction, fantasy can disrupt dominant forms of realism

---

[81]   Paul Coates, *The Gorgon's Gaze: German Cinema, Expressionism, and the Image of Horror* (Cambridge: Cambridge U.P., 1991), p. 1. Coates goes on to argue that *The Student of Prague* and *The Cabinet of Dr. Caligari* are 'the recto and verso' of E.T.A. Hoffman's tale 'The Sandman'; ibid., p. 38.

[82]   Rank, 'Literary Autobiography', op. cit., p. 15.

[83]   Tzvetan Todorov, *The Fantastic: A Structural Approach to a Literary Genre*, trans. Richard Howard (Cleveland: Press of Cape Western Reserve University, 1973), p. 31.

[84]   Rosemary Jackson, *Fantasy: The Literature of Subversion* (London: Methuen, 1981), p. 34.

[85]   Ibid., p. 63.

by highlighting or defamiliarizing habitual modes of interpretation.[86] This is not to devalue the importance that James places on the mechanisms of habit, but Jackson foregrounds the presence of a threshold between two narrative modes which helps to illuminate the devices at work in conventional modes of writing: in other words, 'to introduce the fantastic is to replace familiarity, comfort, *das Heimlich*, with estrangement, unease, the uncanny . . . to introduce dark areas, of something completely other and unseen'.[87]

In 'The Uncanny', Freud discusses Hoffman's short story, 'The Sandman' (1815) and follows the work of German physician Ernst Jentsch by discussing the notion that the *unheimlich* (the foreign and strange) is to be found within the *heimlich* (the everyday and familiar): in other words, 'the uncanny is that class of the frightening which leads back to what is known of old and long familiar'.[88] He stresses that the uncanny cannot be equated fully with the unfamiliar because the *heimlich* develops 'in the direction of ambivalence, until it finally coincides with its opposite'.[89] Whereas the uncanny circumscribes a topological space for Freud, Jackson suggests that it should not only be understood as a place 'one does not know one's way about in'.[90] Rather, in line with Rank's interest in artistic production, the uncanny is primarily a technique of representation, which dramatizes a sense of ambivalence. According to Jentsch:

> in story-telling, one of the most reliable artistic devices for producing uncanny effects easily is to leave the reader in uncertainty as to whether he has a human person or rather an automaton before him . . . This is done in such a way that the uncertainty does not appear directly at the focal point of his attention, so that he is not given the occasion to investigate and clarify the matter straight away.[91]

Whereas Freud wishes to force a distinction between the uncanny as it is manifest in human experience and a literary uncanny, Jentsch emphasizes the 'devices' and 'effects' of representation, but argues that the affect of the uncanny cannot be easily reduced to these 'devices'; rather than trying to codify the uncanny as a specific reading practice, motivated by the attempt 'to discover a central structure or generative device which governs all levels of the text', an uncanny sensation often derives from the indeterminacies of a text.[92] On this model, the uncanny has

---

[86]   See Victor Shklovsky, 'Art as Technique', *Russian Formalist Criticism: Four Essays*, eds and trans. Lee T. Lemon and Marion J. Reis (Lincoln: University of Nebraska Press, 1965), pp. 5–22.

[87]   Ibid., p. 179.

[88]   Freud, 'The Uncanny' (1919), in *Art and Literature*, op. cit., p. 340.

[89]   Ibid., p. 347.

[90]   Ibid., p. 341.

[91]   Ernst Jentsch, 'On the Psychology of the Uncanny', trans. Roy Sellars, *Angelaki*, 2(1) (1995), 13.

[92]   Jonathan Culler, *Structuralist Poetics* (London: Routledge, 1975), p. 172.

an experiential dimension which cannot be wholly explained by reference to the formal technique used to produce it: as Paul de Man comments, 'hermeneutic understanding . . . is always, by its nature, lagging behind' and can never fully encapsulate the 'mystery' of the story.[93]

In line with these ideas, Rank implies that psychic life lies on the borders between ineffable phenomena (which cannot be explained by Freud's model of libidinal drives) and material reality (which, although providing the self with its anchor in the world, disturbs the self with the reality of finitude). In other words, Rank argues that the boundaries between self and non-self are inherently unstable and permeable: there is something of the non-self (*unheimlich*) which disturbs the self (*heimlich*) as an automatous entity separated from the world. The neurotic surrenders him or herself to these schisms and becomes the victim of uncertainty, whereas the artist acknowledges this precarious existence and resolves to make something out of ontological uncertainty, by projecting these conflicts into his or her art-work. In this way, narcissism can either undermine the individual's sense that the world is full of other people who may be undergoing a similar set of experiences, or it can protect the self from threats to its ego boundaries. Throughout his early work Rank suggests that a significant degree of narcissism is necessary to ensure the self remains intact in the face of internal and external pressures, but excessive narcissism may spill back into solipsism and neurosis. This is the fate of Balduin in *The Student of Prague* whose own sense of self-worth is perversely undermined by a phantom *Doppelgänger* of his imagination: his soul-sickness results from his inability to lead a creative life or to forge a union with his beloved. While Balduin's uncanny fears finally cause his undoing, Rank suggests his demise could be checked with a little more self-knowledge and the resolution to inhabit the liminal cracks between different dimensions of psychic existence. But, while knowledge and pragmatism both have places in Rank's work, when the two are in conflict Rank, like James, errs on the side of pragmatic action.

The Sundering of Wills

Although certain aspects of Rank's early work are clear developments of Freud's theory of psychic drives, not until the publication of Rank's most notorious work *The Trauma of Birth* in 1923 did the tensions appear which were later to sever Rank's association from the IPA. Not only did Rank begin to break away from the central tenets of Freud's metapsychology, but he believed psychoanalysis could never fulfil Freud's desire to develop a pure science of the mind. As early as 1916 Rank had asserted that psychoanalysis was the 'bastard' offspring of nineteenth-century natural science, and he proclaimed Freud was not even its father: this

---

[93]   de Man, *Blindness and Insight*, op. cit., p. 32.

credit he gave to Josef Breuer, with Freud as its 'loving foster mother'.[94] The real difference between the two analysts is that, with no formal medical training Rank's theory of mind was essentially polymathic, while Freud tried to forge a unidirectional path for his new science. Moreover, Freudian analysis was a means to reorient his patients by imposing social prohibitions on them, whereas Rankian therapy attempted to unlock potentially creative traits which would provide an outlet for neurotic energy. Artistic fulfilment could only be brought about for Rank with the partial surrender of the intellect:

> the error lies in the scientific glorification of consciousness, of intellectual knowledge, which even Psychoanalysis worships as its highest god although it calls itself a psychology of the unconscious . . . Intellectual understanding is one thing and the actual working out of our emotional problems another.[95]

Rank's emphasis on emotion over intellect is essentially romantic in orientation and links his early analytic work in *The Artist* and *The Double* with his later humanistic work on patient-centred therapy. His belief that the natural and mental sciences were becoming too 'sharply separated' is the primary reason why his version of dynamic psychology diverged radically from Freud's psychoanalysis in the mid-1920s.[96] By distancing his version of romantic science from Freud's more strictly empirical standpoint in *The Trauma of Birth* and his collaborative work with Sandor Ferenczi in *The Development of Psychoanalysis* (written in 1922 and first published in 1924), he wished to redefine the parameters and redirect the therapeutic goals of psychoanalysis.

Rank's reputation is based mainly on the publication of *The Trauma of Birth*, in which he argues that the primary trauma in the infant's life is not the Oedipal scenario as elaborated by Freud, but the phase in the infant's life which includes the moment of birth, the primal relationship to the mother and the process of weaning from her. Whilst Rank maintained his book represented a direct development of Freud's ideas, it actually countered his teacher's work on some crucial issues and sent shock waves through the ring of analysts Freud had gathered around him, particularly Ernest Jones, who claimed *The Trauma of Birth* was 'badly and obscurely composed . . . written in a hyperbolical vein more

---

[94]   Rank, 'Psychoanalysis as General Psychology', *Mental Hygiene*, 10(1) (1916), 13. The dedication page of *The Trauma of Birth* complicates this distinction slightly, in which Rank accredits Freud as being an 'explorer of the unknown' and the 'creator of psychoanalysis'; see the editorial note to 'Rank's Genetic Psychology', an essay in which he responds to Freud's criticisms of *The Trauma of Birth*; *Journal of the Otto Rank Association*, 4(2) (December 1969), 6.

[95]   Taft, *Otto Rank*, op. cit., pp. 149–50.

[96]   Sandor Ferenczi and Otto Rank, *The Development of Psychoanalysis* (New York: Nervous and Mental Disease Publishing Co., 1925), p. 65.

suitable for the announcement of a new religious gospel'.[97] Rank later regretted the publication of *The Trauma of Birth* and placed increasingly less emphasis on the traumatic birth experience as the underpinning principle of his metapsychology. Nevertheless, the volume does display some key ideas about the implications of 'becoming human'.[98] To balance the aesthetic bias in his early writing, *The Trauma of Birth* marks Rank's attempt to expound a biological principle which underpins, rather than completely determines, psychic activity. Although he claims that the birth trauma should be viewed as 'the ultimate biological basis of the psychical' which provides a material 'anchor' for mental life, he does not resort to a fatalistic account of the self directed by uncontrollable forces from the past.[99]

The study begins by attending to both the biological cause and the mental productions of the unconscious which 'harmoniously supplement one another'.[100] As such, his theory of the birth trauma grounds all cultural understanding in an awareness of bodily life in a manner reminiscent of James' work in *Principles* and *Varieties*. For example, he argues that afflictions such as bronchial asthma, migraines, hysterical attacks and infantile convulsions are all expressions of an 'organic language' which can be traced to primal intrauterine conditions.[101] He claims that many of his patients reveal signs of birth symbolism in conversation and the recalling of dreams (especially towards the end of the period of therapy), suggesting analysis itself replicates a certain stage in the individual's early life. He garners these ideas from 'a mass of heterogeneous material', including dream analysis and his reading of literature, in which he repeatedly detects the motifs of pregnancy and the separation of individuals from situations of security.[102] Rank resists interpreting this kind of 'rebirth-phantasy' on a mystical or 'anagogic' level (which he attributes to Jung) 'to the neglect of its libidinal tendencies', but he does suggest that such symbolism can be read as corresponding allegorically to the infant's 'earliest physiological relation to the mother's womb'.[103] He suggests that the analyst must play the role of the mother (as opposed to the judicial paternal role which Freud adopted), not so that the neurotic patient can enter a phase of intrauterine bliss, but to 'sever this primal fixation on the mother, which the patient was unable to accomplish alone'.[104] The anxiety which accompanies such intrauterine desires needs to be acted out in the analytic

---

[97]   Jones, *The Life and Work of Sigmund Freud*, op. cit., p. 523.

[98]   Rank, *The Trauma of Birth* (New York: Dover, 1993), p. xxi.

[99]   Ibid., pp. xxiii, 187.

[100]   Ibid., p. xxiv.

[101]   He also argues that popular cures for infantile convulsions such as 'putting the child in warm water' replicate the desire to return to the womb; Rank, 'Psychoanalysis of Organic Conditions', *Medical Journal and Record*, 120(3) (August 1924), xxxiv.

[102]   Rank, *The Trauma of Birth*, op. cit., p. 5.

[103]   Ibid., pp. 3, 4.

[104]   Ibid., p. 9.

situation in order for it to become controllable, otherwise the uncanny dread of being both human and foetal can paralyse the individual's life. On this model, the identification of death is associated with a return to the mother as a peaceful repose, rather than a fear of the unknown: in other words, a retreat from death rather than a resolute confrontation with it.

Rank takes the biological experience of birth as the underpinning moment in an individual's life, heralding the separation of child from mother and characterizing the infant's helplessness in the world. He finds evidence of birth imagery throughout literary, religious and philosophical writing, but often confuses the actual experience of birth with the psychological impact which primal separation causes to the developing child, in much the same way that Freud confuses the child witnessing his or her parents copulating with the psychological function of such an experience.[105] Nevertheless, the goal of therapy for Rank is to enable the individual to come to terms with these feelings of separation in order to diffuse the fixation on such a traumatic moment. The dynamic of birth as the primary act of creation can be deployed in therapy to counteract the individual's anxiety about separation and fear of loneliness and to heal potential divisions in the self. Not only does an acting out of the birth experience within the relative safety of the therapeutic session (with the therapist as 'experienced midwife') enable patients to free themselves from the neurotic paralysis which birth trauma may induce, but it provides a platform for the individual taking control of his or her own life and becoming self-creators.[106] The individual must repeat the early experience in order that unconscious material can be made conscious in an act of self-becoming.[107] This does not comprise a compulsive-repetitive acting out of the experience as a fixation on the past, but an ability to harness the creative moment in a forward-oriented direction. Rank's practice of active therapy (*Activitat*) with a set short-term time limit (*Terminsetzung*) was devised to equip his patients with the skills to acknowledge their birth trauma and deal with it themselves outside therapy, although he was not such an idealist to believe every period of analysis can be successful.

These therapeutic ideas are outlined in *The Development of Psychoanalysis*, in which Ferenczi and Rank set out a new agenda for psychoanalysis which might allow them to go beyond the scientific constraints of psychology and make therapy more productive for the patient. A Hungarian physician from Budapest, Ferenczi had undergone some early work on hypnotism and made his acquaintance with Freud in 1908. Like Rank in Vienna and Binswanger in Zürich, Ferenczi was to become one of Freud favoured 'sons' whose speculative turn of

---

[105] Esther Menaker argues this point convincingly in her chapter on 'The Trauma of Birth Reexamined'; Esther Menaker, *Otto Rank: A Rediscovered Legacy* (New York: Columbia U.P., 1982), pp. 62–75.

[106] Rank, *The Trauma of Birth*, op. cit., p. 204.

[107] Ibid., p. 183.

mind both stimulated and worried him.[108] Freud's regulative mode of therapy contrasted directly with the other three analysts who recommended that individuals should expansively exert themselves in the face of a deterministic environment. Significantly, both Rank and Ferenczi were drawn to American culture (as was Jung), whereas Freud's 1909 visit to Clark University in Worcester, Massachusetts was plagued with self-doubt and misgivings.[109] This Central European-American axis in part explains Rank's philosophical differences from Freud: if Ibsen was Rank's European model of the 'scientific artist' in his rigorous naturalistic analysis of psychological types, then Mark Twain was his American literary model of the expansive pioneering individual who asserts himself in the face of a hostile environment.[110]

Ferenczi's friendship with Freud continued into the 1910s, but a trip to Sicily in 1911 led to the beginning of a rift between the two, which Ernest Jones claims derived from Ferenczi's 'inordinate and insatiable longing for his father's [that is, Freud's] love' and which later led to the breakdown of Ferenczi's mental 'stability'.[111] However, Ferenczi remained loyal to Freud and with the publication of *The Trauma of Birth* he began to regret his collaboration with Rank, even though he discerned Freud was playing the role of 'castrating god' over the future direction of psychoanalysis. Jones' portrait of Ferenczi is as distorted as his depiction of Rank, but it is interesting that both analysts are caricatured as mentally unstable and having deeply neurotic 'father complexes', especially in the light of Rank's comment that Freud was more 'foster mother' than father. Rather than viewing *The Development of Psychoanalysis* as staging an Oedipal struggle for the right to define the future direction for psychoanalysis, the book better represents a joint attempt to reorient a set of practically motivated theories, which the pair believed were becoming increasingly divorced from patient care. As such, *The Development of Psychoanalysis* lays the groundwork for Rank's

---

[108] Adam Phillips suggests that Ferenczi was Freud's 'repressed unconscious – the prodigal son who keeps coming back for more', even after he had written to him in 1911 to argue that the leader's 'fight against occultism' and mysticism was 'premature'; Adam Phillips, *Terrors and Experts* (London: Faber & Faber, 1995), pp. 20, 19.

[109] When Freud invited Ferenczi to accompany him to America, the Hungarian showed his eagerness by learning English and ordering books on America 'to get a proper orientation on that mysterious country'; Jones, *The Life and Work of Sigmund Freud*, op. cit., p. 342. Freud's misgivings cannot be explained by his aversion to travel. In 1916 he recommended that Rank should visit Troy after his own stimulating trips to Rome and Athens; see Martin Grotjahn, 'Otto Rank on Homer and Two Unknown Letters from Freud to Rank in 1916', *Journal of the Otto Rank Association*, 4(1) (June 1969), 77.

[110] Rank, *The Artist*, op. cit., p. 59. Rank read Twain voraciously as a young boy and granted Huckleberry Finn the same heroic status as the mythical protagonists in *The Myth of the Birth of the Hero*. As Virginia Robinson notes in 'The Double Soul: Mark Twain and Otto Rank', they both changed their names and shared an interest in the double; *Journal of the Otto Rank Association*, 6(1) (June 1971), 32–9.

[111] Ibid., pp. 360, 361.

humanistic psychology and his primary focus on emotional expression as the basis of therapy.[112]

In the opening chapter Ferenczi claims that the notion of 'active therapy' does not actually mark a break with Freud's technique of working through repressed psychic material by an act of remembering, but is actually an extension of it.[113] Referring to Freud's essay 'Remembering, Repeating and Working Through' (1914), Ferenczi and Rank place more emphasis on repetition of the neurosis complex which would lead to a 'gradual *transformation of the reproduced material into actual remembering*'.[114] This notion of acting out key stages of neurotic behaviour in analysis is a development of Rank's earlier work on *The Artist* in which the art-work is a restaging of an uncomfortable episode in the individual's life. Because personal experience is subject to temporal flux, the restaging of a potentially traumatic episode may serve to transform the individual's psychic material into a new set of interpersonal experiences in line with Kierkegaard's formula of repetition-with-difference: a matter of 'reexperiencing, of reproducing in contrast to remembering the preconscious material'.[115] This is not to divorce repetition from remembering, but rather to 'transform one . . . kind of repeating, the reproducing, into another mental form of repeating, the remembering'.[116] Less attention is given to knowledge of the traumatic experience and more emphasis placed on devising techniques in which to actively transform psychic material into a new mode of existence. For Rank, the dramatic restaging of the episode in the analytic situation would mirror the trauma of birth, but potentially transform such trauma into a rebirth experience which may reveal a path forward for the individual. Like James, the path which the creative individual takes away from 'the backward striving tendency' does not follow a unidirectional path, but spirals through a series of figurative encounters with early traumatic experience.[117]

Inevitably the analytic situation will involve a series of transferences and resistances on behalf of patient and analyst, the difference being that the Rankian analyst 'not only uses the transference . . . for the purpose of making the process of unwinding the libido easier, but also points out to the patient the transference at every step in order to finally free him from it'.[118] Initially the patient will approach the analyst as an ideal parent or a desired partner, but the analytic

---

[112] Sandor Ferenczi, *The Clinical Diary of Sandor Ferenczi*, ed. J. Dupont, trans. M. Balint and N.Z. Jackson (Cambridge, MA: Harvard U.P., 1988), p. 188.

[113] Similarly, in *The Trauma of Birth*, Rank's maintains his theory of birth trauma is only 'a contribution to the Freudian structured of normal psychology, [or] at best . . . one of its pillars'; Rank, *The Trauma of Birth*, op. cit., p. 210.

[114] Ferenczi and Rank, *The Development of Psychoanalysis*, op. cit., p. 4.

[115] Ibid., p. 11.

[116] Ibid., p. 26.

[117] Rank, *The Trauma of Birth*, op. cit., p. 216.

[118] Rank, *The Development of Psychoanalysis*, op. cit., p. 7.

encounter is designed to challenge such narcissistic identification by challenging the respective roles which the patient and analyst are playing. The notion that the analyst can help 'unwind' the patient's libido implies that, to recall Rank's theatrical allusions in *The Artist*, he or she should act as a catalyst for the patient's own psychic virtuosity, in much the same role as a director will encourage an actor to adopt a split-perspective: both to inhabit a role and to see the role from the outside. As such, the analyst must be both passive and active: passively allowing the patient to inhabit the stage of analysis and actively coaxing or provoking the patient into an acting out of their neuroses. Following Rank's insights in *The Trauma of Birth*, both analysts agree that the infantile fixation is less Oedipal and more primeval: 'in the correctly executed analysis the whole development of the individual is not repeated, but only those phases of development of infantile libido on which the ego . . . has remained fixed'.[119] This notion of the patient actively expressing him or herself serves to transform neurotic material as an 'adjustment to reality' in much the same way as the artist wards off neurosis by transforming it into art.[120] Rather than the analysis representing a special enclave outside the flow of the patient's life, it is believed to symbolize a distinct 'phase of experience'.[121]

In their assessment of the analytic situation Ferenczi and Rank emphasize the process of translating neurotic expression into productive material. This does not merely entail a theoretical knowledge of the patient's trauma but the ability to participate fully in the analytic situation in order to put 'academic interests aside as much as possible in their practical application'.[122] The analytic practitioner should 'essentially' be a 'craftsman', rather than a theorist 'overflooded by theoretical speculation'.[123] The two analysts do not wish to abandon scientific principles wholesale, but develop a more flexible relationship with the patient, allowing time for behavioural observation and encouraging reflection to enhance a fuller interpersonal understanding. In the final chapter of *The Development of Psychoanalysis*, Rank claims such a 'doctor of souls'[124] is both scientist and artist:

> Under the influence of this increase in consciousness the physician, who has developed from the medicine man, sorcerer, charlatan, and magic healer, and who at his best often remains somewhat an artist, will develop increasing knowledge of mental mechanisms, and in this sense, prove the saying that medicine is the oldest art and the youngest science.[125]

---

[119] Ibid., p. 19.
[120] Ibid., p. 27.
[121] Ibid., p. 55.
[122] Ibid., p. 53.
[123] Ibid., pp. 63, 58.
[124] Ibid., p. 65.
[125] Ibid., p. 68.

Neither complete empathy with the world of the patient nor comprehensive technical knowledge of it will equip the analyst with the multiple perspective which Rank believes is central to protecting the self from paralysis or extinction. By initiating role-playing exercises Rank argues that the analyst can awaken the patient's ability to cope with threatening or uncomfortable forces and find within themselves a hidden artistic virtuosity.

At the centre of his model of active therapy Rank places the Jamesian dynamic of willing. Rank claims Freudian analysis represents a throwback to Schopenhauer's denial of the will in viewing any manifestation of the patient's assertiveness as a form of resistance which needs to be eradicated before a resolution can be foreseen, whereas he invests 'therapeutic value' in the conflict of wills between analyst (or, more accurately, therapist) and patient.[126] Rank recommends the therapist should refrain from judging resistance to be bad and actively encourage the patient's own 'insistence on personality'.[127] In a summary of his therapeutic ideas in *Will Therapy* (1936), Rank contrasts psychoanalysis (which he characterizes as a pedagogic technique of conquering resistance) with 'constructive therapy': the goal of which is 'the transformation of the negative will expression', or the patient's 'counter-will', 'into positive and eventually creative expression'.[128] In this scenario, the analyst would assert his or her own will as a means to activate the patient's own sense of self. The end result of such practice is to make the patient aware that neurosis derives largely from a paralysed or negative will and, by actively encouraging the clash of wills in therapy, to enable patients to impose their will positively, or creatively, in interpersonal situations. Drawing much of the impetus of will therapy from his work on *The Trauma of Birth* (but without his earlier emphasis on the biological moment of birth), Rank encourages the patient to realize his or her individuality in an interpersonal experience of creative rebirth. The 'rehabilitation of the will' does not constitute a Nietzschean ethos of will-to-power, nor the transcendence of corporeality, but a realization of hidden abilities which may help to heal internal divisions and expend psychic energy productively.[129] In essence, the patient should learn that he or she is both biological 'creature' and aesthetic 'creator' and utilize such understanding in a life-inducing manner.[130] This double mode of existence does not replicate the internal schisms of the *Doppelgänger* experience, but locates the seat of complementary in the creative refashioning of the self.

---

[126] Rank, *Will Therapy*, trans. Jessie Taft (New York: Norton, 1978), p. 17.
[127] Ibid., p. 8.
[128] Ibid., p. 19.
[129] Ibid., p. 10.
[130] Rank, *Truth and Reality*, trans. Jessie Taft (New York: Norton, 1978), p. 2.

The Artist's Life

By the time Rank was jettisoned from the Vienna circle of analysts he had achieved a number of major goals in his work: outlining the applicability of artistic and mythic analysis to psychoanalysis, providing a biological foundation for psychic understanding and recommending a patient-centred approach to analysis to counter the overly technical side of Freud's work. In the mid to late 1920s Rank spent his time between Paris and New York holding seminars on his analytic technique and practising a form of active therapy in which he set a variable but short-term time limit depending on the individual and the nature of the condition. During analysis he would work extensively with the patient for a few weeks and then began to wean them off analysis in the end phase of treatment by seeing them less frequently. By refining this method he hoped to encourage his patients to widen the repertoire of roles they might adopt in social situations and circumvent potentially neurotic limitations. From the mid-1920s, one of his patients Jessie Taft, an American psychologist who met Rank at an American Psychoanalytic Association conference in Atlantic City in June 1924, did much to translate Rank's work and promote him in the Anglophone world.[131] In her later study, *Otto Rank* (1958), she claimed that at that time she was unaware of the enmity towards Rank and in 1927 she joined a growing group of supporters dedicated to discussing and disseminating Rank's ideas among American psychiatrists and analysts. Such was the friendship between Rank and Taft that she was responsible for translating *Will Therapy* and *Truth and Reality* in the mid-1930s which together summarize Rank's analytic position and his therapeutic emphasis on the will. Taft's commentary and her access to Rank's notebooks after his death are invaluable additions to Rankian scholarship, but the creative writers Henry Miller and Anaïs Nin provide most insight into the development and the aesthetic implications of Rank's thought in the early 1930s.[132]

While in Paris in 1933 Rank began his acquaintance with Henry Miller, the expatriate American writer who had left New York for Europe in 1930 to escape what he perceived to be the shallow progressivism of 'Babbittry' and the moral and creative restrictions of his homeland. As Miller wrote to Nin in March 1933, he came away from his first meeting with Rank

---

[131]   In tribute to Taft who died in 1960, Anita Faatz, a leading member of The Otto Rank Association which ran until December 1982 (with 500 members at its peak and branches in five American cities), called her 'the one person to whom everyone had turned for answers to questions about Rank' after his death; Anita Faatz, 'Individuals in Association', *Journal of the Otto Rank Association*, 1(1) (Fall 1966), 83.

[132]   Much of Rank's correspondence and many of his manuscripts are available for consultation in the Rare Book and Manuscript Library, Columbia University, New York City.

in a state . . . like that which must have been Napoleon's when he came
away from the signing of the great treaty – a mere affixing of the signature
to something long ago inspired, envisaged and destined – the seal of
conviction as it were.[133]

The almost mythological self-image which this initial meeting gave Miller was
accompanied by 'a tremendous and fathomless exultation going on in the depths
of me, a trampling down of many lives in me, of many failures, of many
misgivings'.[134] There is little in Miller's account to suggest more than a
stimulating exchange of ideas and, rather than seeing Rank as a mirror of his own
cultural uprootedness, he chastises him for his 'self-deceptive' optimism 'which
always leads the German mind, in the end, into the bogs of hopeless
mysticism'.[135] Although Miller concludes his letter by claiming flirtatiously that
Nin herself (not Rank) was 'the living example, the guide who conducted me
through the labyrinth of self to unravel the mystery of myself', his fictional work
from the mid-1930s displays the creative dynamic which Rank had explored from
his early daybooks through to his extended study of the artist in *Art and Artist*
(1932):

> A poet is like a sponge. He absorbs himself in an experience, an adventure,
> a pain and then puts out all the accumulated material in a concentrated
> form. There follows . . . a time of thinness, dryness, and emptiness until a
> new life experience stirs him up again.[136]

This life-philosophy of creative energy which afterwards collapses into a period
of 'dryness' derives from Schopenhauer's basic human dynamic of desire
followed by boredom, but, unlike Schopenhauer, Rank shares with Nietzsche the
belief that wilful activity can be asserted in order to make meaningful patterns out
of the chaos of neurotic life.

Miller was attracted to the way in which Rank's emphasis on the will restores
'the magic, the deeply religious quality to his creative faculties'.[137] In the
aftermath of the Great War and in a decade when many American writers were
expressing a loss of cultural bearings, such a philosophy provided a lifeline for
creative production. Although Miller's work is full of despair, out of such
dissolution emerges a creative individual whose life fuses with his art in an act of
spontaneity: his first-person fictional personae are impressionistic in their
outlook and promiscuous in their engagement with life. Miller's fiction in the
1930s explores the romantic themes of growth, freedom and self-creation but

[133] Henry Miller, 'Encounter with Rank' (1933), *Journal of the Otto Rank Association*, 1(1) (Fall 1966), 58.
[134] Ibid., p. 58.
[135] Ibid., p. 59.
[136] Taft, *Otto Rank*, op. cit., p. 26.
[137] Henry Miller, 'Encounter with Rank', op. cit., p. 64.

with none of the noble sentiments of the mid-nineteenth-century American romantic writers Whitman and Emerson. His rejection of America in *Tropic of Cancer* (1934), in which he describes New York as 'a whole city erected over a hollow pit of nothingness', and his persistent nihilism are distinct from Rank's optimistic association with the spirit of American progressivism, but his emphasis on wilful creativity and the special status of the artist are apiece with Rank's philosophy: for Miller, 'an artist is always alone – if he *is* an artist'.[138]

Miller was particularly interested in the way in which Rank developed Nietzsche's insights about the transgression of limitations which leaves the individual morally free to explore his or her own desires. Out of the nothingness which Miller's protagonists have become, divorced from their homeland and without moral bearings, emerges a new being:

> I made up my mind that I would hold on to nothing, that I would expect nothing, that henceforth I would live as an animal, a beast of prey, a rover, a plunderer . . . The dawn is breaking on a new world in which the lean spirits roam with sharp claws.[139]

Such descriptions suggest that Miller's protagonists are both creatures in the biological sense and wilful creators who can adapt to their environment without totally transcending it. Only by adopting a wilful outlook can such a protagonist prevent himself becoming stifled by a moribund environment in which myths are extinguished by the threat of finitude: 'our heroes have killed themselves, or are killing themselves'.[140] Much of Miller's work is characterized by an aggressive masculine stance in which he poses as a predatory beast, but elsewhere he responds sensitively to the fluid 'chaos' of creative life with a more androgynous voice:

> Once you have given up the ghost, everything follow with dead certainty, even in the midst of chaos. From the beginning it was never anything but chaos: it was a fluid which enveloped me, which I breathed through the gills . . . In everything I quickly saw the opposite, the contradiction, and between the real and the unreal, the irony, the paradox.[141]

The beginning of *Tropic of Capricorn* (1939) displays fuller evidence of Rank's theory of the birth trauma and the artist's emergence from the womb as a creative being. Furthermore, Miller's ability to embrace contradictions suggests that a double or multiple life can be embraced in a potentially liberating manner even in the face of adversity. The potent image of the river Seine at the end of *Tropic*

---

138   Henry Miller, *Tropic of Cancer* (London: Flamingo, 1993), pp. 74, 72.
139   Ibid., p. 104.
140   Ibid., p. 9.
141   Miller, *Tropic of Capricorn* (London: Flamingo, 1993), p. 9.

of Cancer suggests a psychic flow which the artist can harness, providing a temporary resting place in the midst of experience.

If Miller's fiction is a masculinist development of Rank's work then Anaïs Nin provides another aesthetic interpretation of the Austrian analyst's work. Nin did more than anyone else to bring Rank's work to the attention of the reading public in her fiction and her accounts of meetings with Rank, published in her multi-volume diary in the 1960s. From the mid-1930s Nin's fiction was heavily influenced by Rank's therapeutic aims, providing a creative outlet for exploring her erotic fantasies.[142] Before she met Rank, Nin was already beginning to style herself as an artist who 'lived out all the themes he wrote about, The Double, Illusion and Reality, Incestuous Loves Through Literature, Creation and Play', but confessed her need for therapy in the face of psychic confusion.[143] The expression of her life as a 'shattered mirror' derived partly from her national rootlessness (the tensions between her Spanish family background, her adolescence in New York and her adult life in Paris) and her fraught relationship with her father 'who had told her she was ugly, beaten her, adored her with his camera lens, and then abandoned the family'.[144] Early in therapy she describes her life as a labyrinth (developing Miller's metaphor), without a discernible shape, claiming there are 'always, in me, two women at least', but describes herself as neurotic (a 'failed work of art') because she can find no way of connecting these split selves.[145] Rather than trying to resolve such conflicts, Rank encouraged her to dramatize them by improvising new disguises and personae, as a means of preventing her from slipping back into psychic chaos and providing these new inventions with direction and purpose. With this reassurance, Nin admits: 'I felt that my lost identity was already being reconstructed with his recognition and vision of me. He had not thrown me back upon a vague ocean of generalities, a cell among a million cells.'[146]

However, Nin's feelings of dependency suggest that Rank soon begins to take the place of her father as the director of her life but his role (as she soon realizes) is to facilitate, or provide a channel for, her own creative rebirth. In this way Rank hopes to utilize the transference of Nin's feelings towards her parents on to him and he encourages her to increase her 'power of creation in order to sustain and

---

[142] Nin treated Rank as a fellow artist corroborating Rank's earlier opinion of himself: 'Otto Rank was a poet, a novelist, a playwright, in short a literary man, so that when he examined the creative personality it was not only as a psychologist, but as an artist'; Anaïs Nin, 'Rank's Art and Artist' (1968), in ed. Philip K. Jason, *Anaïs Nin Reader* (Chicago: Swallow, 1973), p. 280.

[143] Anaïs Nin, *The Journals of Anaïs Nin, Volume 1: 1931–1934*, ed. Gunther Stuhlmann (London: Peter Owen, 1970), p. 269.

[144] Ibid., p. 270; Noel Riley Fitch, *Anaïs: The Erotic Life of Anaïs Nin*, op. cit., p. 11.

[145] Nin, *The Journals of Anaïs Nin, Volume 1*, op. cit., p. 270.

[146] Ibid., p. 273.

balance the power of emotion'.[147] This model of therapy is not primarily retrospective, conjuring up a past self in a moment of calm contemplation, because 'there is no one stable viewpoint . . . that would enable us to explain all phenomena on a single basis'. Nor is the patient encouraged to confront Oedipal trauma as recommended by Freud, because Rank believes it infantilizes the patient's condition, suggesting 'that he [or she] cannot wholly be free from the past'.[148] In order to resist the psychoanalytic emphasis on the past, Rank argues that any retrospective foray should be an attempt to reorient the self in the present and towards the future: 'this is the authentic psychological side of the so-called "reality problem", which is nothing but the problem of the present, in other words, the consciousness of living'.[149] The path-finding which James encourages is echoed in Rank's recommendation that an awareness of lines of convergence between different manifestations of the self should not be replaced by manic recreation. Instead, active creation should be tempered by moments of reflection without leading to either infantile fixation, paralysis of the will, or the fear that the self can never be fully known.

Both Miller and Nin shared Rank's view of aesthetic creation as a liberation from potentially paralysing aspects of the self – in Miller's words, as a progression from 'the *via purgativa* to the *via unitiva*'.[150] The multiple personae which the two writers create in their fiction are coextensive with their living experiences: Nin understands Rank to mean 'the flow of life and the flow of writing must be simultaneous so that they may nourish each other'.[151] Even when the writer is exploring the past, the individual is encouraged to look toward future manifestations of the self: 'henceforth I travel two ways, as sun and as moon. Henceforth I take on two sexes, two hemispheres, two skies, two sets of everything. Henceforth I shall be double-jointed and double sexed.'[152] Here, doubleness represents a positive splitting of the self into complementary fragments instead of replicating Balduin's fatally divided life in *The Student of Prague*. This notion of *dédoublement* echoes James' description of the sick-minded individual who perceives (usually overlooked) correspondences between self and world, suggesting that ambiguity and paradox (as well as existential suffering) are central aspects of human existence. Rank's chief therapeutic aim is to facilitate links between the different aspects of personality, rather than embarking on the futile attempt to fully know the self. In this way, Nin claims she learns to 'describe fragmentation without the disintegration which usually

---

147 Ibid., p. 283.
148 Rank, *Will Therapy*, op. cit., p. 30; p. 36.
149 Ibid., p. 41.
150 Jonathan Cott, 'Reflections of a Cosmic Tourist' (1975), in Frank L. Kersnowski and Alice Hughes, eds, *Conversations with Henry Miller* (Jackson: University Press of Mississippi, 1994), p. 190.
151 Nin, *The Journals of Anaïs Nin, Volume 1*, op. cit., p. 283.
152 Ibid., p. 191.

accompanies it. Each fragment had a life of its own.'[153] Both Miller and Nin can be seen to construct fictional selves in order to proliferate future possibilities which may not otherwise be available; this explains why they share Rank's fondness for the metaphors of 'journey' and 'path' to describe this creative process.[154]

Nin's diaries were especially useful for recording her daily experiences and enabling her to enlarge the boundaries of objectively verifiable reality. She distinguishes diary-writing from fiction in that the former represents a discursive 'laboratory' (echoing Rank's 'microscope' in his *Tagebücher*), in which she 'checked realities and illusions', retained 'psychological authenticity' and provided her with the freedom to 'fictionalize only externals, situations, places'.[155] In her long essay *The Novel of the Future* (1968) she reflects on her growing awareness of the 'mutual influence' of diary and fiction as two aspects of the same creative continuum; diary writing is not as laborious as fiction but, according to Nin, maintains 'impetus and the exhilaration born of freedom'.[156] Moreover, the diary enables the writer to keep touch with the self behind the roles which society induces the individual to play: 'I had to find one place of truth, one dialogue without falsity.'[157] This does not imply that Rank or Nin believed there is an essential self behind all the masks, but merely that the process of reflection which diary writing can facilitate may enable the individual to trace patterns between and behind these social fragments. Indeed, Rank warns that the dominant self which manifests itself in the diary should also be resisted as potentially delimiting. Consequently, late in 1933, Rank suggested that Nin abandon her intimate diaristic self-analysis (because he feared her condition was narrowing and she was 'being kept' by the diary) and revert to the form of a 'sketchbook' in order to participate in wider cultural myths and stories which may give more definite shape to her life.[158] On this model, switching between different written forms prevents the multifarious self from rigidifying into a particular type, while keeping creative outlets open.

In essence, Nin links creative development with sexual liberation from fears and preventative barriers in the self. This view is very similar to James' notion of potential energies which usually remain untapped, but which may provide a resource for contending with psychic and social obstacles. The emphasis on kineticism in James is echoed in the sensuality dramatized in Nin's writing. Just like Miller's expression of doubleness, Nin's mature writing is full of bisexual expressions which enable her to transgress the conventional sphere of femininity

---

[153] Nin, *Anaïs Nin Reader*, op. cit., p. 301.

[154] Ibid., p. 301; Kernowski and Hughes (eds), *Conversations with Henry Miller*, op. cit., p. 190.

[155] Nin, *Anaïs Nin Reader*, op. cit., p. 296.

[156] Ibid., p. 299.

[157] Nin, *The Journals of Anaïs Nin, Volume 1*, op. cit., p. 286.

[158] Ibid., pp. 289, 298.

without the threat of social prohibition. Unlike the therapeutic limitations which the American poet H.D. (Hilda Doolittle) detected in Freud's analysis of her (she characterizes him as a cold and rational scientist), Nin's relationship with Rank is much more sensual.[159] Although their relationship in the mid-1930s developed outside strictly analytic boundaries, Rank continued to act primarily as a facilitator of Nin's exploration of herself. She characterizes him in a later story, 'The Voice' (1948), as a 'man without identity . . . the Voice of the man who was helping her to be born again'.[160] In this semi-fictional piece, 'The Voice' helps the character Djuna discover her 'larger' self which pushes 'the smaller one to act and speak greatness, not smallness or doubt or fear'.[161] The dynamic of repetition-with-difference which Rank promulgates throughout his work suggests that the self is in flux but also self-sustaining: 'The blood must pass. There must be change. [But] this was only a tide, and the self turning, was feeding the rotation of desire.'[162] Such creative 'rotation' is not only to be found in formal therapy or fictional explorations: for example, Nin's role as dancer in the American avant-garde film-maker Maya Deren's *Ritual in Transfixed Time* (1946) and masquerade character in Kenneth Anger's *Inauguration of the Pleasure Dome* (1954) enabled her to enact fantasies in an aesthetically heightened dreamscape in which new performative selves could emerge from sensual interaction with others.

   Both Miller and Nin develop Rank's insights in their work without reverting to the nineteenth-century aesthetic of self-development. By embarking on a middle path between the linear form of the *Bildungsroman* and the modernist novel of psychic disintegration, the two late modernist writers imbue their fictional selves with complex patterns. Rank does not apply a universal template because he realizes these patterns take many forms and evolve through time and with new experiences. But he does allude to possible patterns in his reference in *The Double* to Adalbert von Chamisso's quasi-picaresque tale *Peter Schlemihl's Remarkable Story* (1814), in which the eponymous hero is dogged by misfortune but lives an essentially good, if wayward and haphazard, life. Similarly, Rank shares with James an interest in the figure of the spiral which suggests an open circle from which productive life can issue, without the certainty of linear development. The spiral also recalls Rank's earlier work on myth which corresponds with the French anthropologist Claude Lévi-Strauss' comments in 'The Structural Study of Myth' (1955), in which he argues that, although there is an repetitive structural pattern to myths, their growth and development develops through time and cumulative retellings: 'myths grow spiral-wise until the

---

[159] For an account of H.D.'s literary exploration of Freudian ideas see Claire Buck, *H.D. and Freud: Bisexuality as a Feminine Discourse* (New York: Harvester Wheatsheaf, 1991).

[160] Anaïs Nin, *Winter of Artifice* (London: Peter Owen, 1991), p. 122.

[161] Ibid.

[162] Ibid., p. 124.

intellectual impulse which has produced it is exhausted'.[163] On this view, each retelling transforms the myth in an unfolding spiralling process, just as a particular speech act (*parole*) works within the structures of language (*langue*), but is unique in terms of timbre and semiotic significance. For Rank, such an understanding bridges the scientific, artistic and moral concepts of the self: a spiralling life entails an acceptance of the limitations of biology as the structuring template, a commitment to live with others as an acknowledgement of intersecting myths and an individualistic enactment of the artist's 'will to form'.[164] However, while Miller and Nin explore these ideas in their stories and diaries, their creative individualism often characterizes the excessive narcissism which Rank warned against in his early work. The commitment to others which recurs as an ideal throughout Rank's work remains a frustrated promise in the prodigal fictional adventures of Miller and Nin.

Beyond Psychology

In a paper delivered for the First International Congress on Mental Hygiene, Washington DC in May 1930, and in the presence of detractors like Franz Alexander, Rank discussed what he called the 'vital' human element of individuality which cannot be 'measured and checked and controlled' by scientific analysis of it.[165] He argued that although psychoanalysis had recognized 'the high importance' of this irreducible 'human side', it tended to concentrate on the 'transference phenomenon' at the expense of 'the most essential aspect of personality'.[166] He also emphasized the individual's personal experience and self-knowledge (both introspective and practical knowledge), over the 'intellectual knowledge' which provides the cornerstone of psychoanalysis.[167] Far from suggesting that these 'two world views' in which humans are either creature or creator are intrinsically incompatible, Rank suggested that they only diverge at their extremes when theory is divorced from therapy or when analysis loses its practical applicability. In order to find a middle ground between the self-explorations of his early daybooks and the biologically grounded work of *The Trauma of Birth*, Rank's work in the 1930s is representative of an almost Nietzschean attempt to surpass the limits and restrictions of rationalistic psychology. He continued to fuse the theory and practice of analysis in his writing and began to develop an ethically oriented therapeutic discourse which checks the heavy emphasis on individuality displayed in his interaction with Nin and Miller.

---

[163] Claude Lévi-Strauss, *Structural Anthropology, Volume 1*, trans. Claire Jacobson and Broke Grundfest Schoef (New York: Basic Books, 1963), p. 229.

[164] Rank, *Art and Artist*, op. cit., p. 4.

[165] Taft, *Otto Rank*, op. cit., p. 148.

[166] Ibid., pp. 148–9.

[167] Ibid, p. 149.

Rank's work in the 1930s developed his understanding of the self in three different, but complementary, ways. Firstly, in *Truth and Reality* and *Will Therapy* (both 1936) he clarified his therapeutic goals by explaining the role of the will and outlining the advantages of short-term sessions with his patients. Secondly, in *Psychology and the Soul* (1930) and *Art and Artist* he developed another way of understanding individual psychology: what he calls 'soul science' (*Seelenkunde*) is commensurate with James' desire to radicalize the sphere of empirical observation in its rejection of the limitations of rationalistic and 'soulless' psychology.[168] The third strain of thought is most evident in his posthumously published book *Beyond Psychology* (1939), in which he widens the field of inquiry of orthodox psychology in order to locate creative individuality within a broader sphere of cultural activity. The remainder of this chapter discusses the type of romantic science which Rank outlined in the last phase of his career, focusing particularly on this third trend in his work. However, because aspects of his socially oriented understanding of the self are only partially formed, especially the emergent theory of ethical commitment to others, I will draw on Stanley Cavell's discussion of 'moral perfectionism' to identify a central dimension of romantic science which Rank started to articulate in his later work before his early death in 1939.

In *Psychology and the Soul* Rank asserts that the 'soular' dimension of human beings is usually overlooked by empirical science because of its intangibility.[169] The problem with Freudian psychoanalysis is that it does not embrace the philosophical or speculative discourses which are necessary to conceptualize the soul. For Rank, psychoanalysis should be 'a unique marriage' of science and philosophy 'without a clear separation or differentiation of the two spheres'.[170] The fact that most rival versions of psychoanalysis are one-sided (Rank cites the dissidents Jung and Adler) suggests that this 'two-sided' marriage is extremely precarious to maintain.[171] Rank does not want to develop a dualistic science but one that creatively blends opposites into a therapeutic manifold. Scientific psychology denies the existence of the soul because it only understands a narrow version of human activity; for Rank, psychoanalysis should deal not only with facts and causal explanations but 'ideas created by soul-belief'.[172] James' claim

---

[168] Rank, *Psychology and the Soul: A Study of the Origin, Conceptual Evolution, and Nature of the Soul*, trans. Gregory C. Richter and E. James Lieberman (Baltimore: John Hopkins U.P., 1998), pp. 1, 2.

[169] Lieberman notes that *seelisch* is best translated as 'soular' in order to free it from the religious implications of 'spirit'; Lieberman, *Acts of Will*, op. cit., p. xxxii.

[170] Ibid., p. 6.

[171] Anita Faatz, 'The Summing Up', *Journal of the Otto Rank Association*, 7(1) (June 1972), 42. In 1929, Rank argued that Jung emphasized the 'racial factor' too heavily and Adler went too far in grounding an understanding of the individual within a social context; Rank, 'The Psychological Approach to Personal Problems' (1929), *Journal of the Otto Rank Association*, 1(1) (Fall 1966), 14.

[172] Rank, *Psychology and the Soul*, op. cit., p. 125.

that the will is a valid hypothesis as long as an individual believes in its existence also holds for this conception of the soul as a creative dimension of the self. Rank's major claim in *Psychology and the Soul* is that psychoanalysis (and psychology in general) should deal with 'attitudes of the self' which can only be detected through interaction with others and not by means of passive observation.[173]

Beyond Psychology develops some of the key issues introduced in *Psychology and the Soul* and represents Rank's fullest attempt to articulate his desire to transgress the limitations of psychoanalysis, a term he includes under the broader study of 'rationalistic' and explanatory psychology. He argues that the historical crisis and 'bewilderment' of the late 1930s in part derives from the collective attempt to explain irrational behaviour from within a rational system of thought, in order to make it 'intelligible' and 'acceptable'.[174] What is lost in such a system is the 'irrational basis of human nature' which psychoanalysis cannot address without translating it into the order of reason and mastery.[175] As an antidote to rational discourse, Rank portends the French Feminist revolt against patriarchal analysis in the 1970s, by suggesting that psychology needs to develop a 'new vocabulary' for understanding and articulating irrationality: in short, 'a psychology of difference'.[176] Partly an individualistic and rebellious act of the will 'to create beyond' the 'given natural self' and partly an acquiescence to unknown (and unknowable) dimensions of the self, this understanding promises to break free from the shackles of natural science, but without offering any ultimate 'solution to our human problems'.[177] The balance between giving forth and yielding is directly comparable to James' recommendation in *Varieties* that the individual should strive to modify delimiting aspects of his or her life-world, without, on the one hand, giving up a notion of agency or, on the other, becoming a ruthless Nietzschean individualist. James and Rank share the belief in the inherent adaptability of human beings to psychic and environmental circumstances, arguing that internal schisms need not diminish the self, but can 'operate as a force of balance and not only as a source of conflict'.[178] Because clean empirical science cannot do justice to these ideas, the psychology that Rank recommends is 'not an objective instrument, like a telescope or microscope' (developing his earlier metaphor of 'a human soul under the microscope'), but actually dirties itself in the confusion of modernity. Here, Rank's earlier idealism

---

[173] Ibid., p. 128.

[174] Rank, *Beyond Psychology* (New York: Dover, 1941), p. 11.

[175] Ibid., p. 12.

[176] Ibid., pp. 12, 29. The masculine bias to his early work is replaced not only by his attention to the role of the mother for individual development but also, in *Beyond Psychology*, to the plight of women in general who have suffered from 'suppression, slavery, confinement and subsequent persecution'; ibid., pp. 287–8.

[177] Ibid., pp. 11, 16.

[178] Ibid., p. 21.

is tempered by a more 'realistic' or 'living' psychology which he hopes can contend with the darker irrational forces at large in Europe on the brink of another war.[179]

Rank foreshadows the Foucauldian notion that the self regulates and polices socially unacceptable impulses in order to maintain a unified personality, but often at the risk of debilitating neurosis. Indeed, such techniques of 'self-observation' and surveillance are at odds with Rank's 'introspective self-therapy' as a tool for broadening the sphere of personal experience.[180] In *Beyond Psychology* Rank bemoans the fact that American democratic principles have led to a levelling of individuality, rather than creating a fertile environment of diversity and cultural difference. Rather than democracy embodying a political principle, following Kant he maintains that the democratic ideal is primarily an ethical imperative. As such, the individual's experience of irrational forces which are normally held in abeyance need not lead to excessive narcissism nor a life of capricious impulse, but actually 'permits the individual to accept his inner limitations or outer restrictions in his own terms and on his own free volition'.[181] Rank attempts to reconcile the opposing forces of determinism and free will by suggesting a future-oriented philosophy which harnesses the will as a 'spontaneous organising force', partly determined and partly within the sway of human volition.[182] This intermediary zone does not frustrate creativity, but provides a environment in which creativity can flourish as a striving towards a yet-to-be-achieved goal, either as a 'protest against denied liberty' or as temporary release from self-captivity.[183] More than anything else, such a philosophy gives the individual a set of therapeutic techniques which ensures the maintenance and development of the self and helps bridge rifts in personal, spiritual and social worlds.

Both *Psychology and the Soul* and *Beyond Psychology* gesture towards an ethical imperative as a 'science of relationships', which Rank argues should be the individual's primary motivating force for looking beyond the narcissistic limitations of the self, but he does not explore these ideas as far as he might.[184]

---

[179] Ibid., p. 27.
[180] Ibid., p. 44.
[181] Ibid., p. 47.
[182] Ibid., p. 50.
[183] Ibid., p. 57.
[184] Rank, *Psychology and the Soul*, op. cit., p. 2. In a joint review of *Truth and Reality* and *Will Therapy*, the American social theorist Kenneth Burke argues that Rank fails to offer a meaningful 'bridge' between creative individualism and the 'collectivistically "communicative" aspect' of social and political existence; Kenneth Burke, 'Without Benefit of Politics', *The Nation* (18 July 1936), 78. However, in an Advanced Curriculum Course at the Pennsylvania School of Social Work in 'The Symbols of Government' taught between 1937 and 1938 (on which Burke's *Permanence and Change* (1935) was recommended reading), Rank began to outline some of the productive links between irrationality and social law.

Here, Stanley Cavell's lectures *Conditions Handsome and Unhandsome* (1990) provide a means of developing some of the romantic-scientific implications of Rank's late work, especially his Jamesian conception of the interdependence of yielding and action. Cavell takes the title of his lectures from Emerson's essay 'Experience' (1844): 'I take the evanescence and lubricity of all objects, which lets them slip through our fingers then when we clutch hardest, to be the most unhandsome part of our condition.'[185] To grasp or clutch at the world implies that the 'unhandsome' self is one who acts out of desperation in the face of doubt and uncertainty.[186] Despite the urgency of which Cavell and Emerson speak, only by allowing the world 'to be' can one be committed or resolute in relation to it. Not only does this stimulate a renewed set of relationships with and in the world, but it enables the individual to devise techniques and use what is 'to hand' to maximize the possibility of transforming it.[187] A secure self cannot be rendered through passive conformity or a quietist world-view, but neither can change be brought about by furiously grasping at the world. Cavell thus understands the 'handsome' part of the human condition to be 'what Emerson calls being drawn and what Heidegger calls getting in the draw, or the draft, of thinking'.[188] The 'draft' of thinking is brought about through interaction and provides 'the conditions for my recognizing my difference from others as a function of my recognizing my difference from myself'.[189] What makes Cavell's reading of selfhood interesting for a consideration of romantic science and Rank's psychology of difference is that, as the 'knotted' self sides with the next potential self, it must also side 'against my attained perfection'.[190] In the knots of the self the individual is obliged to turn away from stasis and solipsism to recognize the existence of 'an other' and to accept and adjust to that Other (as Rank sees it), whether it is within the context of family, community, or a personal relationship with the physician.

One of the most problematic areas of thought which Cavell has developed over recent years is his discussions of 'moral perfectionism'.[191] He defines this term variously as 'being true to oneself' (often by a romantic act of defiance), a 'tradition of the moral life . . . called the state of one's soul' and the neoplatonic 'transformation of self through self-education'.[192] Cavell acknowledges finally that not only has he 'no complete list of necessary and sufficient conditions for

[185] Emerson, *Lectures and Essays*, op. cit., p. 473.
[186] In *Psychology and the Soul*, Rank argues that most scientific psychologies understand the self and the world by 'grasping' at them, rather than interacting in a reciprocal manner; Rank, *Psychology and the Soul*, op. cit., p. 125.
[187] Cavell, *Conditions Handsome and Unhandsome*, op. cit., p. 38.
[188] Ibid., p. 41.
[189] Ibid., p. 53.
[190] Ibid., p. 31.
[191] Ibid., pp. 1, 3.
[192] Ibid., pp. 3, 7.

using the term', but he has 'no theory in which a definition of perfectionism would play a useful role'.[193] His understanding of genre as open to a number of possible permutations (as discussed in the Introduction) is comparable to the 'open-ended thematics . . . of perfectionism' which can be redefined to suit the needs of the particular individual.[194] The criticism that Cavell endorses a form of individual exceptionalism (in the sense of a perfected or perfectible self) can be defended when one considers that for neither Rank nor Cavell does the self imply perfectibility. For example, Rank is attracted to Nietzsche's 'lifelong' glorification of his illness, 'because he discovered through his own experience that *becoming* well is of greater value than being well'.[195] Just as there are no absolute grounds for positing the self (at least epistemologically speaking), neither is there a *telos* at which point one can know the self through reflection. Instead, Cavell understands that '*each* state of the self is, so to speak, final'; it is possible to understand the self in a provisional sense, but the potential 'next self' suggests the impossibility of final understanding.[196] That is not to say that the self cannot maintain identity, but only that the interpretative process leaves traces of unknowability, or a residue of mystery, which defies the ability to arrive at final understanding. In this sense, Rank's will therapy concurs with Jamesian path-finding: the individual does not get 'well', but becomes 'creative'.[197]

Cavell's hermeneutics develop Rank's belief that there is no truth as such to be understood, but a set of possibilities, or pragmatic paths, which can lead one through the kind of dense network of possible selves dramatized in the fiction of Miller and Nin. Just as the self is not to be understood by extrapolating backwards from a *telos*, neither can the journey along an interpretative path be understood 'beyond the way of the journey itself'.[198] Cavell concludes that to recognize the 'unattained but attainable self' is actually 'a step in attaining it': in other words, '"having" "a" self is a process of moving to, and from, nexts . . . we are from the beginning, that is from the time we can be described as having a self, a next, knotted'.[199] The implication (which Rank does not develop much further than his early work on myths) is that the self must be positioned within a narrative framework to lend it spatial and temporal coordinates. Although stages of the narrative of self may suggest closure, the plotting of a master-narrative, or final denouement of a unifying narrative, is deferred because there is always the possibility of a 'next' narrative to be told: in other words, the self is always actively under way and is never finally attained. Cavell's work extends Rank's claims in *Beyond Psychology* that cultural production 'serve the purposes of

---

[193] Ibid., p. 4.
[194] Ibid.
[195] Rank, 'Self-Inflicted Illness', op. cit., p. 14.
[196] Cavell, *Conditions Handsome and Unhandsome*, op. cit., pp. 3, 9.
[197] Rank, 'Self-Inflicted Illness', op. cit., p. 14.
[198] Cavell, *Conditions Handsome and Unhandsome*, op. cit., p. 10.
[199] Ibid., p. 12.

strengthening the self' within a social framework of ethical commitment, and propels Rank's interest in creative individuality in the narrative direction which the later romantic scientists, Erikson and Sacks, would develop in more depth. Rank's lasting legacy, however, is his attempt to move beyond the normative psychology of Freudian psychoanalysis and to embody a set of therapeutic ideas deriving from romanticism and articulated as a psychology of difference.

Chapter 3

Ludwig Binswanger: The Existential Romantic

Despite James' attempts in *Principles* to resolve the dichotomy between the physical and psychological dimensions of the self and Rank's desire to study human beings as both creature and creator, their redescriptions of the embodied self remain troubled by metaphysics. Overlapping with the major phase of Rank's work, the Swiss psychiatrist and existential analyst Ludwig Binswanger (1881-1966) tackled a similar set of issues between the 1910s and 1950s, in order to reconceptualize the sphere of human existence without retaining the dualisms implicit in the work of James and Rank.[1] Binswanger radically questioned the 'Magna Charta' of psychiatry established in the mid-nineteenth century, which, like Freud's early work, based the psychological study of humans on a model taken from natural science. Binswanger's extensive correspondence with Freud in the 1910s revolved around a disagreement over the purpose of analytic inquiry, echoing the concerns of Adler, Jung and Rank about the authority Freud was beginning to exert over European psychoanalysis. However, although he was careful to distance himself from psychoanalytic terminology, Binswanger's emphasis on Being (*Sein*) was not devised as an outright assault on Freudian psychoanalysis, but an attempt to complement it by introducing an dimension of inquiry hitherto neglected in the human sciences.

Whereas James' attention was directed primarily towards either himself as experiencer or the abstract subject of *Principles*, Binswanger shared with Rank a focus on the idiosyncratic life-world (*Lebenswelt*) of individuals in order to check theoretical speculation in and for itself with the practical application of analytic insights. This chapter outlines the major theoretical implications of his writings, especially his relation to German existential philosophy, before considering two important case studies, 'Lola Voss' and 'Ellen West', published in *Schizophrenie* (1957). Binswanger's work not only influenced the spread of existential therapy across Europe and American in the mid-century (influencing the likes of Medard Boss, Ronald Laing and Michel Foucault in Europe and Rollo May, Carl Rogers and Erich Fromm in America), but significantly contributed to the emerging twentieth-century tradition of romantic science.

---

[1]   Binswanger's only direct link with James is a reference to his 'endogenous depressive desperation and anxiety' (suffered by the French correspondent in *Varieties*) in *Über Ideenflucht* (Zürich, 1933), p. 138. However, as this chapter makes clear, there are a number of conceptual similarities which link key areas of their thought.

Existential Analysis

Swiss psychiatry made important advances in the early years of the twentieth century, especially in the German-speaking parts of the North, where Binswanger studied at the Burghölzli hospital under the direction of Eugen Bleuler and Carl Jung, becoming a volunteer physician in 1906. Burghölzli had served as a training clinic for Zürich University since 1860 and became an important focal point for Swiss psychiatry after Jung's arrival in 1900. The development of psychoanalysis in Zürich led to the formation of a Swiss 'Freud Group' which included Jung, Bleuler and Binswanger and in 1910 Jung was made head of the newly-formed International Psychoanalytic Association, much to the annoyance of the Viennese psychoanalysts. Binswanger first met Freud on a trip to Vienna in 1907 which soon developed into a close friendship after his move to Kreuzlingen, where he took up a post at Bellevue sanatorium, founded by his grandfather Otto Binswanger (who wrote the standard Swiss textbook on hysteria in 1904) and continued by his father Robert. Freud was acquainted with the Binswangers and was impressed by their work, even referring some of his own patients to Bellevue.[2]

Although Jung, Rank and Binswanger disagreed about similar aspects of Freudian orthodoxy, unlike the other two the younger Swiss retained his personal friendship with Freud long after their initial meeting. The friendship is documented in *Erinnerungen an Freud* (1956),[3] in which Binswanger comments on the paternal role Freud took in relation to him (an attitude which Jung had rebelled against and which later soured the relationship between Freud and Rank).[4] In his diary Binswanger comments that he cared more for Freud as a 'human being' for his 'scope and depth of humanity' than he respected him as a scientist.[5] This expression of emotion does not merely illustrate Binswanger's depth of feeling for Freud, but is central to his analytic project to develop an empathic relationship with his patients. His belief in the need to empathize

---

[2]   Jung's own apprenticeship at Berghölzli is discussed in Frank McLynn, *Carl Gustav Jung* (London: Bantam, 1996), pp. 55–75.

[3]   Translated as *Sigmund Freud: Reminiscences of a Friendship* (New York: Grune & Stratton, 1957). This book provides an alternative view to the standard biographies of Freud by Ernest Jones and Peter Gay, who both underplay the private and professional relationship between Binswanger and Freud. See also Herausgegeben von Gerhard Fichtner, *Sigmund Freud/Ludwig Binswanger, Briefwechsel 1908–1938* (Germany: Fischer, 1992).

[4]   Illustrated in the letters Jung sent to Freud in the years preceding his resignation as president of the IPA in 1914; reprinted in *The Freud/Jung Letters*, ed. William McGuire (London: Penguin, 1991), p. 256. Unlike Rank, Jung and Binswanger were both Swiss and non-Jewish which created further tensions in their relationships with Freud.

[5]   Fritz Schmidl, 'Sigmund Freud and Ludwig Binswanger', *Psychoanalytic Quarterly*, 28 (1959), 47.

(*einfühlen*) with the individual prior to scientific study creates problems of transference in therapy; nevertheless, he wished to link empathic understanding with a sensitivity to aspects of the individual's psychological world which the taxonomies of natural science usually fail to include. Like James and Rank, Binswanger addressed crucial aspects of psychic life (especially spiritual and aesthetic impulses) not directly available to empirical investigation. In short, all three thinkers wished to radicalize empiricism by developing a phenomenological approach to the self.

Many of Binswanger's theoretical essays and public addresses discuss and criticize aspects of Freud's thought he had inherited from a nineteenth-century tradition of biological science.[6] In a late essay on 'Freud and the Magna Charta of Clinical Psychiatry' (1955), Binswanger states that psychoanalysis had inherited its 'basic conceptual categories' from one of the founding moments of clinical psychiatry in 1861.[7] The second edition of Wilhelm Griesinger's *Pathologie und Therapie der psychischen Krankheiten* appeared in this year (originally published in 1845), in which the first director at Burghölzli claims that because psychic phenomena are organically rooted in brain processes, they 'ought only to be "interpreted" by natural scientists'.[8] Binswanger is critical of Griesinger's comparison of the psyche to physiological reflex actions and his proposal to investigate it through similar methodological procedures; for Griesinger, subjectively felt experiences are thought to derive from 'elementary processes in neural matter'.[9] In a later case history, Binswanger declares his reservations about such reductionism: 'we immediately realise how radical is the process of reduction which the natural-historically oriented clinical method must use in order to be able to speak of a disease-process and to project it upon the "organism" and the structure and modes of functioning of the brain'.[10] Although Griesinger did not rule out the possibility of leaving 'the door open for a "descriptive and analytical" or *verstehende* psychology', Binswanger claims that 'today the charter shows signs of having become so dogmatically rigidified that many of its advocates deem any measure proper that would condemn and excommunicate those scientists who seem to hold opposing views'.[11]

If the Magna Charta proclaimed that Western psychiatry should be understood as a branch of natural science, then Freudian theory 'fills in a very pronounced

---

6   Freud's attempt to forge psychoanalysis as a natural science is evident from the abandoned 'Project for a Scientific Psychology' in the 1890s to his metapsychological papers written in the mid-1910s and his lectures on psychoanalysis in the 1930s.

7   Jacob Needleman (ed.), *Being-in-the-World: Selected Papers of Ludwig Binswanger* (New York: Basic Books, 1963), p. 186.

8   Ibid., p. 186.

9   Ibid., p. 187.

10   Rollo May et al. (eds), *Existence: A New Dimension in Psychiatry and Psychology* (New York: Basic Books, 1958), p. 329.

11   Needleman (ed.), *Being-in-the-World*, op. cit., pp. 187, 188.

gap in the charter', but also 'deepens the very ideas contained in it by casting light upon things that could never have been seen from within the charter alone'.[12] By strengthening the status of psychiatry as a branch of natural science, Binswanger accuses Freud of translating the 'pictorial language of psychology' into 'more uniform, coarser, chemical and physical "images" that, for him, seemed closer to reality'.[13] He does not take exception to Freud's understanding of empiricism *per se*, but criticizes the materialistic and epistemological presuppositions underpinning his conception of psychic life. Binswanger's basic point is that psychoanalysis cannot fulfil itself as a general theory of psychology if it is conceived only as a branch of natural science, whereas, in 'A Philosophy of Life' (1932), Freud argued that psychoanalysis is not 'in a position to create a *Weltanschauung* of its own'.[14] Freud claimed a *Weltanschauung*, as a 'philosophy of life', is 'an intellectual construction which gives a unified solution of all the problems of existence . . . a construction, therefore, in which no question is left open and in which everything in which we are interested finds a place'.[15] Like James at the beginning of *Principles*, rather than offering a 'unified solution' Freud suggests that psychoanalysis 'does not take everything into its scope, it is incomplete and it makes no claim to being comprehensive or to constituting a system'.[16]

Binswanger's disagreement with Freud rests upon what he understands to be a misconception of *Weltanschauung* as a totalizing philosophy of life. Freud claims that psychoanalysis 'is quite unsuited to form a *Weltanschauung* of its own; it must accept that of science in general', whereas Binswanger wishes to develop Martin Heidegger's existential philosophy by examining the basic structures of existence.[17] He argues that an existential branch of psychiatry would enable the analyst to address the limits of the natural-scientific view and contribute to 'the effort of psychiatry *to understand itself as science*'.[18] Binswanger's disagreement is essentially with Freud's resistance to the 'encroachment' of speculative philosophy into the realm of natural science (from which psychology had broken free in the late nineteenth century), against his assertion that such modification would provide science with the means to

---

[12]   Needleman (ed.), *Being-in-the-World*, op. cit., p. 204.

[13]   Ibid., p. 200. Much of Freud's work is contaminated with the language of natural science, but, as Paul Ricoeur argues, he sought to extend the scope of medical psychiatry by developing psychoanalysis as a specifically interpretative practice; Paul Ricoeur, *Freud and Philosophy: An Essay on Interpretation* (New Haven: Yale U.P., 1970), pp. 20–36.

[14]   Sigmund Freud, *New Introductory Lectures on Psycho-analysis*, 3rd edn, trans. W.J.H. Sprott (London: Hogarth, 1946), pp. 229, 232.

[15]   Ibid., p. 202.

[16]   Ibid., p. 232.

[17]   Ibid., p. 203.

[18]   Needleman (ed.), *Being-in-the-World*, op. cit., pp. 207, 208.

'understand itself'.[19] He clarifies this point in his claim that an 'analytic of existence . . . can indicate to psychiatry the limits within which it may inquire and expect and answer and can, as well, indicate the general horizon within which answers, as such, are to be found'.[20] In this way, Binswanger seeks to rescue the spirit of *Weltanschauung* from Freud's criticisms and to forge an analytic method which addresses the 'total phenomena' of psychic life.[21]

Stimulated by the defections of Adler and Jung in the 1910s and Rank in the mid-1920s, Binswanger claims that the founding suppositions of psychoanalysis had long been under threat. He argues that the deepening crisis can only be resolved by carrying out a form of 'empirical existential-analytic research' to reassess the implications of psychic disturbance.[22] Most psychoanalysts tended to view patients within the 'natural-scientific' horizon of understanding, in which the object of study is 'the "sick" organism', whereas he wishes to attend to the sick 'human being' and to view the way illness manifests itself in the individual psyche.[23] Although this comment does not seem very different from reading Freud's case studies as close investigations into psychic disturbance, Binswanger emphasizes the importance of being receptive to the psychological requirements of patients in order to prioritize individual experience over the cataloguing of symptoms.

Despite his desire to broaden the 'horizon of understanding' of psychiatry, Binswanger does not view scientific and anthropological investigations as mutually exclusive: 'in actual practice, these two conceptual orientations of psychiatry usually overlap'.[24] However, he concurs with Rank in his claim that the analyst's empathy for the patient must occur prior to any systematic account of the patient as a specific type. This point corresponds to Heidegger's division between ontological and ontic understanding: the former addresses 'the idea of being' and the structures of existence and the latter refers to the scientific study of behaviour.[25] Binswanger argues that only if ontological understanding provides the focus for analysis can ontic understanding be rescued from a partial understanding of psychic disturbance. He is not worried about Freud's rigorous scientific approach, but he is concerned with the dismissal and reduction of dimensions of psychic life that are not directly discernible from a strictly empirical perspective. The aim of Binswanger's analytic of existence is to indicate what is withdrawn, or lost, when humans are understood only with

---

19   Ibid., p. 208.
20   Ibid., p. 211.
21   May et al. (eds), *Existence*, op. cit., p. 329
22   Needleman (ed.), *Being-in-the-World*, op. cit., p. 211.
23   Ibid., pp. 208–9.
24   Ibid., pp. 207, 209.
25   Ludwig Binswanger, 'On the Relationship between Husserl's Phenomenology and Psychological Insight', *Phenomenology and Phenomenological Research*, 2 (1941), 201.

reference to the ontic sphere of understanding. As such, like James, Binswanger's phenomenological approach to the self is radically empirical.

Freud and Binswanger crucially differ over their respective understanding of what an empirical approach to the self entails. In 'The Existential Analysis School of Thought' (1946), Binswanger characterizes these two types of knowledge: firstly, *'discursive inductive* knowledge' which entails 'describing, explaining, and controlling "natural events"' and, secondly, *'phenomenological empirical* knowledge in the sense of a methodical, critical exploration or interpretation of phenomenal contents'.[26] In the hermeneutic thought of Paul Ricoeur (also influenced by the phenomenology of Heidegger and Gabriel Marcel), the former provides the basis for an explanatory project which seeks to *'expose and to abolish idols* which are merely the projections of the human will', whereas the latter stresses understanding which 'requires *a willingness to listen with openness* to symbols and to "indirect" language'.[27] As a close reader of Freud, Ricoeur wishes to combine explanation with understanding to compensate for the limitations of each: the reductive tendency of the former and the uncritical nature of the latter.[28] In a similar manner, Binswanger actually seeks to redress the explanatory bias of Freudian writing, not to deliberately oppose it.

Binswanger agrees with Freud's belief that analytic concepts should not derive from a 'speculative system of ideas', but should result from experience, 'being founded either on direct observations or on conclusions drawn from observations', but he is unhappy with Freud's reductive views of religion and art.[29] Freud claims both are the epiphenomena of instincts which themselves lie 'on the frontier between the mental and physical'; as such, spiritual and aesthetic impulses are sublimations of instinctual drives which form the basic dynamics of psychic life.[30] Because imagination and spirit are impalpable Freud perceives them to be illusions: art is 'almost always harmless and beneficent, it does not seek to be anything else but an illusion' and the irrational affirmation of religious life is seen to be diametrically opposed to rational inquiry.[31] He acknowledges the rival claims of religion, but emphasizes that 'truth' can be unearthed through scientific methods alone. In 'A Philosophy of Life' and *The Future of an Illusion* (1927), he subjects religion to the 'critical examination' of science and concludes that, unlike art, religion is not merely an illusion, but a dangerous delusion which

---

[26]   May et al. (eds), *Existence*, op. cit., p. 192.

[27]   Anthony Thisleton, *New Horizons in Hermeneutics* (London: HarperCollins, 1992), p. 344.

[28]   Although it is an oversimplification to suggest that Freud adheres only to the explanatory pole and Binswanger to the ideal of understanding, the general drift of their respective writings can be characterized as such.

[29]   Ibid., p. 208.

[30]   Freud, *On Metapsychology: The Theory of Psychoanalysis*, ed. Angela Richards, trans. James Strachey (London: Penguin, 1984), p. 108.

[31]   Freud, *New Introductory Lectures*, op. cit., p. 205.

leads to the kind of pacification which Nietzsche claims to derive from deeply entrenched habits.[32] Binswanger reacts strongly to these views, believing that such conclusions derive from Freud's strict adherence to a limited model of empirical inquiry, simplifying 'the life of the psyche' to 'crude natural-scientific schema governed by a few principles'.[33] Closely reflecting James' theoretical move from epistemology to radical empiricism, Binswanger believes an empirical method should entail close description of psychic phenomena as experienced by the individual. Religious experiences are to be understood and investigated as they subjectively arise, to recall James, as the fruits of experience, rather than translated into a set of metapsychological principles. Unlike Rank, Binswanger does not reintroduce the language of soul (*das Seele*) into analytic vocabulary,[34] but preserves a phenomenological language of spirituality:

> In phenomenological experience, the discursive taking apart of natural objects into characteristics or qualities and their inductive elaboration into types, concepts, judgments, conclusions, and theories is replaced by giving expression to the content of what is purely phenomenally given.[35]

Here the opposition between 'taking part' and 'giving expression to' stresses the creative and collaborative aspects of therapy over the type of analytic study which reduces phenomena to the level of instinctual drives. This does not mean that phenomenology cannot dispel illusions, only that Binswanger wishes to shift his emphasis from determinate explanations to descriptions of structures which characterize the individual's life-world.

Three major criticisms of Binswanger's phenomenological approach need to be addressed to defend his project against his detractors. Firstly, behaviourists often criticize phenomenology for being an uncritical and purely descriptive mode of inquiry. However, in an essay 'On the Nature and Aims of Phenomenology' (1943), the critic John Wild asserts that phenomenology should be distinguished from phenomenalism, the latter 'which sets for itself the really hopeless task of transcribing the infinitely variable succession of appearances as they occur'.[36] Wild argues that phenomenology is actually radically empirical by directing 'its attention to that formal or eidetic structure of the phenomena, which is implicit in them, but which requires painstaking and repeated observation'.[37]

---

[32]   Ibid., p. 213.

[33]   Needleman (ed.), *Being-in-the-World*, op. cit., p. 203

[34]   Binswanger claims the soul 'is a religious, metaphysical, and ethical concept' which can be admitted 'only as a theoretical, auxiliary construct within a specific field' like psychopathology; Needleman (ed.), *Existence*, op. cit., p. 231.

[35]   Ibid., p. 192.

[36]   John Wild, 'On the Nature and Aims of Phenomenology', *Philosophy and Phenomenological Research*, 3 (1943), 85.

[37]   Ibid.

He stresses the 'laborious' effort involved in such description, but asserts that this kind of study is preferable to any 'premature attempts at synthesis'.[38]

The second major criticism of phenomenology is that it preserves the object-subject dichotomy and does nothing to break from the Cartesian tradition of rational inquiry. Here, Heidegger's rejection of the ambition of Husserl's phenomenology helps to dissolve this dichotomy. Whereas Husserl sought to extract the essences which give form to phenomena, Heidegger sees no possibility of objectively dismantling these structures. As such, he moves away from Kant's transcendental phenomenology, which attempts to rescue the things-in-themselves lying behind phenomena, towards a hermeneutic phenomenology which recognizes that the inquirer is always already embedded in the world: the subject of consciousness does not face the object, but each is implicated by, and partially constitutive of, the other. This idea enables Heidegger to reject both idealism and materialism, without reverting back to dualism. By dispensing with the terminology of subject and object, he stresses the interconnected nature of each and replaces the dualism with the term *Dasein* or Being-in-the-world. This does not represent a return to the transcendental phenomenology of Husserl, but implies Being-in-the-world is a condition of existence anterior to the subject-object split.[39] In this way, Heidegger's notion of 'thrownness' into existence transforms Kant's transcendental ego and Husserl's essences into an exploration of the 'temporality of human existence'.[40] By adopting this Heideggerian position, Binswanger makes temporal and narrative understanding central to his existential project.

The third, broadly post-structuralist, criticism of phenomenology is that it seeks to disclose the 'original pre-analytic data' of phenomena.[41] Because, as Derrida and de Man argue, interpretation is always already the domain of any data one wishes to examine. If 'pre-analytic data' is held to be synonymous with pre-linguistic or culturally unmediated material then Binswanger's phenomenological method reaches a serious theoretical impasse. However, he defends his position in 'The Existential School of Thought' by claiming that 'the phenomena to be interpreted are largely language phenomena'.[42] The existential structures which he seeks to disclose are fundamentally linguistic, through which the structures 'actually ensconce and articulate themselves and where, therefore, they can be ascertained and communicated'.[43] Although his comments on language are not a dominant focus of his writings (he worked briefly on aphasia in 1926), instead of attempting to discover Husserl's essences stripped of

---

[38]   Ibid., pp. 90, 91.

[39]   For Heidegger 'the existential analytic of *Dasein* comes *before* any psychology or anthropology, and certainly before any biology'; Martin Heidegger, *Being and Time*, trans. John Macquarrie and Edward Robinson (Oxford: Blackwell, 1962), p. 71.

[40]   Ibid., p. 206.

[41]   Wild, 'On the Nature and Aims of Phenomenology', op. cit., p. 86.

[42]   May et al. (eds), *Existence*, op. cit., p. 200.

[43]   Ibid.

language, he realizes that the structures themselves are part of a wider semiotic and cultural matrix.

The question remains how Heidegger's existential interpretation of pheno-menology enables Binswanger to move away from Freud's natural-scientific explanations. Fundamentally, this rests on Binswanger's attempt to redescribe *Weltanschauung*, not as a totalizing picture of the world, but as an holistic inquiry into Being. Even if one rejects Freud's reductive criticisms of spirituality and art, Binswanger argues that the scientific conception of man 'as a physical-psychological-spiritual unity does not say enough' because it fails to address what it means to *be* in the world, whereas 'existential analysis undertakes to work out being human in all its existential forms and their worlds, in its being-able-to-be (existence), being-allowed-to-be (love), and having-to-be (thrownness)'.[44] He claims that psychoanalysis is equipped to deal only with 'having-to-be' (a form of psychic determination), whereas existential analysis enables Binswanger to affirm the spiritual dimension of human psychic life and furnishes him with an analytic vocabulary which transcends the biological basis of Freud's project.

The Question of Being

A decade after the publication of Binswanger's most important analytic papers, a younger Swiss analyst, Medard Boss, also sought to appropriate Heideggerian thought to compensate for the limitations of Freudian orthodoxy. Like Binswanger, in *Psychoanalysis and Daseinanalysis* (1963) Boss takes issue with the 'prescientific presuppositions' which result in splitting up 'the unity of man's "Being-in-the-world" into three primordially separated particles: the "psyche," the human body, and the external world'.[45] Reacting against these 'theoretically abstracted parts of man's world', Boss appears to agree with Binswanger's affirmation of *Dasein* or Being-in-the-world as the primary and irreducible condition of human life. However, he seeks to determine whether 'there is any connection at all' between *psychoanalyse* and *Daseinanalyse*, by distinguishing his Heideggerian analysis from 'the shocking intellectual confusion' which he claims 'has come to prevail under the blanket term "existentialism"'.[46] While Binswanger's name is only mentioned twice in *Psychoanalysis and Dasienanalysis*, the writings of the elder Swiss analyst represent one strain of 'fashionable' existentialism from which Boss wished to disassociate himself.[47]

---

[44]   Ibid., p. 315.

[45]   Medard Boss, *Psychoanalysis and Daseinanalysis*, trans. Ludwig B. Lefebre (New York: Da Capo Press, 1982), p. 78.

[46]   Ibid., pp. 2–3.

[47]   The criticisms of Binswanger are more evident in the original German version of *Psychoanalyse und Daseinsanalytiker*, the manuscript of which was approved by Heidegger.

Boss identifies several misconceptions of *Dasein* before embarking on his own reading of Heidegger. His primary concern is to prevent existential analysis from falling back into an idealistic conception of self, which he detects in Binswanger's 'subjectivistic revision' of Heidegger's *Being and Time*.[48] According to Boss, this entails the recuperation of a pre-existentialist notion of subjectivity and pushes 'aside the real meaning of Being-in-the-world'.[49] Far from resolving the subject-object dichotomy, 'Being-in-the-world is then pictured as a property, or as a character trait' and 'turns out to be merely a somewhat wider and more useful version of the concept of subjectivity' which 'gains access to the unity of human existence only at the price of losing all psychotherapeutic possibilities'.[50] As such, Boss criticizes Binswanger (along with Sartre and Merleau-Ponty) for retaining traces of the Cartesian subject. But, whilst his criticisms seem valid for French existentialism, he ignores Binswanger's insistence on 'Being-in-the-world-with-others' as the crucial starting point for his redescription of self. Indeed, although in *Being and Time* Heidegger argues that man is fundamentally 'Being-with', both he and Boss underplay the implications of such an assertion.[51]

Heidegger's conception of Being-in-the-world implies an unbreakable connection with a shared world, within which *Dasein* is co-determined. The pronoun 'I' does not begin 'by starting out and isolating' the self against the backdrop of others, but implicates those others 'from whom, for the most part, one does *not* distinguish oneself – those among whom one is too'.[52] Heidegger implicitly attacks the Cartesian subject for conceptually dividing the world into a private self and those other shadowy objects and entities which inhabit the world. This split rests upon Descartes' association of the mind with the true self and the body with the contingent world in which the mind is suspended. Whilst Husserl and Sartre conform to a neo-Cartesian model of the self, Heidegger's claim that *Dasein* is anterior to the subject-object split challenges this model and posits that 'with' and 'too' should be 'understood *existentially*' as relational terms linking the

---

[48] Ibid., p. 51.

[49] Ibid., p. 53.

[50] Ibid., pp. 54, 53.

[51] Martin Heidegger, *Being and Time*, trans. John Macquarrie and Edward Robinson (Oxford: Blackwell, 1962), p. 153. William Sadler cites a personal letter from Binswanger in which the Swiss analyst notes 'he had in fact misunderstood the nature of Heidegger's ontology; however, he hoped that his own development of a Heideggerian type of existential analysis would be considered a fruitful misunderstanding'; William Sadler, *Existence and Love: A New Approach in Existential Phenomenology* (New York: Scribners, 1969), p. 118. Boss first corresponded with Heidegger in 1946, when he had already 'turned' from the concerns of *Being and Time* towards a more explicit interest in language. Heidegger continued to stress the chasm between ontology and psychology, but in this later phase, during which time he gave seminars in Switzerland to Boss' medical students, he began to outline some common ground between the two disciplines.

[52] Heidegger, *Being and Time*, op. cit., p. 154.

self to others: 'by reason of this *with-like* (*mithaften*) . . . the world is always the one that I share with Others. The world of *Dasein* is a *with-world* (*Mitwelt*)'.[53]

Heidegger argues that 'I' is not an empty indexical term of self-reference but a 'locative personal designation' which confirms the coexistence of spatial and temporal planes: 'I-here' and 'I-now'.[54] Binswanger develops this understanding by claiming the relational model of *Dasein* reveals 'the world-design or designs in which the speaker lives or has lived' and circumscribes a social dimension which patient and analyst share and in which they can communicate.[55] On this model the existential analyst should attend closely to the behaviour and expressions of the patient in order to understand the way in which 'world-designs' structure the individual's life-world.[56] Dialogic exchange not only bridges the epistemological gap between behavioural study and introspective inquiry (third-person and first-person), but provides an interpersonal exchange of 'Being-with' in which alterations in the patient's life-world can be detected through collaborative analysis. Binswanger stresses the communal dimension of language in his description of the primacy of 'certain phenomenal, intentional, and preintentional modes of Being-together (*Mitseinandersein*) and co-being (*Mitsein*)'.[57] He also subdivides James' description of habitat into 'three world-regions': the natural world (*Umwelt*), the social world (*Mitwelt*), and personal world of self-referentiality (*Eigenwelt*). By emphasizing the interrelated nature of these three worlds, Binswanger claims that any singular understanding of 'being-oneself' (*Eigenwelt*) is only one expression of *Dasein* which needs to be understood in terms of the other two world-regions (*Mitwelt*, *Umwelt*). He characterizes his patients' inability, or the diminution of their ability, to understand their world designs from such a multiple perspective as being closely associated with psychosis; only the reaffirmation of a coexisting world can rectify the turning inward often associated with psychiatric disorder.

However, Boss regards Binswanger's concept of world-design with a particularly critical eye. In 'The Existential Analysis School of Thought', Binswanger argues that the world-design is peculiar to human beings (as distinct from animals) because it is 'rooted in this multifold potentiality of being': 'the *what* of the respective world-design always furnishes information about the *how* of the Being-in-the-world and the *how* of being oneself'.[58] In other words, Binswanger claims that, unlike genetically determined animals, on an ontic level humans are capable of taking up a variety of roles within society and on an

---

53   Ibid., p. 155.

54   Ibid.

55   May et al. (eds), *Existence*, op. cit., p. 201.

56   Binswanger expands most thoroughly on the forms which underlie behavioural patterns in *Gründformen und Erkenntnis Menschlichen Daseins*, 2nd edn (Zürich, 1953).

57   May et al. (eds), *Existence*, op. cit., p. 226.

58   Ibid., pp. 197, 195.

ontological level they are capable of motivating, or creatively 'designing', themselves 'toward the most different potentialities of being'.[59] This claim has similarities with James' and Rank's recommendations that the limitations of the self can be overcome by exertion of the will, but Binswanger is careful to emphasize that world-designs are anterior to consciousness and thus limit the range of 'potential modes of being for the self'.[60]

He asserts that animals do not have reciprocal worlds in the same way as humans because they do not possess a notion of selfhood: that is, the inability to say 'I-you-we'.[61] Selfhood is thus defined not merely as the capacity to say the words 'I-you-we', but also the ability to position the self as a subject within a shared discourse. Moreover, humans have the unique potential to transcend their determining environment into which each one is thrown, by redesigning, or opening up, their world. Boss' criticisms seem valid if Binswanger means that the pre-Heideggerian self has the potential to separate itself from the world to 'decide independently in and for a situation'; however, he asserts that a world-design is not a conscious act of fashioning, or will-to-power, by which the individual imposes him or herself on the world.[62] Indeed, he claims that 'world' does not 'refer to anything psychological but to something which only makes possible the psychic fact'.[63] One cannot change a world-design in its entirety by an act of will, but neither does this prohibit the self from devising alternative modes of Being within the parameters of the world-design.

There are obvious links between Heidegger's exposition of Being (*Sein*) and Binswanger's conception of world-design. In his introduction to *Being and Time*, Heidegger claims that Being is 'the most universal and the emptiest of concepts', which eludes definition but 'does not eliminate the question of its meaning'.[64] Binswanger's project to disclose world-designs adheres closely to Heidegger's assertion that 'what is sought in the question of Being is not something completely unfamiliar, although it is at first totally ungraspable'.[65] He argues that a world-design is discernible in each of the utterances and acts of an individual, but is not wholly deducible from them. Similarly, Heidegger asserts that '*Dasein* is ontically "closest" to itself, while ontologically farthest away': in other words, only by addressing the ontic level can the ontological level be approached.[66] However, the problem of how to move from the ontic (*Seiende*) to the ontological (*Sein*) haunts the writings of Heidegger and Binswanger. Indeed, Jacob

---

[59]   Ibid., p. 198.
[60]   Ibid.
[61]   Ibid.
[62]   Ibid.
[63]   Ibid., p. 204.
[64]   Heidegger, *Being and Time*, op. cit., pp. 21, 23.
[65]   Ibid., p. 46.
[66]   Ibid., p. 58.

Needleman argues that Binswanger's project to disclose world-designs never actually breaches the ontological level, but instead is wholly of the ontic.

Needleman translates Binswanger's notion of world-design as a 'meaning-matrix' (*Bedeutungsrichtung*) through which the self experiences itself in relation to the world and to others.[67] This does not constitute an innate propensity to view the world in a certain manner (as a type of genetic determinism), but conditions aptitudes and provides a frame of reference in which an individual can understand him or herself in terms of the threefold world. An individual's awareness of the structure of his or her world-design would represent the possibility of finding meaningful or authentic language with which to express the self. To deploy structuralist vocabulary, Binswanger claims a world-design revolves around a set of binary oppositions, one side of which, under certain conditions, is limiting or debilitating and the other which is enabling or freeing. He outlines a number of oppositions (full-empty, order-chaos, movement-stasis, continuity-discontinuity, rising-falling) but argues that the exact manifestation and combination of these oppositions is peculiar to the individual. Ideally, existential therapy should not only encourage the individual to invest new meaning in his or her world-design, but also generate new techniques for increasing the scope of experience.

Throughout Binswanger's writing there is an ambiguity as to whether there is one world-design (one axis of meaning) for each individual, or whether a number of complementary world-designs condition a particular set of activities or responses to stimuli. At different stages of his work he seems to endorse both readings, or, at least, he suggests the possibility that several world-designs structure the life of an individual. But he does distinguish between what he means by world-design and the 'categories' of understanding through which the analyst is made aware of it:

> the emptier, more simplified, and more constricted the world-design to which an existence has committed itself, the sooner will anxiety appear and the more severe will it be. The 'world' of the healthy with its tremendously varied contexture of reference and compounds of circumstance can never become entirely shaky or sink. If it is threatened in one region, other regions will emerge and offer a foothold. But where the 'world' . . . is so greatly dominated by one or a few categories, naturally the threat to the preservation of that one or those few categories must result in more intensified anxiety.[68]

On one level, as a romantic scientist, Binswanger associates health with plurality and illness with constriction, but he seems uncertain as to whether 'categories' are the interpretations of universal world-designs, or whether they constitute a

---

67   Needleman (ed.), *Being-in-the-World*, op. cit., p. 29.
68   May et al. (eds), *Existence*, op. cit., p. 205.

personal set of world-designs which influence how an individual comes to cognitively understand, and act upon, experience.

A brief glance at one of his shorter case studies clarifies this issue. In 'The Existential Analysis School of Thought' Binswanger describes 'a young girl who . . . experienced a puzzling attack of anxiety and faint when her heel got stuck in her skate'.[69] These fits continue throughout the woman's life 'whenever a heel of one of her shoes appeared to loosen' or when she saw 'a loose button hanging on a thread or a break in the thread of saliva'.[70] By closely observing these symptoms, Binswanger discerns a 'depletion' of Being-in-the-world of the woman, eventually narrowing 'to include only the category of continuity'.[71] This case indicates that 'categories' are really different ways of conceiving the world within the framework of a world-design, which, in the case of this woman, is limited to one mode of interpretation. A world-design is understood as a formal structure which facilitates the development of a habitual set of responses which, in turn, helps to mould particular categories of understanding. Shock or trauma serves to rigidify this multiplicity into one pattern of meaning. For example, the world-design of another patient is shown to be polarized around the axis of 'urgency' and 'narrowness', which allows for 'no steadiness' and in which 'everything occurs by jerks and starts'.[72] But, only when a world-design becomes contracted into one manner of interpretation does psychosis manifest itself.

The question remains as to whether these existential structures are actually ontological or 'meta-ontic'.[73] According to Needleman, 'any discipline that concerns itself with the transcendentally *a priori* essential structures and possibilities of concrete human existence is, strictly speaking, neither ontological nor ontic, but lies, rather, somewhere in between'.[74] From this premise, Needleman suggests that world-designs 'are the universal forms that stand to the experience of *each* human being'.[75] Boss is concerned with Binswanger's assertion that existential analysis provides the means to expose the 'universal forms' which underlie *Dasein* and thereby overcome the blockages of psychosis. Boss shares Heidegger's understanding that authenticity emerges from the attempt to disclose the question of Being from which humans are under most conditions estranged.[76] The main concern for Heidegger and Boss is that uncovering a meta-ontic structure does not disclose Being and may actually

---

[69]   Ibid., p. 203.

[70]   Ibid., p. 205.

[71]   Ibid.

[72]   Ibid. pp. 206, 207.

[73]   Needleman (ed.), *Being-in-the-World*, op. cit., p. 26.

[74]   Ibid., p. 27.

[75]   Ibid.

[76]   Heidegger suggests humans are homeless in two ways: they are estranged both from Being and from an authentic mode of dwelling in the world. It is the task of philosophy to work through this double estrangement.

constitute another obscuring factor in the pursuit of Being. But this is the point at which Binswanger's project of *Daseinanalyse* departs from Heidegger's *Daseinanalytik*, distinguishing the phenomenological anthropology of the Swiss analyst from the phenomenological ontology of the German philosopher. While Heidegger outlines 'a phenomenological hermeneutic of Being understood as existence' on an ontological level, Binswanger proposes 'a hermeneutic exegesis on the ontic-anthropological level, a phenomenological analysis of actual human existence'.[77] Because Binswanger's project is practical by nature, the 'ontic-anthropological level' necessarily encroaches on the ontological realm. Paul de Man, in his essay, 'Ludwig Binswanger and the Sublimation of the Self' (1966), understands that a 'certain degree of confusion arises when this knowledge is interpreted as a *means* to act upon the destiny that the knowledge reveals. At this moment ontological inquiry is abandoned for empirical concerns that are bound to lead it astray.'[78] But de Man also realizes that such a tension arises in all studies that forsake 'the barren world of ontological reduction for the wealth of lived experience'.[79]

Binswanger's central argument is that humans have 'the possibility of transcending' ontic being (*Seiende*, not *Sein*), 'climbing above it in care and swinging beyond it in love'.[80] Although this appears comparable to Nietzsche's description of overcoming, the terms 'care' and 'love' are crucial limiting factors in such an activity. Both terms involve giving over the self: 'care' (*Sorge*) is directed toward the Otherness of the self (*Eigenwelt*) and 'love' (*Liebe*) is a display of concern for others with which one inhabits the world (*Mitwelt*). For Heidegger, 'care' is fundamentally ontological: 'it lies "before" . . . every factical "attitude" and "situation" of *Dasein*'.[81] On this account, care is a looking after, and a looking out for, the self: 'because Being-in-the-world is essentially care, Being-alongside the ready-to-hand could be taken in our previous analyses as *concern*, and being with the *Dasein*-with of Others as we encounter it within-the-world could be taken as *solicitude*'.[82] What Heidegger terms 'solicitude', Binswanger (following the existential work of Max Scheler, Gabriel Marcel and Martin Buber) calls 'love': a way of relating to another human which enables a limiting or depleting situation to be transcended in a joint venture.[83] Like Scheler,

---

77   May et al. (eds), *Existence*, op. cit., pp. 269–70.

78   de Man, *Blindness and Insight*, op. cit., p. 48.

79   Ibid., p. 49.

80   May et al. (eds), *Existence*, op. cit., p. 198

81   Heidegger, *Being and Time*, op. cit., p. 238.

82   Ibid., p. 237. Heidegger defines the 'Being of the ready-to-hand (involvement) as a context of assignments or references'; Heidegger, *Being and Time*, op. cit., p. 121. Hubert Dreyfus translates *Zuhandenheit* as 'availableness': those 'available' techniques for coping or dealing with situations.

83   Despite Boss' criticisms of Binswanger, their projects are fundamentally similar in the search for 'that highest form of humanness in the relation to others, namely, of that selfless, loving, "vorspringende" caring which frees the other to his own selfhood';

Binswanger cultivates the importance of an ethic of genuine sympathy (*Mitgefühl*) in his therapy: 'sympathy in its true sense is the way in which we transcend mere external stimuli and passing emotions so as to see the other not as an example of organic life classified as *homo sapiens*, but as a unique, spiritual-historical being'.[84] While care is a technique of looking after the self, only sympathetic love can induce a creative analytic encounter which combines solidarity with independence, simultaneously affirming Being-with and disclosing Otherness.

By appropriating this idea of sympathetic love from Scheler, Binswanger clarifies his development of Heidegger's position: 'Being-in-the-world as being of the existence for the sake of *myself* (designated by Heidegger as "care") has been juxtaposed with "being-beyond-the-world" as being of the existence for the sake of *myself* (designated by me as "love").'[85] Binswanger stresses intersubjectivity (the 'overswing' of love) over and above the care of the self (a Nietzschean 'overclimb' of care) and claims that only through reciprocity can the self overcome its existential limitations.[86] He sees love as possessing a two-fold potential: firstly, as an ideal of eternity (*Ewigkeit*) to be striven towards and, secondly, as a home or 'haven' (*Heimat*) which the sympathetic encounter with the analyst offers.[87] This kind of reciprocity does not mean a renunciation of the limitations of finitude for an unobtainable ideal, but an acquiescence to the conditions of existence without surrendering completely to them. Authenticity for Binswanger thereby represents a striving to disclose *Dasein* with the care and support of another being. By stressing 'Being-with' as the central concern of existential analysis and therapy, Binswanger emphasizes the patient's (and the analyst's) embeddedness in the world of others.

In the case of the woman who gets her heel stuck in her skate as a child, the anxiety Binswanger describes is appropriated from Kierkegaard (via Heidegger) as a subliminal dread of existence which cannot be associated with an identifiable object or a particular set of psychological demands. In a critical article on 'Dread and Authenticity' (1989), Julius Heuscher distinguishes between fear which 'is linked to "actual" dangers, whereas "dread" and "anxiety" are viewed as having "imaginary," "uncanny," "incomprehensible," and "irrational" elements'.[88] Although Binswanger often associates the locus of anxiety with a traumatic moment in childhood, this understanding does not contradict such a distinction.

Boss, 'Martin Heidegger's Zollikon Seminars', in ed. Keith Hoeller, *Heidegger amd Psychology* (New Jersey: Humanities Press, 1988), p. 10.

[84]   Max Scheler, *The Nature of Sympathy*, trans. Peter Heath (London: Routledge & Kegan Paul, 1962), p. 58.

[85]   May et al. (eds), *Existence*, op. cit., p. 195

[86]   Ibid., p. 195.

[87]   Ibid., p. 201.

[88]   Julius Heuscher, 'Dread and Authenticity', *The American Journal of Psychoanalysis*, 49(2) (1989), 140.

The actual moment of trauma seems less important than its subsequent limiting of experiential categories. For Binswanger, therapy would not constitute a coming to terms with the incident in order to purge the patient's feelings of anxiety: this would represent an inauthentic 'covering up'. Instead, the therapeutic goal seeks, firstly, to disclose anxiety as a vital response to authentic existence and, secondly, to disturb a repetitive pattern of compulsion by nurturing a mutual relationship of care and love. Heuscher later agrees with one of the key commitments of existential analysis when he claims that 'dread is the shadow-side of love: not of an ethereal, fantastic love, but of the committed love by an embodied human being for another mortal individual'.[89]

Dreaming through Existence

Before considering two of Binswanger's most illuminating case studies, it is important to address his influential paper *Dream and Existence* (1930), in which he restates his central arguments concerning existential analysis and emphasizes the close association between imaginative creation and dreaming. It is useful to read this essay alongside Michel Foucault's commentary on Binswanger's essay 'Dream, Imagination and Existence' (1954) and Foucault's later work on the aesthetics of existence, by combining certain ideas which emerge in Binswanger's thought with the French theorist's more rigorous expression of self-creation. Here, Nietzsche's work provides the point of convergence between Binswanger and Foucault, as well as raising many of the ideas already discussed in the previous chapters on James and Rank.

Freud's seminal work *The Interpretation of Dreams* (1900), followed by his own synopsis of it, 'On Dreams' (1901), set the tenor for the central role of dreams in psychoanalytic theory. Although Freud's emphasis on interpretation in these works is more akin to the theoretical drift of Binswanger's project than the biological model which underpins his metapsychology, there are fundamental differences between their respective views of dreams. Indeed, Freud's model of the unconscious is alien to the phenomenological method which Binswanger adopts. For Freud the unconscious is a repository for repressed libidinal feelings only discernible through dreams or psychic lapses, such as slips of the tongue, jokes or repetitive behaviour. Binswanger agrees about the influence of dreams upon waking life, but rejects Freud's notion of a structural unconscious. As Foucault notes, Freud 'restored a psychological dimension to the dream, but he did not succeed in understanding it as a specific form of experience'.[90] Although

    [89] Ibid., p. 142.
    [90] Michel Foucault, 'Dream, Imagination and Existence', in ed. Keith Hoeller, trans. Forrest Williams, *Dream and Existence* (New Jersey: Humanities Press, 1993), p. 43.

Binswanger describes dreaming as having a status distinct from waking thoughts (agreeing with Freud on the logic of condensation and displacement which characterize dream states), for him, the conscious and unconscious are joined in the plenum of existence, on a model comparable to James' stream of thought. As Foucault corroborates: 'if dreams are so weighty for determining existential meanings, it is because they trace in their fundamental coordinates the trajectory of existence itself'.[91]

The unconscious has a place in Binswanger's thought only as a region of consciousness which the experiencer is unable to reflect upon at any one particular moment: in his words, as a 'sense of nonattention or forgetting'.[92] Thus, the repression of unconscious desires is not the primary cause of neurosis, but only an aspect of it. Referring to the case of the patient who caught her heel in the ice-skate, Binswanger claims:

> We should . . . not explain the emergence of the phobia by an overly strong 'pre-oedipal' tie to the mother, but rather realize that such overly strong filial tie is only possible on the premise of a world-design exclusively based on connectedness, cohesiveness, continuity. Such a way of experiencing 'world' – which always implies such a 'key' (*Gestimmtheit*) – does not have to be 'conscious'; but neither must we call it 'unconscious' in the psychoanalytical sense, since it is outside the contrast of these opposites. Indeed, it does not refer to anything psychological but to something which only makes possible the psychic fact.[93]

Binswanger rejects Freud's psychic apparatus because he argues that 'an unconscious id is not in the world in the sense of existence', rather, it is a scientific construct which 'objectifies existence'.[94] The primacy of *Dasein* does not permit him to conceive of an unconscious as a 'second person' existing in a negative realm 'behind' consciousness, because Being-in-the-world 'always means to be in the world as I-myself, He-himself, We-ourselves, or anonymous oneself; and least of all does the id know anything of "home," as its true of the dual We, of the I and Thou'.[95]

Binswanger claims that the Freudian unconscious is merely an extension of his biological model of psychic instincts, with which existential analysis, its emphasis being on an anthropological approach to humans, cannot concur. Moreover, the interrelation of space (the 'home' of *Dasein*) and time (Heidegger's 'thrownness' into being) implies a need for existential analysts to describe the unconscious world of dreams in terms of a multi-dimensional

[91]   Ibid., p. 60.
[92]   May et al. (eds), *Existence*, op. cit., p. 326.
[93]   Ibid., p. 204.
[94]   Ibid., pp. 326, 327.
[95]   Ibid., p. 327.

locative region or landscape. Borrowing a term from Heraclitus, Foucault reads Binswanger's description of the dream-world to represent an '*idios kosmos*': 'a world of its own . . . in the sense that it is constituted in the original mode of a world which belongs to me, while at the same time exhibiting my solitude'.[96] Thus, the dream-world is shown to be coextensive with the waking-world by providing a stage for the imaginative dramatization of unresolved psychic problems.

Like Rank, Binswanger frequently elides the verbal (or written) accounts of dreams with passages from literature (most often from German romantic literature) which are characterized by similar elevated moods (*Stimmung*). Indeed, in a paragraph omitted from the 1947 version of 'Dream and Existence' (and thereafter), Binswanger claims that to understand dreams 'we must even today still turn primarily to individual creators of language, to the poets'.[97] There are obvious problems here concerning the craft and artifice associated with the creation of art versus the, at least supposed, immediacy of dream recreation. However, following the romantic tradition of Blake, Hölderlin and Coleridge, the visionary transformation of dreams suggest a particular kind of elevated consciousness which is aesthetic in character. Thus, for example, Nietzsche speaks of 'elevated moods' (*hohe Stimmungen*) in *The Gay Science*, which he understands as transitory (but necessary) goals which can be striven towards as permanent possibilities. Similarly, in the first volume of *The World as Will and Representation*, Schopenhauer claims that in such aesthetic states 'man relinquishes the common way of looking at things'; one 'can no longer separate the perceiver from the perception' because 'both have become one, because the whole of consciousness is filled and occupied with one single sensuous picture'.[98] In this way, Binswanger's understanding of dreams show marked similarities with the heightened moods characteristic of the symbolist poetry of Stephane Mallarmé, in which the perceiver and perception are fused together.

If one takes as given the inevitable creative element at work in the recollection of dreams, it is possible to view both dream-reflection and literary creation on a *re*-constructive level, whilst accounting for their respective contexts of production or utterance. Because Freud dismissed aesthetic production as a sublimation of libidinal instincts, Binswanger's model can be seen to be

    [96]   Foucault, 'Dream, Imagination and Existence', op. cit., p. 54. For Heraclitus *kosmos* does not just signify universe (*idios kosmos* as a private universe), but also, as his translator indicates, 'an orderly arrangement' and 'something which beautifies and is pleasant to contemplate'; Jonathan Barnes, *Early Greek Philosophy* (London: Penguin, 1987), p. 19. This second dimension to *kosmos* links up to Binswanger's and Foucault's notion that individual's possess a capacity to make (*poetria*) or aesthetically create a personal world.

    [97]   Binswanger, 'Dream and Existence', in ed. Keith Hoeller, *Dream and Existence* (New Jersey: Humanities Press, 1993), p. 104.

    [98]   Arthur Schopenhauer, *The World as Will and Idea, Volume 1*, 6th edn, trans. R.B. Haldane and J. Kemp (London: Kegan Paul, Trench, Trübner & Co., 1907), p. 231.

essentially very different: he considers autobiographical accounts, diary entries, poems and testimonies by others to explore the 'human individuality' of his patients and expand on his claim that writing and dreams are imaginative visions of the existential structures which constitute an individual life.[99] In essence, the relative value which the two analysts assign to imagination provides the key to understanding their central differences.[100] As such, Binswanger's anthropological approach counters the ontic limitations of Freudian inquiry and demands 'a new way of conceiving how meanings are manifested' in dreams.[101]

Whilst Freud subscribes to the nineteenth-century understanding of a dream as 'a rhapsody of images', Binswanger and Foucault both argue that there is a textual quality to dreams which 'has a content all the richer to the degree that it is irreducible to the psychological determinations to which one tries to adapt it'.[102] This view shifts the interpretative emphasis from image to symbol, which is inexhaustible on the level of meaning and irreducible to any one mode of inquiry. As James detects in *Principles*, objects are 'imagined by a "cluster" or "gang" of ideas', but these cannot be broken down into their constitutive parts in terms of the phenomenological world of the individual: 'an imagined object, however complex, is at any one moment thought in one idea, which is aware of all its qualities together'.[103] For Binswanger, the imaginative impulse of dreams stimulates an elevation in the mood of the dreamer, demanding the analyst to respond with particular sensitivity to, and qualitative appreciation of, the language and form in which the dream is expressed.

Like Freud, Binswanger studies dreams in order to understand the psychic world of the individual and the structures of existence which underpin it, but the emphasis upon these evaluations is very different. Binswanger emphasizes the 'manifest content of the dream' (the form of its expression and its remembered content) which he claims 'has in modern times receded all too far into the background', in contrast to the 'latent dream thoughts' which Freud believed could be uncovered through dream-work.[104] Binswanger does not lapse into a naive form of dream interpretation, but attempts to understand the meaning of

---

[99]   May et al. (eds), *Existence*, op. cit., p. 267.

[100]   James' chapter on 'Imagination' in *Principles* emphasizes the romantic scientist's interest in the imaginative and creative world of the individual. James defines imagination as rooted in lived experience: 'sensations, once experienced, modify the nervous organism, so that copies of them rise again the mind after the original outward stimulus is gone'; James, *Principles*, vol. 2, op. cit., p. 44.

[101]   Foucault, 'Dream, Imagination and Existence', op. cit., p. 33.

[102]   Ibid., pp. 43, 44.

[103]   James, *Principles*, vol. 2, op. cit., p. 45.

[104]   Binswanger, 'Dream and Existence', op. cit., p. 88. In his two books on dreams, *The Analysis of Dreams* (1958) and *'I dreamt last night'* (1977), Medard Boss dispenses entirely with the language of dream-work. He argues that Freud's causal relation between latent thoughts and manifest content undermines his and, in this case, Binswanger's focus upon the individual's subjective response to dreams.

dreams in terms of the life-world of the individual: by 'steeping oneself in the manifest content of the dream . . . one learns the proper evaluation of the primal and strict interdependence of feeling and image'.[105] Aligning himself with the aims of the French symbolists, Binswanger interprets image and mood as reflecting 'the larger and deeper rhythms of normal and pathologically manic and depressive "disattunement" (Verstimmung)'.[106] He understands dreams to be distinct from waking life by virtue of the logic of image-condensation, but they are continuous in the moods they evoke in the individual's consciousness. By grasping the double displacement of dreams, in which the dreamer takes both the roles of participant and observer, these moods can be identified and rearticulated in an analytic encounter. Binswanger presents an example of such a 'double movement' in dreams in which 'I' (the experiencer) 'keep my feet on the ground even as I fall and introspectively observe my own falling'.[107] He focuses particularly on the moods associated with rising and falling, but similar moods are encountered for other categories which have existential significance for the individual: analysis should be directed towards the mode in which 'the pulse of Dasein, its systole and diastole, its expansion and depression, its ascension and sinking' is expressed.[108] The task of the therapist is to encourage patients to remember their dreams, to stimulate the re-enactment of this double movement of observer and participant in waking life and thereby open up the possibilities of new modes of being. Binswanger claims that the individual needs language as a precondition for recreating dreams: 'if the individuality is what its world is . . . and if its world is only affirmed in language . . . then we cannot speak of individuality where language is not yet language, that is, communication and meaningful expression'.[109] The collaborative endeavour of existential analysis may enable both therapist and patient, in Foucault's words, 'to arrive at a comprehension of existential structures' which are anterior to psychic phenomena.[110]

Freud describes three kinds of dream: firstly, 'intelligible' dreams which 'can be inserted without further difficulty into the context of our mental life'; secondly, dreams which 'have a bewildering effect, because we cannot see how to fit that sense into our mental life'; and, thirdly, 'dreams which are without either sense or intelligibility, which seem disconnected, confused and meaningless'.[111] Freud directs his analytic attention particularly to the latter two types in his attempt to establish coherent meaning from the fragments and discontinuities of dreams and overcome 'the obscurity of the dream-content and

105   Ibid., p. 88.
106   Ibid.
107   Binswanger, 'Dream and Existence', op. cit., p. 85.
108   Ibid., p. 88.
109   May et al. (eds), Existence, op. cit., p. 326.
110   Foucault, 'Dream, Imagination and Existence', op. cit., p. 33.
111   Freud, The Essentials of Psychoanalysis, ed. Anna Freud, trans. James Strachey (London: Penguin, 1986), pp. 89–90.

the state of repression . . . of certain of the dream-thoughts'.[112] He closely attends to the puzzling character of dreams that are modified by repetition and claims that 'faithful and straight-forward reproductions of real scenes only rarely appear in dreams'.[113] As Slavoj Žižek discerns, the real mystery which dream interpretation seeks to clarify is not the latent content in itself, but the nature of dream-work 'which confers on it the form of a dream'.[114] Latent dream-thoughts cannot be retranslated into 'the "normal", everyday common language of inter-subjective communication', because that would serve only to ignore the *form* of the "dream" in which unconscious desire [is] articulated'.[115] As Freud asserts, 'the dream-work . . . creates that form, and it alone is the essence of dreaming – the explanation of its peculiar nature'.[116]

For Binswanger there is no transformative logic to unravel, only the need to recapture the heightening of mood and intensification of dream-image. Nevertheless, his emphasis on the form of dreams is comparable to Freud's. Whereas Freud speaks of dream-work, Binswanger identifies the intensity of mood as the dynamic factor of dreams: '*Dasein* moves within the meaning matrix (*Bedeutungsrichtung*) of stumbling, sinking and falling . . . in this case form and content are *one*': in dreams literally 'we are completely uprooted, and we lose our footing in the world'.[117] Thus, the mood of the dream is encapsulated both by the images of rising and falling and the affect induced in the '*dramatis persona*' of the dreamer.[118] The persona of the dreamer in no way need resemble 'the individual body in its outward form', but refers 'to something that can serve as the subject of the particular structural moment . . . this subject may well be, in its sensory aspect, an alien, external subject. It is, nevertheless, *I* who remain the primal subject of that which rises and falls.'[119] The 'structural moment' in which 'I' is assigned a particular perspective is the central concern here. Binswanger goes on to suggest that 'falling itself . . . and its opposite, rising, are not themselves derivable from anything else. Here we strike bottom ontologically'.[120]

---

[112] Ibid., p. 114.

[113] Ibid., p. 103.

[114] Slavoj Žižek, *The Sublime Object of Ideology* (London: Verso, 1989), p. 12.

[115] Ibid., p. 13.

[116] Freud, *The Interpretation of Dreams* (Harmondsworth: Penguin, 1977), p. 650.

[117] Binswanger, 'Dream and Existence', op. cit., pp. 81, 82.

[118] Ibid., p. 85.

[119] Ibid., p. 84. The claim that the 'isolated form' of the body is 'unessential' in itself and may be transmuted into a 'thousand' persona in dreams, leaves Binswanger vulnerable to the criticism that he is establishing an argument for idealism; ibid. If he is understood to equate the essential part of the self with the disembodied 'I' then such a reading is tenable. But his point that the 'essential ontological structure' is conceived prior to any dualistic split implies that the self is always already embodied; ibid. In this way, he can be aligned with Merleau-Ponty in his stress upon the meaning which 'I' assigns to 'my' embodiment.

[120] Ibid., p. 83.

In many of the examples in 'Dream and Existence' the rhythm of rise and fall is dramatized by dreams in which birds are the symbolic focus. Binswanger's examples suggest two main types of 'bird dream': firstly, the awe evoked by the sight of a majestic bird (he gives examples of an eagle, described in Eduard Mörike's *Painter Nolten* (1832), and a kite, taken from the journals of the Swiss writer Gottfried Keller); and, secondly, scenes in which birds of prey attack smaller birds symbolic of beauty or frailty, such as a dove or a pigeon.[121] He implies these symbols are universals (in the first case, perhaps, a variation on the Icarus myth and, in the second, a generic narrative of anxiety), but there is a particularly Germanic feel to his examples in the overreaching archetypal figure of Faust, caught between his soaring desire for the heavens and a gravitational attraction to the physical world.[122] Although Binswanger concentrates primarily on bird imagery, he does provide other examples of rising and falling moods, such as in a passage describing a 'boat dream' in Goethe's *Italian Journey* (1816), where a 'sudden change of a victoriously happy vital current into one that is fraught with anxiety is expressed by the fading or disappearance of brilliantly lit colors and by the obscuring of light and vision'.[123] The passage from Goethe is significant not only as an example of Binswanger's romantic-scientific interest in optics, but also in its description of dreams and visions, which 'have a charm, for while they spring from our inner self, they possess more or less of an analogy with the rest of our lives and fortunes'.[124] The 'charm' of the dream (inviting the reader/analyst to feel himself into world of the writer/patient) and its function in waking life are two aspects of dreaming in which he shows particular interest.[125]

Binswanger focuses on those dreams which fall into Freud's first two categories: those which are intelligible, but may also induce bewilderment or unease in the dreamer. These are the types of dream in which the experiencer

---

[121]   These are not just isolated cases of dreams in which birds feature as important symbols. His patient Ellen West writes an early poem entitled 'Spring Moods' in which she expresses an ecstatic longing for death: 'I'd like to die just as the birdling does/That splits his throat in highest jubilation'; May et al. (eds), *Existence*, op. cit., p. 246. Such visions are also prevalent in romantic literature: in the second part of Coleridge's 'Christabel' (1801) Bard Bracy's dreams of a snake coiled around a dove and Georg Trakl's autumnal wanderer follows migrating birds in 'The Wanderer' and 'Song of the Departed' (1914).

[122]   As Johannes Pfeiffer discerns in his reading of *Faust*, Goethe's hero is divided between a self-serving drive to master the secrets of nature and a search for 'authentic being as the highest value'; Vernon Gras, *European Literary Theory and Practice: From Existential Phenomenology to Structuralism* (New York: Dell, 1973), p. 218. The two modes of being – the single mode of egotism and the dual mode of love – represent the two extremes between which many of Binswanger's patients shuttle.

[123]   Ibid., p. 89.

[124]   Ibid., p. 90.

[125]   Ibid., p. 88. An alternative translation of this sentence is equally revealing: 'such fantastic images give us great delight, and, since they are created by us, they undoubtedly have a symbolic relation to our lives and destinies'; J.W. von Goethe, *Italian Journey*, trans. W.H. Auden and Elizabeth Mayer (London: Penguin, 1962), p. 112.

stages, or works through, an emotional crisis or an affective response to an event in an imaginative recreation of it. Although Binswanger does not overlook fragmentary and contradictory dreams, his Rankian conception of dramatically staging unresolved, or troublesome, feelings is important because it implies some level of meaningful activity in the dream-world of the individual. This does not mean the individual is in control of his or her dream as the conscious and lucid maker of it (the individual is 'the one for whom . . . the dream occurs'); rather, the creative moment oscillates between the remembering and interpretation of the dream in waking life and the dream-world itself:

> the dreamer awakens in that unfathomable moment when he decides not only to seek to know 'what hit him', but seeks also to strike into and take hold of the dynamics in these events, 'himself' – the moment, that is, when he resolves to bring continuity or consequence into a life that rises and falls, falls and rises. Only then does he *make* something.[126]

If the dream is seen as a vignette of waking life, then by attending to the meaning of the symbols as experienced in the dream and situating them within the context of the individual's threefold world, one can recreate the self through an act of narration. In this way, Binswanger departs from Freud's wholly epistemological model of dream interpretation to one which approximates to James' belief that pragmatic activity and self-creation should supplement self-knowledge.

Binswanger positions the self in an ongoing narrative, the retentive (backward-looking) and potential (forward-oriented) elements of which characterize Heidegger's description of the temporality of human existence. The spatial coordinates of dreams (the 'overlapping of journeys, paths crossing, roads which converge to the same place on the horizon') are given an explicit temporal meaning within the life-world of the experiencer through an act of narration.[127] The portentous and revelatory aspects of dreams express the individual's movement towards infinity and death: 'just as we do not know where life and dream begin, so we are, in the course of our lives, ever again reminded that it lies beyond man's powers 'to be "individual" in the highest sense'.[128] Binswanger maintains that a concerted attempt to understand dreams is not a futile activity merely because one can never reach such a transcendent moment (although it may be experienced fleetingly in the rhythm of dreams); nor should a fully coherent or complete understanding of dreams ever hope to be elicited. In an essay unpublished in his lifetime, 'On a Quote from Hofmannsthal' (1981), he states: 'the poet's [or dreamer's] quest never arrives at its destination. Every poem [or dream] signifies . . . a being-at-home and a breaking-up simultaneously'.[129] As

---

[126] Binswanger, 'Dream and Existence', op. cit., p. 102.

[127] Ibid., p. 61.

[128] Ibid., p. 103.

[129] Binswanger, 'On a Quote from Hofmannstahl: 'What Spirit is Only the Oppressed Can Grasp', trans. Vernon Gras and Irmgard Hobson, *Boundary* 2, 9(2) (1981), 192.

Foucault indicates, 'the ultimate in all those dreams that are haunted by the anguish of death' may, in fact, 'paradoxically' disclose 'the movement of freedom toward the world, the point of origin from which freedom makes itself world'.[130]

Foucault understands the responsibility of self-interpretation to be a vital element of an individual's *idios kosmos*. Despite turning away from the phenomenological approach to self during his career, he returned to examine practices which are constitutive of the self in *The History of Sexuality* (1976–84). Here he outlines 'etho-poetic' rules of conduct which can transform the self, but retain an emphasis on 'the way a human turns him- or herself into a subject'.[131] Foucault shuns what he considers to be the oppressive subjugation of self in recent history, and instead favours the Stoic technique of *askesis*: 'an exercise of oneself in the activity of thought'.[132] In his analysis of 'the relation between care and self-knowledge', he explains that *askesis* is 'not a disclosure of the secret self' (the laying bare of an unconscious), but an active 'remembering'.[133] He interprets *askesis* as:

> [the] progressive consideration of self, or mastery over oneself, obtained not through the renunciation of reality but through the acquisition and assimilation of truth. It has as its final aim not preparation for another reality but access to the reality of this world. The Greek word for this is *paraskeuazo* ('to get prepared'). It is a set of practices by which one can acquire, assimilate, and transform truth into a permanent principle of action. *Aletheia* becomes *ethos*. It is a process of becoming more subjective.[134]

While it is not necessary to assimilate the entire Stoic doctrine to appropriate the movement from *aletheia* to *ethos* in existential therapy, Foucault importantly indicates that one Stoic technique of *premeditatio mallorum* ('an ethical, imaginary experience') dictates that 'one shouldn't envisage things as possibly taking place in the distant future but as already actual and in the process of taking place'.[135] The tenor of this practice is central to Binswanger's conception of dream interpretation, which combines self-understanding and the active moral commitment to exercise insights in waking life.

Binswanger also takes his lead from pre-Socratic versions of a 'dynamistic

---

[130] Foucault, 'Dream, Imagination and Existence', op. cit., p. 54; p. 51.

[131] Hubert Dreyfus and Paul Rabinow, *Michel Foucault: Beyond Structuralism and Hermeneutics* (Chicago: University of Chicago Press, 1982), p. 208.

[132] Foucault, *The History of Sexuality Volume 2: The Uses of Pleasure*, trans. Robert Hurley (London: Penguin, 1992), p. 9.

[133] Foucault, *Technologies of the Self: A Seminar with Michel Foucault*, eds Luther Martin et al. (London: Tavistock, 1988), pp. 21, 35.

[134] Ibid., p. 35.

[135] Ibid., p. 36.

world-view', as contrasted to a modern 'explicative (*erklärich*)' science.[136] His writings are caught between Nietzsche's stress on the ceaseless flux of becoming and Heidegger's emphasis on Being revealed through hermeneutic disclosure. But, as the political theorist Leslie Thiele discerns, if Being is understood not as 'stable and unchanging', but as 'a coming to presence, less a *what* than a *how*, then becoming is not only compatible with Being but constitutes its very essence'.[137] Binswanger illustrates this reconception by quoting a remark made by the Roman philosopher Gaius Petronius, 'Each man creates his own!' (*sed sibi quisque facit!*), which becomes his central maxim for considering the relation between dream and existence. In pondering the *Quisque* of Petronius, Binswanger fixes upon the 'problem of man's moral responsibility for his dreams', which leads him away from the *idios kosmos* of the isolated dreamer (and the morally dangerous overreaching which is implicit in Nietzsche's doctrine of self-overcoming) towards a responsible attitude to the '*koinos cosmos*', or the larger community of humans.[138] By shifting from the private world of dreaming to the collaborative venture of interpretation, Binswanger affirms his emphasis on love and care in order to help the patient understand his or her structures of existence. He draws a distinction between the temporary state of dreaming and permanent delusion which represents a withdrawal into 'private opinion' or '*doxa*'.[139] From Heraclitus he understands genuine 'awakeness' to result only if the individual struggles to comprehend the whole plenum of existence and 'live in awareness' of the 'interconnection' between individual, natural and social worlds.[140] In the analytic encounter this would be represented by the patient,

> [who] must decide whether, in pride and defiance, to cling to his private opinion . . . or whether to place himself in the hands of a physician viewed as the wise mediator between the private and the communal world, between deception and truth . . . None can attain to genuine health in his innermost being unless the physician succeeds in awakening in him that little spark of spirituality that must be awake in order for such a spirit to feel the slightest breath.[141]

---

136 Binswanger, 'Dream and Existence', op. cit., p. 96. Here Binswanger's comments can be aligned with the thought of Heidegger, who, in 'The Question Concerning Technology' (1953), favourably contrasts the *teckne* of the pre-Socratics (particularly *poiesis*) with the menace of modern technology.

137 Anthony Thiele, *Timely Meditations: Martin Heidegger and Postmodern Politics* (Princeton, NJ: Princeton U.P., 1995), p. 26.

138 Binswanger, 'Dream and Existence', op. cit., p. 99.

139 Ibid., p. 98.

140 Ibid.

141 Ibid., p. 99.

The analyst is not equipped with miraculous powers of health-giving, but may be able to isolate existential structures which the patient overlooks or cannot acknowledge. The analyst can only achieve this if he or she is aware of the meaning of the dream symbolism for the patient. Appropriately, the collaborative activity of analyst and patient is the key to Binswanger's dream analysis. However, as the remaining two sections demonstrate, the realities of dialogic communication often renders this therapeutic goal as optimistic in the extreme.

Lola Voss and the Fear of Living

Published in 1957, Binswanger's series of case studies *Schizophrenie* provides a useful testing ground for his theoretical premises.[142] Eugen Bleuler, his former teacher at the Burghölzli, coined the term 'schizophrenia' to describe patients who display a particular form of dementia, characterized by 'a loosening of the tension of associations, in a manner more or less similar to what happens in dreams or in daydreams'.[143] As described in *Dementia Praecox, oder Gruppe der Schizophrenien* (1911), Bleuler's organo-dynamic theory broke with the strictly organicist theories of nineteenth-century psychiatry and gave rise to extensive research into schizophrenia in the ensuing years, especially in Zürich. Just as James addresses the *fin-de-siècle* urban illness of neurasthenia, so Binswanger follows the work of Karl Jaspers and Eugene Minkowski by focusing closely on the various mental and behavioural phenomena which characterize schizophrenia.[144] Whilst descriptive phenomenology appears inadequate for patients who cannot remember or articulate their experiences to the analyst, Binswanger's extensive study of existential forms complements the patient's incomplete phenomenological descriptions of his or her life-world, by filling in gaps and disclosing 'connections and interrelations' between data.[145]

In the introduction to *Schizophrenie*, Binswanger states that his primary aim is to approach each patient without theoretical preconceptions in order to glean

---

[142] The collection consists of an introduction (translated by Needleman in *Being in the World*, op. cit., pp. 249–64) and five case studies which had been printed separately in journals over the previous decade: the cases of 'Ilse', 'Ellen West', 'Jürg Zünd', 'Lola Voss' and 'Suzanne Urban'. 'Lola Voss' is translated by Needleman (pp. 266–341) and the first two cases are available in May et al. (eds), *Existence*, op. cit., pp. 214–36 and pp. 237–364.

[143] Henri Ellenberger, *The Discovery of the Unconscious: The History and Evolution of Dynamic Psychiatry* (New York: Basic Books, 1970), p. 287

[144] See Jasper's *General Psychopathology* (Berlin, 1923) and Minkowski's *La Schizophrénie* (Paris, 1927). One of Minkowski's earlier papers, 'Findings in a Case of Schizophrenic Depression', is available in translation; May et al. (eds), *Existence*, op. cit., pp. 127–38.

[145] May et al. (eds), *Existence*, op. cit., pp. 99–100.

'insights into the *specific* ontological structure of our cases'.[146] He outlines his hostility to the clinical tendency to 'transform' data into 'symptoms of illness' because this implies the patient is conceived as a 'disease unit'.[147] Instead, he wishes to view patients as individuals who reveal 'a *unity* of definite existential structures and processes'.[148] Although schizophrenia is varied and cannot be reduced to one definable set of symptoms, for Binswanger it represents a disproportionate mode of Being-in-the-world characterized by delusions and grandiose ideals. This enables him to dispense with the normative taxonomies 'of healthy and sick, normal and abnormal' in order to concentrate on the 'distinct modes of existence' which schizophrenic patients display.[149] For example, in 'The Case of Lola Voss' (1949), Binswanger claims the analyst must 'focus upon the "world" in which Lola presents herself to us already as a very sick person . . . without bothering in the least about the purposive biological judgment that pronounces Lola as "sick".'[150] Similarly, in an article published two years before *Schizophrenie*, Manfred Bleuler, the son of Eugen and also a professor at the Burghölzli, outlines this kind of approach to schizophrenia:

> Existential analysis refuses absolutely to examine pathological expres-
> sions with a view to seeing whether they are bizarre, absurd, illogical or
> otherwise defective; rather it attempts to understand the particular world of
> experience to which these experiences point and how this world is formed
> and how it falls apart . . . The remarkable result of existential analytical
> research in schizophrenia lies in the discovery that even in schizophrenia
> the human spirit is not split into fragments . . . If the mental life of a
> schizophrenic . . . is not merely a field strewn with ruins but has retained
> a certain structure, then it becomes evident that it must be described not as
> an agglomeration of symptoms, but as a whole and as a *Gestalt*.[151]

The attempt to explicate these structures forms the basis of Binswanger's theory and implies there is an underlying world-design (or a combination of world-designs) which can be mapped despite disorderly, and often contradictory, behaviour, but the question remains whether personal expression is reducible to a structural principle. Actually, he claims that temporal experience is not wholly reducible to a synchronic pattern: following a romantic precedent, he emphasizes the transformative nature of expression and suggests a world of flux and instability which is constantly remade through time. Accordingly, the cases in *Schizophrenie* introduce individuals whose behavioural patterns change through the course of their illness, rather than just intensifying in degree.

---

146 Needleman (ed.), *Being-in-the-World*, op. cit., p. 251.
147 Ibid., pp. 250–51.
148 Ibid., p. 250.
149 Ibid.
150 Ibid., p. 289.
151 May et al. (eds), *Existence*, op. cit., p. 124.

It is interesting to compare the cases of Lola Voss and Ellen West, not only because they share similar schizophrenic characteristics, but also because they illustrate the role of the body in Binswanger's thought: Lola displays an outward phobia towards clothes and Ellen incorporates her anxiety into her eating habits. Both cases indicate that, while *Dasein* is primarily concerned with meaning structures, because *Dasein* is embedded in the world it seeks expression through bodily activity. However, compared with the emphasis on the material body in the work of Merleau-Ponty, Binswanger can be accused of focusing on ideals and meanings at the expense of corporeality. He explicitly defends himself from those who 'accuse *Daseinanalyse* of "neglecting the body"', but is keen to stress that 'if physical needs are given authority over the whole of man's being, then the image of man becomes one-sidedly distorted and ontologically falsified': as such, the spatial and temporal trajectories of corporeality should be 'accommodated within the totality of man's knowledge of himself'.[152] Although the body is sometimes underplayed in his work, it reappears in metaphors relating to the world designs of depth/surface and rising/falling which characterize patients suffering from schizophrenia. The acute anxiety experienced by Lola Voss and Ellen West is manifested by their inability to come to terms with the Otherness of embodied life; they find themselves in a state of exile or, to recall Heidegger, a condition of double estrangement from themselves.

In the introduction to *Schizophrenie* Binswanger outlines four characteristic stages of schizophrenia. Firstly, the patient undergoes a 'breakdown in the consistency of natural experience' accompanied by the 'inability to "let things be" in the immediate encounter with them, the inability . . . to reside serenely among things'.[153] The Heideggerian notion to 'let be' (*sein lassen*) is characterized by composure of mind and equanimity: the 'mood springs from resoluteness, which, in a moment of *vision*, *looks* at those Situations which are possible in one's potentiality-for-Being-a-whole as disclosed' in the anticipation of death.[154] Far from being a 'quietist' philosophy, the resoluteness (*Entschlossenheit*) of 'letting things be' is a positive encounter with the limits of existence. In this way *Dasein* is rendered 'ahead of itself in *resolve*', which is comparable to Foucault's understanding of the Stoic preparatory technique of *askesis*.[155] Unable to bear such anxiety, Binswanger's patients also seem unable 'to come to terms with the inconsistency and disorder of their experience' and so

---

[152] Ibid., p. 212; Needleman (ed.), *Being-in-the-World*, op. cit., p. 160.

[153] Ibid., p. 252.

[154] Heidegger, *Being and Time*, op. cit., p. 396.

[155] In his interpretation of the first division of *Being and Time*, Hubert Dreyfus indicates that *Entschlossenheit* means both resoluteness and, when written with a hyphen, 'unclosedness' or 'openness'; Hubert Dreyfus, *Being-in-the-World* (Cambridge, MA: MIT Press, 1991), p. 318. Dreyfus understands that by embracing openness one acknowledges the 'impotence' of the self for making absolutely rational and 'lucid' decisions; ibid.

they construct delusions in their search 'for a *way out* so that order can be re-established'.[156] In *Schizophrenie* these delusions take the form of each patient choosing, or compulsively following, an 'inappropriate' course of action; their world-design 'manifests itself without exception in the formation of Extravagant (*verstiegene*) ideals that masquerade as a life-stance, and in the hopeless struggle to pursue and maintain these ideals'.[157]

The second quality of schizophrenia is a consequence of the first: the individual displays a 'splitting off of experiential consistency into alternatives' which take the form of a 'rigid *either-or*' logic.[158] Such disproportionate logic leads back to Binswanger's theory of the diminution of the plurality of world-designs which limit the individual into one fixed pattern of response:

> the complete submersion of the *Dasein* in the particular pair of alternatives also means that the existence can, in general, temporalize itself only in the mode of 'deficiency' – in the mode, namely, that we have come to know as *fallenness* to the world, or in short, as *'mundanization'* (*Verweltlichung*).[159]

Thirdly, in attempting to avoid facing the alternative modes of Being-in-the-world the individual often conceals, or inauthentically 'covers', those choices which are 'unbearable to the *Dasein* so that the Extravagant ideal might thereby be buttressed'.[160] The fourth stage is the feeling of 'no longer being able to find a way out or in', a self-surrender 'to existential powers alien to itself' (either in the form of 'the will of others' or compulsive behaviour) and a final *'resignation* or a *renunciation* of the whole antinomic problem' which 'takes the form of an existential *retreat*'.[161]

The case studies in *Schizophrenie* are each divided into roughly three parts: the first is an orthodox case history of the patient; the second forms Binswanger's

---

[156] Ibid., p. 253.

[157] The concept of 'Extravagance' is evident in all the cases in *Schizophrenie*, a translation of *Verstiegenheit* to which William Sadler objects. He prefers to translate the term as 'high-flown' which 'suggests a climber who has gone too high up the mountain and has become stuck on a ridge; he can neither rise higher nor can he come back down onto common ground', as embodied in the character of Halvard Solness in Ibsen's *The Master Builder*; Sadler, *Existence and Love*, op. cit., p. 292. The extravagance of the master builder is an important image for Binswanger which recurs in his essay on *Verstiegenheit* (Needleman (ed.), *Being-in-the-World*, op. cit., pp. 342–49), originally the first chapter of *Drei Formen Missglückten Daseins* (Tübingen, 1956), and in his longer work *Henrik Ibsen und das Problem der Selbstrealisation in der Kunst* (Heidelberg, 1949), an extract of which is translated as 'Ibsen's Masterbuilder'; Gras, *European Literary Theory and Practice*, op. cit., pp. 185–216.

[158] Needleman (ed.), Being-in-the-World, op. cit., p. 254.

[159] Ibid., p. 257.

[160] Ibid., p. 258.

[161] Ibid., pp. 258, 259, 263, 258.

existential analysis; and the third, a clinical analysis in which he attends closely to the particular form of psychopathology displayed by the patient. This format is similar to Freud's case studies and serves to contextualize the forms of psychopathology within the fuller life-history of the patient. Binswanger is forced to attend scrupulously to individual testimony (verbal and written accounts of letters, diaries, dreams and hallucinations) and also information garnered from families and associates of the patient, as well as from medical records. His primary interest is not in the final stage of schizophrenia *per se*, but rather 'the existential process' which leads to a retreat from the communal life of Being-in-the-world-with-others.[162] The crucial factor is 'the particular nature of the *resignation* or final capitulation of the *Dasein* culminating in the withdrawal from the *Dasein*'s decisional frame of reference . . . the *Dasein*'s surrendering of itself to the will of "alien" forces or "alien" persons'.[163] In his investigation of the 'existential process' Binswanger focuses his analysis on the personality of the patient and the unique situation in which they find themselves.

The data available to Binswanger concerning the life of Lola Voss before she enters Kreuzlingen at the age of twenty-four is limited. He is forced to rely on her father to fill in the biographical details of her life, details later confirmed by her mother. Lola was born in South America to a German father and Spanish-American mother. Spoilt as a child, she suffered from typhoid fever at the age of twelve and attended a German boarding-school at thirteen where she showed a propensity for needlework and painting. Not until she fell in love with a Spanish doctor at the age of twenty and her father expressed concern about the suitability of the match did she show signs of fasting, appearing 'joyless and depressed'.[164] At twenty-two she displayed the first signs of 'clothes-phobia' on a trip with her mother to a German spa, refusing to 'go aboard the boat unless a certain dress was removed from the luggage'.[165] When the Spanish doctor delayed his proposal to Lola she became 'melancholic and peculiarly superstitious', displaying an aversion to particular objects as well as to hunchbacked women ('hunchbacked men she considered lucky, however, and even tried to touch them').[166] While spending time with her aunts in Germany she turned against her mother, considering anything associated with her 'bewitched', particularly personal items such as clothes and underwear.

In the early days in the sanatorium no systematic study was attempted, although her first physician noticed that she was 'verbally astute', particularly in 'the art of lying'.[167] From this early stage it seems unlikely that Binswanger's collaborative ideal is a viable mode of therapy to help Lola contend with her

---

[162] Ibid., p. 249.
[163] Ibid., p. 264.
[164] Ibid., p. 267.
[165] Ibid.
[166] Ibid., p. 268.
[167] Ibid., p. 269.

growing anxiety, which began to manifest itself in extreme superstition and a refusal to change her clothing. A new physician soon took over supervision of Lola, with whom she was more inclined to communicate. Although she continued to resist talking about herself, this physician eventually learnt that her superstition regarding clothes derived from a stay in New York when she began to fear that 'something might happen to her friend if she wrote to him while wearing a particular dress'.[168] Her fear of hunchbacks is also thought to be linked closely to a letter expressing bad feeling between her and a male friend. During this trip, she visited a department store to buy a new dress, but she ran out 'in horror' when she noticed that 'the saleslady' was 'squint-eyed'.[169] These 'obsessive ideas' become more intense during the period of her engagement, at which time her superstition begins to take on different guises: she 'explained that it was the compulsion to *"read"* *something into everything* that made her so exhausted, and the more so, the more she was among people'.[170] This hyper-analytic state worsened as Lola became obsessed with objects whose names she associates with good or bad luck.

One particular instance, which she initially claims was 'so terrible that she could not possible talk about it', concerned an umbrella, which in German (*Schirm*) contains the word *si*, 'an affirmation'.[171] She associated this instance with a previous traumatic moment when 'her father had bought a new umbrella' and she saw a hunchback woman: 'now all the bad luck emanating from the hunchback was displaced onto the umbrella' which is 'confirmed through the meaning of the *si*'.[172] Other instances are documented in which an umbrella symbolizes for Lola the transferral of bad luck onto people (her mother and nurse), or the infection of objects in contact with it. She expressed her suffering in a letter to her physician and the unrelenting terror and sadness which accompanied the 'horrible possibility' of bad luck.[173] Her residence at Bellevue was soon terminated (after fourteen months) when she received news that her family 'was now ready to permit her to marry the Spanish physician'.[174] To supplement this information, the next year of Lola's life in Paris and, later, in South America is documented in letters to her physician, in which she expresses her ongoing fears and anxieties. The first section of the case study ends here.

In his existential analysis Binswanger outlines definite signs of schizophrenia in Lola's behaviour. He discerns a shift from the early stages in which 'the self was still able to preserve itself to some modest degree', towards the more extreme second stage of the schizophrenic process in which Lola 'is completely delivered

---

[168] Ibid., p. 271.
[169] Ibid., p. 275.
[170] Ibid., p. 271.
[171] Ibid., p. 272.
[172] Ibid.
[173] Ibid., p. 276.
[174] Ibid., p. 277.

to the superior power of the world, and benumbed by it'.[175] This is characterized by:

> what we call mundanization . . . a process in which the *Dasein* is abandoning itself in its actual, free potentiality of being-itself, and is giving itself over to a specific world-design . . . the *Dasein* can no longer freely allow the world to be, but is, rather, increasingly surrendered over to one particular world-design, possessed by it, overwhelmed by it.[176]

Binswanger calls the state in which *Dasein* surrenders itself to powers external to itself 'thrownness' (*Geworfenheit*), as distinct from Heidegger's use of the term to describe existence in general. Thrownness for Binswanger describes the gradual erosion and 'absorption' of existence by those objects and people that populate the world of *Dasein*.[177] If *Dasein* is overwhelmed by the Otherness of relations it often leads to delusion: in Lola's case, a belief in malevolent fate and her fear of paranoia and persecution.

Lola's extravagant ideal is understood to protect herself from the anxiety she associates with human relations and secure her existence by 'being left alone by the world . . . to let no one and nothing come close to her'.[178] As a consequence, 'when it retreats from the world of fellowmen, from its coexistors, the *Dasein* also forgoes itself, or rather forgoes itself as a self'.[179] This kind of retreat represents an inability to be 'oneself' or of 'achieving the self': in other words, seeking to avoid others actually results in submitting to her carers or investing magical controlling properties in objects.[180] Such a loss of selfhood is a form of existential weakness, as the individual 'does not take his existence upon himself but trusts himself to alien powers . . . he makes alien powers "responsible" for his fate instead of himself'.[181] In James' rhetoric this weakening of will constitutes an inability, or a refusal, to be socialized in an appropriate manner. This claim appears to reimpose those normative structures of health/sickness, sane/insane which Binswanger wishes to discard, by covertly criticizing the schizophrenic patient for not living up to social expectations. However, rather than resisting the constraints of expectation, a Foucauldian reading of Binswanger suggests that *Dasein* should first accept these limitations in order to transcend them in an act of self-transformation. Paradoxically, for Binswanger, 'existential richness' and resolve spring from letting things be, rather than surrendering to the will of others or retreating from the wider world of human relations.[182] The significant others in

---

[175]   Ibid., p. 312.
[176]   Ibid., p. 284.
[177]   Ibid., p. 285.
[178]   Ibid.
[179]   Ibid., p. 288.
[180]   Ibid., pp. 284, 296.
[181]   Ibid., p. 290.
[182]   Ibid., p. 289. Here, the similarity between Binswanger's resolute self, Rank's creative self and James' strenuous self is striking.

Lola's life (for the most part her mother and nurse) do not impose this condition on her, but make manifest to her 'a superior, uncanny, even dreadful *it*, confronted with which the *Dasein* feels completely forlorn'.[183] Although Lola feels 'abandoned by the others' (and here Binswanger makes the physician the only exception as 'the post to which existence clings while adrift in the whirlpool'), she surrenders herself over to the constant threat of 'the Uncanny and the Dreadful'.[184]

Binswanger suggests two responses through which Lola experiences 'the Uncanny': she attempts to '"capture" the Dreadful and anticipate its "intentions" with the help of words and playing on words' and she also tries 'to put spatial distance between [herself] and the persons and objects struck by the Dreadful's curse'.[185] He understands Lola's tendency to read profound significance into phonetic associations and her retreat from 'spatial closeness' to derive from a feeling of being overpowered by each of the spheres of existence (*Umwelt*, *Mitwelt* and *Eigenwelt*), leading to a 'narrowing of her life-space'.[186] Her superstitious belief in fate throws her into a state of 'anxiety without any possibility of regaining herself, or even becoming aware of herself'.[187] This does not necessarily mean that Lola is unaware of her anxiety, rather that she cannot see that her belief in fate as 'an uncanny *objective* power' is unjustifiable.[188] Binswanger draws the analogy between this state of hyper-analysis and 'astrological superstition'.[189] He claims that 'what is common to both is the clinging to alleged, blindly operating, power and evasion of the opportunity to retrieve *oneself* from thrownness and return to being one's real self or to accept genuine religious faith'.[190] This 'return' is not to some essential, or foundational, selfhood. Although memory is a retrospective activity, it acts primarily as a means for self-transformation.

In a manner similar to the grotesque vision experienced by James' French correspondent, Binswanger understands Lola's fear of female hunchbacks to be a potential staging of her future, a staging which she believes to be confirmed by the hunchback's association with the umbrella. The symbol of the female hunchback represents an 'ominous portent (as does the squinting salesgirl) from the "abnormality" of these life-phenomena, an abnormality in the sense of a

---

[183] Ibid., p. 292.

[184] Binswanger admits that he favours Schelling's definition of the uncanny ('anything which ought to remain in secrecy and obscurity and has become manifest is known as uncanny') to Freud's discussion in his 1919 paper 'The Uncanny'; ibid., p. 306. Binswanger supplements Schelling's understanding by stating 'what was supposed to remain in secrecy and hiding is the original anxiety, which now "has emerged"'; ibid.

[185] Ibid., p. 292.

[186] Ibid., pp. 293, 296.

[187] Ibid., p. 297.

[188] Ibid.

[189] Ibid., p. 298.

[190] Ibid.

"descending life" . . . namely, of deformation, crippling, disfigurement'.[191] Whereas Heidegger describes authenticity as a Being-towards-death, Lola retreats from a symbol which she associates with her 'descending' life. In doing so, her anxiety 'makes the world appear ever more insignificant, ever more simple, because it "petrifies" existence, narrows its openness, its "here", down to ever smaller circles'.[192] Contrasted to the optimistic plenitude of Emerson's 'self-evolving circle', the image of decreasing circles is characteristic of a restricted and petrified life without escape.[193] The resulting 'loss of freedom and compulsive entanglement in the net of external circumstances' gives way to a mundane present which repeats itself *ad infinitum*.[194] This 'covering' strategy and her retreat from the *Mitwelt* represent Lola's attempt to stave off her Being-towards-death, but only serve to compound her anxiety.

Closely connected to Lola's fear of female disfigurement is her peculiar relation to clothing.[195] Binswanger suggests that clothing for Lola, particularly dresses, 'become actual representatives . . . of the mother'.[196] In contrast to another of the patients in *Schizophrenie*, Jürg Zünd, who refuses to remove his coat for fear of others thinking he is degenerate, Lola actually fears 'the world of garments'.[197] Ellen West is anxious about what lies beneath the surface of the body, whereas Binswanger claims that Lola is concerned 'only with the cover provided by the clothes'.[198] But Lola's refusal to wear clothes only associated with good omens can be read as an attempt to cover the *depth* of her body. Because Binswanger sees Lola to have lost her grasp on her 'physical and psychic world', the surface world of clothing 'assumes prime importance' even in her dream-world.[199] Furthermore, because Lola cannot express her 'physical and psychic world' except through the metonymy of clothing, when she cuts them up she enacts a displaced masochistic violence to her body.

In Lola's wish to literally wear good luck she attempts to wrest 'from the intangible, uncanny Dreadful a personlike character, namely the personification of a fate that proceeds according to predictable intentions'.[200] Rather than

[191]   Ibid., p. 298.
[192]   Ibid., p. 299.
[193]   Emerson, *Essays and Lectures*, op. cit., p. 147.
[194]   Needleman (ed.), *Being-in-the-World*, op. cit., p. 300.
[195]   Binswanger admits that to address this issue fully, 'we would have to conduct a biographical investigation. Unfortunately, we do not have any historical points of reference at our disposal'; ibid., p. 300. Details such as the dresses Lola dislikes and what other items of clothing mean to her are presumably unavailable. Such limiting factors are common in this kind of analysis and check the therapeutic ideal stressed in 'Dream and Existence'. Binswanger is more successful at collecting biographical information in the case of Jürg Zünd.
[196]   Ibid., p. 300.
[197]   Ibid., p. 301.
[198]   Ibid.
[199]   Ibid.
[200]   Ibid., p. 302.

associating her 'worn-out dress' with the 'descending life' signified by the female hunchback, Binswanger understands Lola's reluctance to buy new clothes as an sign of the imminent 'catastrophe' she feels to be 'implicit in the new dress', as she accepts the 'external mundane continuity' of the old dress rather than risking the unforeseen consequences of acquiring a new one.[201] The habit of wearing worn-out clothes becomes a compulsive-repetitive pattern which offers no apparent escape for Lola. Binswanger understands her clothes-phobia to be closely related to delusions of persecution, not so much a symptom of her underlying fear of the future, but the way in which inexpressible uncanniness is articulated in terms of her world-design.

In an essay on 'The Loneliness of Contemporary Man' (1961) the American analyst Carl Rogers outlines 'two elements to the sense of aloneness' characteristic of schizophrenic patients.[202] Both elements are indicative of Lola's inability to express her condition: firstly, Rogers describes 'the estrangement of man from himself, from his experiencing organism . . . we find man lonely because of an inability to communicate freely within himself'; and, secondly, 'in our loneliness is the lack of any relationship in which we communicate our real experiencing – and hence our real self – to another'.[203] Even if one ignores the problematics of 'reality' for Rogers, this view of a self divided from itself and from others whose language it shares, but with whom it consistently fails to communicate, illustrates not only the condition of schizophrenia but the central isolating experience of modernity. Binswanger's ideal of empathic understanding and collaborative overcoming is less an antidote to such loneliness than a belief in future possibility. For him, a case is rarely closed but open to self-transformative reworkings. However, as the next section discusses, the optimism of medical care faces its sternest test when faced with terminable cases in which the 'potentially fatal division' which Rogers describes moves inexorably towards its conclusion.

Ellen West and Spiritual Longing

'The Case of Ellen West' (1944–45) is the one single piece of Binswanger's work which has received most critical attention. Not only is Ellen West presented as the most articulate of Binswanger's patients, but the denouement of her case provides one of the most dramatic moments in Western analytic history. The case adds a further dimension to Binswanger's theory of the active and embodied self and also indicates how schizophrenia can delineate a certain imaginative space, at the

---

[201]   Ibid., p. 303.
[202]   Carl Rogers, 'The Loneliness of Contemporary Man', *Review of Existential Psychology and Psychiatry*, 1(2) (1961), 94.
[203]   Ibid.

same time as the illness diminishes the possibility of participating in the world of others. The insights gleaned from Ellen's dreams, poems, diaries and letters are invaluable sources for Binswanger's analysis of her. Moreover, these personal accounts introduce elements which are either overlooked or cannot be assimilated into the anthropological schema of existential analysis. Appropriately, this section indicates the implications, as well as some of the limits, of Binswanger's project.

The background to Ellen's adult condition is full of incident, with her extended family displaying various instances of mental instability. She has a Jewish father with 'suicidal ideas'; a nervous mother; a mentally ill aunt; an uncle who shot himself; a grandmother who died of 'dementia senilis'; and a manic-depressive great-grandmother.[204] Ellen is characterized as a 'headstrong and violent child', prone to excessive mood swings and expressing a wish to be a boy (a desire which she later renounces). At the age of seventeen she read the Danish writer Jens Peter Jacobsen's novel *Niels Lyhne* (1880) which stimulated her transition 'from a deeply religious person . . . to a complete atheist'.[205] The characteristics of 'rigorous esthetic individualism and religious nihilism' displayed by Jacobsen's protagonist are reflected in Ellen's own spiritual struggle and existential anxiety, which she attempts to stave off by thinking less and working more.[206] Her early adult life is characterized by 'happiness, yearning, and hopes' (as she look forward to an overseas journey) and the feeling that she is 'small and wholly forsaken in a world which she cannot understand' (when she goes to nurse her brother in Sicily).[207]

As a child Ellen displayed a liking for meat and a 'resistance' to desserts and sweets. At twenty-one her anxiety manifested itself with worries over the size of her body and by being 'constantly tormented by the idea she is getting fat'.[208] In her diary she wrote: 'my inner self is so closely connected with my body that the two form a unity and together constitute my "I", my unlogical, nervous, individual "I"'.[209] This close identification of her 'inner self' and body becomes a central characteristic of her eating disorder, manifested in her poems by the figure of death as a imaginative and seductive force: 'not a man with a scythe but "a glorious woman, white asters in her hair, large eyes, dream-deep and gray"'.[210]

In the following year she married her cousin, but experienced a worsening struggle between physical mortification ('I long to be violated – *and indeed I do violence to myself every hour*') and spiritual aspiration ('I still will not give up my "ideal"').[211] She writes powerfully of her feelings of imprisonment: 'I am in a

---

[204] Ibid., pp. 237, 238.
[205] Ibid., p. 239.
[206] Ibid., p. 272.
[207] Ibid., pp. 241, 242.
[208] Ibid., pp. 237, 238.
[209] Ibid., p. 238.
[210] Ibid.
[211] Ibid., pp. 255, 250, 249.

prison caught in a net from which I cannot free myself. I am a prisoner within myself; I get more and more entangled, and every day is a new, useless struggle; the meshes tighten more and more.'[212] As a response to her condition, she was advised to rest and avoid over-stimulation, but such advice was to little avail. Nevertheless, she suffered a miscarriage at the age of twenty-nine and began to imbibe laxatives leading to a general decline in strength. Beginning psychoanalysis in her early thirties she showed little response. At the age of thirty-three she attempted suicide for the first time 'by taking fifty-six tablets of Somnacetin' (which she vomited up) and soon after she made her second suicide attempt 'by taking twenty tablets of a barbiturate compound'.[213]

Soon after her first suicide attempt she joined the Bellevue clinic, but continued to display eating disorders, suicidal tendencies, increased agitation and profound depression. Much of the information concerning the turmoil of Ellen's life-world is documented during this time as diary entries and dreams which express her longing for death. In the light of her 'increasing risk of suicide' and her failure to respond to therapy at Bellevue, the three physicians (Binswanger, Eugen Bleuler and 'a foreign psychiatrist') eventually decided either to move her into a 'closed ward' or to release her; the latter option they concluded would mean 'certain suicide'.[214] After lengthy consultation with Ellen and her husband the doctors decided to 'give in to the patient's demand for discharge'.[215] On returning home initially her symptoms did not alter, but 'on the third day of being home she is as if transformed'.[216] In 'festive mood', she ate healthily, walked with her husband, read Rilke, Storm, Goethe and Tennyson (writers of high passions and imaginative exuberance) and wrote a letter to a patient-friend (a transcript of this letter is not provided). The report closes with a statement, presumably offered by her husband: 'in the evening she takes a lethal dose of poison, and on the following morning she is dead. "She looked as she had never looked in life – calm happy and peaceful".'[217]

In one extreme response to the publication of the full case history, David Lester argues that not only are the facts 'distorted by omission in the report of the case' (there are undeniable absences, particularly a full report by Ellen's husband) in order that Binswanger could use Ellen's 'life and death . . . as a vehicle to demonstrate the technique of existential analysis', but the details of the case conceal the possibility of 'psychic homicide'.[218] Lester argues that this phenomenon, in which 'suicide victims may be propelled into unconscious or

---

[212] Ibid., p. 258.
[213] Ibid., p. 252.
[214] Ibid., p. 266.
[215] Ibid.
[216] Ibid., p. 267.
[217] Ibid.
[218] David Lester, 'Ellen West's Suicide as a Case of Psychic Homicide', *Psychoanalytic Review*, 58 (1971), 251.

partly conscious acting out of the wishes of others', takes the form of 'hostile neglect' on the part of the husband.[219] Together with the gross irresponsibility which he attributes to Binswanger (by agreeing to release Ellen from Bellevue and by orchestrating 'an intellectual exercise in abstract thought with no implications for behavior change or psychotherapy'), Lester argues that the case expresses Binswanger's 'need to acquit himself for responsibility for her death'.[220] While such a reading overlooks the sensitivity with which Binswanger addresses Ellen's death wishes, Lester's comments indicate serious weaknesses when reading this case from Binswanger's anthropological perspective. Similarly, in *The Voice of Experience* (1982), R.D. Laing claims the case is 'generally taken to be a standard work in its field, an exemplary model of its kind', but he criticizes Binswanger for his diagnosis of Ellen as schizophrenic and for evidence of central contradictions in his methodology.[221]

The 'dual mode of love' outlined in Binswanger's theoretical work is present only as an unfulfilled ideal in the cases of Lola Voss and Ellen West. Having established this ideal it is surprising that he claims 'the foregoing account summarizes what we know, on the basis of credible autobiographical and biographical documents and testimonies, about the human individuality to whom we have given the name Ellen West'.[222] Already the reader has the sense that Binswanger does not know Ellen as a 'whole' person, since he never interacts with her at length nor analyses her dreams using her terms. He goes on to claim: 'this knowledge is of a purely historical sort . . . her specific name loses its function of a mere verbal label for a human individuality'.[223] Moreover, Binswanger attempts to suspend clinical judgements ('be they moral, esthetic, social, medical, or in any other way derived from a prior point of view, and most of all our own judgment'), even at the expense of personal contact with Ellen.[224] Of course, Binswanger cannot be held wholly responsible for attending to the patient for the full duration of her stay at Bellevue, but, as Laing scathingly indicates:

> in view of [his] theoretical reflections it is surprising that Binswanger writes . . . that conditions were particularly favourable to existential analysis, just because he did not know her [Ellen] personally. Better than that, he has at his disposal an abundance of written material. Usually, in such cases of deteriorated schizophrenia, material for existential analysis

---

[219]   Ibid., p. 252.

[220]   Ibid., p. 262.

[221]   R.D. Laing, *The Voice of Experience* (London: Allen Lane, 1982), pp. 53, 59–60. Although Ellen displays traces of the four stages of schizophrenia, the other analysts dealing with the case had variously diagnosed hysteria, severe obsessional neurosis with manic-depression, melancholia and psychasthenia.

[222]   May et al. (eds), *Existence*, op. cit., p. 267.

[223]   Ibid.

[224]   Ibid., p. 268.

can only be obtained by persistent and systematic exploration of patients over months and years. Evidently, the attempt to establish a 'dual' relationship with such patients is only a waste of time. In her case, he has pages and pages of useful material. He can spread it out before him, all at once, and look at it. No need to spend time in the presence of a person whose presence in the world is so totally unfortunate and miserable. The existential Gestalt that is Ellen West is unable to 'relate'. His study exemplifies exactly what he attacks.[225]

Laing argues that Binswanger contradicts his emphasis on love (although he does retain it as an unfulfilled possibility) and, despite his comments on empathy, his professional distance amplifies Ellen's feelings of loneliness.[226] However, given the terms of the case, he can (at the stage of writing it) only hope to analyse the material he and his colleagues have collected over the previous years 'in as many details as it is at all possible'.[227] Furthermore, by adhering to the central tenets of romantic science, Binswanger asserts that 'Extravagance can never be understood solely from a subjective point of view, but only from the combined perspective of subjectivity and objectivity.'[228]

To his credit, Binswanger writes of his 'uncertain, fluctuating, and incomplete' judgements and he expresses a need to respond to the details sympathetically: 'love alone, and the imagination originating from it, can rise above this single point of regard'.[229] Here, though, he has shifted his stance from analyst-therapist of patients to analyst-hermeneutician of the text of Ellen's life. If the conflation of life and text is all that can be hoped for after Ellen's death, then the 'dual mode' of intersubjectivity is transformed into a mode of textual interpretation. Binswanger seems to have shifted from his therapeutic practice to reorient the patient in her world towards an understanding of world-design as an existential structure which, at least in this case, can be fully discerned only after the event with the hindsight of narrative closure. The resulting form of posthumous literary criticism (similar to Binswanger's interpretation of Ibsen's *The Masterbuilder*, Rank's reading of myths and Erikson's psychohistories) retains the search for an underlying world-design, both as a structuring pattern and a temporal dynamic that undergoes transformations during the course of Ellen's life. However, Ellen's text generates an excess of meaning which resists Binswanger's desire to uncover *the* central pattern of her life.

The very opening words of the case, 'Ellen West, a non-Swiss . . . ', suggest an Otherness that cannot be contained within the framework of the case history, and, perhaps, indicate that Binswanger's description of world-designs is, at best,

225   Laing, *The Voice of Experience*, op. cit., p. 61.
226   May et al. (eds), *Existence*, op. cit., p. 273.
227   Ibid., p. 268.
228   Needleman (ed.), *Being-in-the-World*, op. cit., p. 349
229   May et al. (eds), *Existence*, op. cit., p. 268.

culturally specific (she is also a Jew in war-torn Europe). Otherness in this case is twofold: firstly, the difficulty which Binswanger experiences in appraising Ellen's 'death-history' and, secondly, Ellen's constant battle with her split self: 'I confront myself as a strange person.'[230] Binswanger is not insensitive to the strange nature of the case, but his ideal of empathy is inadequate to Ellen's internalized experiences. However, it is productive to see Ellen's case and Binswanger's analysis of it as extending and problematizing the complex world of *Dasein* which his comments on love brush over. This section will suspend commentary on Binswanger's medical ethics in agreeing to release Ellen, noting only that, following Heidegger, he stresses that Ellen's existence is authentic 'only when she faces death', a comment which does not necessarily mean he is in agreement with her final act.[231] When he writes 'only in her decision for death did she find herself and choose herself', he interprets her decision from the perspective of her own symbolic world rather than making a moral judgement.[232]

Ellen's 'becoming' can be seen as an attempt to discover a self-reliant mode of life; her early religious distance from her parents indicates a lasting wish to take responsibility for herself. From one perspective it is commendable that 'in no respect does she care about the judgment of the world', but her later renunciation of the worldly (rejecting the significant others in her life) and the corporeal (the wasting of her body) displays how extravagantly she seeks her independence.[233] Beneath Ellen's desire for independence lies a profound unhappiness with her earthly condition: her will to self-reliance conceals her resignation to those alien forces within herself. Because she aligns her identity so closely with her body, it is not surprising that the constricted world-design manifests itself in a compulsion to eat followed by bodily mortification. Rather than becoming herself through starvation, her attempts to empty her body perversely turn into 'nothing but a metamorphosis of freedom into compulsion'.[234] As such, she can be seen to abandon a future of 'definite possibilities' for one which is empty and 'hangs in the air'.[235]

Just as the hero of Jacobsen's *Niels Lyhne* renounces the promises of conventional religion, Ellen has 'no star' to follow to guide her in her life:

> she suddenly discards her faith, which she had cherished in opposition to her father, and feels confirmed, indeed strengthened, in her individualism.

---

[230] Ibid., pp. 254, 296.

[231] Ibid., p. 310. For another response to the (in)authenticity of Ellen's actions see Foucault's 'Dream, Imagination and Existence' and James Miller's interpretation of Foucault's response to Ellen's transgressive 'limit experience'; James Miller, *The Passion of Michel Foucault* (London: Flamingo, 1994), pp. 73–8.

[232] May et al. (eds), *Existence*, op. cit., p. 298.

[233] Ibid., p. 239

[234] Ibid., p. 341.

[235] Ibid., p. 303.

Feeling no longer any trust in, or obligation to a deity, 'nowhere caring' again about the judgement of the *Mitwelt*, she is now completely reliant on herself, determining the guide-lines and goals of her actions, in the words of Niels Lyhne, entirely 'as a solitary individual,' 'by what she in her best moments ranks highest according to what there is in her.'[236]

Ellen introjects all her yearning into the world of the *Eigenwelt*, but, according to Binswanger, 'that also means that the self remains limited to passionately wishing and dreaming'.[237] The 'moral responsibility' which Binswanger discusses in 'Dream and Existence' should lead the individual away from the *idios kosmos* of the dreamer towards a participation in the communal *'koinos cosmos'*.[238] Rather than self-reliance leading to freedom and optimistic creativity, Ellen's poems and dreams reveal a morbid sensibility and a desire to escape the confines of worldly existence (*Mitwelt* and *Umwelt*) for a realm of poetic and ethereal isolation.

Four dreams are recorded which link Ellen's desire to eat with encroaching images of death: dreams in which she experiences 'the joyous expectation that I shall soon die'; a scene in which she asks a man 'to get a revolver and shoot them both'; when 'she jumped into the water through a porthole'; and when she attempts to 'set herself on fire in the forest'.[239] Obviously, the dreams have been selected by Binswanger for their reflection of her anorexic schizophrenia and their anticipation of her final suicide: he claims that, taken together, the dreams form 'the expression of one and the same anthropological fact, namely, the intertwined, inner connectedness of the motif of gluttony and death'.[240] On this interpretation, only in her Being-towards-death is Ellen able to extricate herself from a repetitive pattern which invades both her dreaming and waking life. However, to read Ellen's final act into her earlier dream sets up a form of psychic determinism. A fairer reading would be to suggest the images of death are expressions of Ellen's deep unhappiness about her condition. Indeed, the variety of death-experiences in her dreams connote death as an empty signifier which may suggest a figurative death and rebirth instead of an actual one. As Binswanger's extensive analysis of the third 'sea dream' makes clear, the peculiar combination of fire and water of the last two dreams is atypical of Ellen's *Dasein*

---

[236] Ibid., pp. 272–3.

[237] Ibid., p. 273.

[238] Binswanger, 'Dream and Existence', op. cit., p. 99.

[239] Ibid., p. 263.

[240] May et al. (eds), *Existence*, op. cit., pp. 291–2. Ellen's eating disorder is certainly a form of both 'intake-restricting (or abstinent) anorexia and bulimia/anorexia (characterized by alternating bouts of gorging and starving and/or gorging and vomiting)'; Susan Bordo, 'Anorexia Nervosa: Psychopathology as the Crystallization of Culture', *The Philosophical Forum*, 17(2) (1985), 94. Binswanger uses the term anorexia nervosa sparingly (Ellen's case is one of the first extended case studies of anorexia) in order not to obscure the personal dynamics of the case.

which usually passes between opposing images of earth and air.[241] Binswanger interprets the deep water of Ellen's past to be contrasted, in the fourth dream, with the fire of purification which later manifests itself in her resolution to commit suicide. The upward movement of fire and air combine to portend a revolt against, and a flight away from, time (water) and body (earth) which is as true of her bulimia as it is of her suicide.

As Binswanger comments in his essay on 'Extravagance', it is 'possible for human existence to go *too* far' by enacting a 'disharmony in the relation between rising upward and striding forth'.[242] In Ellen's case, such an 'anthropological disproportion' is characterized by an imbalance in the structure of her empty/full world-design. Emptying her body of food (an ascetic starving, or a gorging followed by the imbibing of laxatives) provides an extravagant manner of 'rising upward' from her earthly state.[243] Binswanger speaks of the desire to overcome 'earth's gravity', as a desire to 'gain a "higher" perspective' upon that which surrounds and limits the self.[244] This desire is realized in Ellen's case by a choice to rise '*above* the particular worldly situation and thus *above* the ambit of the known and seen'.[245] While Ellen's renunciation of the world ('the worm on earth') may seem like a form of asceticism, in a letter to her husband she speaks of the 'greed to realize my ideal' and the 'hatred of the surrounding world which wants to make this impossible'.[246] Characterized by modes of excessive hardship, asceticism shows the same kind of extravagance and disproportion as greed.[247] The ascetic starving of Ellen's body ('the birdling . . . that splits his throat in highest jubilation') becomes a negative choice, rather than a positive acceptance, or a letting things be. In rising above or 'being-beyond-the-world' she robs herself of '*communio* and *communicatio*' and 'can no longer widen, revise, or examine' her existential condition.[248]

Binswanger interprets the images of the worm and the jubilant bird as characteristic of the two poles of Ellen's world-design between which she shuttles. On the one pole accumulate the terms fat, earthy, burdened, passive, lazy, withered and aged, and on the other pole gather images of slenderness, airiness, freedom, activity, flowering and youth. Ellen's wish to renounce the former pole is evident in her early hatred of indolence and conformity:

---

[241] Appropriating the elemental symbolism of Gaston Bachelard, Binswanger suggests these elements are universal properties. See Bachelard's *La Psychoanalyse du Feu* (Paris, 1938) and *L'Air et les songes* (Paris, 1943).

[242] Needleman (ed.), *Being-in-the-World*, op. cit., pp. 342–3.

[243] Ibid., p. 343.

[244] Ibid., p. 345.

[245] Ibid.

[246] Ibid., p. 251.

[247] This interdependence of asceticism and greed is scientifically analysed in the American writer Frank Norris' naturalist novel *McTeague* (1899).

[248] Ibid., p. 343.

I am twenty-one years old and am supposed to be silent and grin like a puppet. I am no puppet. I am a human being with red blood and a woman with a quivering heart. And I cannot breathe in this atmosphere of hypocrisy and cowardice, and I mean to do something great and must get a little closer to my ideal, my proud ideal . . . I am not thinking of the liberation of the soul; I mean the real, tangible liberation of the people from the chains of their oppressors . . . Call it unsatisfied urge to action . . . To me it is as if this boiling in my blood were something better. Oh I am choking in this petty, commonplace life. Bloated self-satisfaction or egotistical greed, joyless submissiveness or crude indifference; those are the plants which thrive in the sunshine of the commonplace. They grow and proliferate, and like weeds they smother the flower of longing which germinates among them . . . The morning must come after this siege of nightmares.[249]

It is worth quoting the patient at length because Binswanger's interpretation of a single world-design (empty/full) is here complicated by a third term: an expression of an intense existence ('boiling' blood) which is both of a bodily and ethereal order. Her fear of being reduced to an uncanny puppet (neither entirely human nor non-human) is a central image of Ellen's desire to fulfil this third way excluded within the rigid either-or dichotomy of her world-design.

To deploy Jamesian rhetoric, this third way would enable Ellen to fuse her romantic yearning with a celebration of her earthly existence in a strenuous drive to meaningful action. However, Ellen's poems and dreams seem to suggest this path is barred: 'all your projects/ . . . all of them lie buried,/Scattered in wind and storm,/And you've become a nothing/A timid earthy worm.'[250] Later on she does not even compare her existence with that of the worm, but 'to lifeless, worthless material': a 'discarded husk, cracked, useless, worthless'.[251] As Ellen's youthful romanticism is quashed by the mundanity of her existence, so the possibility of becoming herself through this third way diminishes: 'she now hates her body too and beats it with her fist.' As such, the rest of her life dramatizes a desperate wish 'not to be oneself'.[252] The retreat from herself is played out in the negative cycles of gorging and fasting which eventually lead to her suicide: according to Binswanger, 'in her death we perceive . . . the existential meaning, or more accurately, contra-meaning, of her life. This meaning was not that of being herself, but rather that of being not herself.'[253] He is sensitive to the early possibility of the third way as an 'attempt at harmonizing the ethereal world ideal with the world of practical action', but he concludes that 'in view of the powers

---

[249] Ibid., pp. 243–4.
[250] Ibid., p. 244.
[251] Ibid., p. 286.
[252] Ibid., p. 297.
[253] Ibid.

available to her this attempt is so high-flown as to make us dizzy'.[254] Ellen is thought to lack the power to achieve her dreams after she severs herself from the shared worlds of *Mitwelt* and *Umwelt* and introjects all her desires into *Eigenwelt*. The suggestion here is that isolated existence leads to unrealistic or extravagant ideals: only in communion with the natural and social worlds could her goals be achieved.

It is interesting to note that one limitation of Binswanger's perspective reveals itself in the pronoun 'us' in the above quotation, implying that it is the analyst (as reader of Ellen's life) who is as liable to fall as the patient herself. Such an admission complicates Binswanger's otherwise sound analysis of Ellen's condition and indicates the fragile basis on which his analytic interpretation is based. By setting up a coherent narrative pattern which structures Ellen's existence and actions, Binswanger's analysis falls short of addressing the discrete experiences of Ellen's life: the complex marriage to her cousin, her recurrent wish to violate her body and the letter to the patient-friend at the end of her life. Perhaps lacking vital information for some of these incidents, the third-person analytic narrative nevertheless reveals its own inadequacy in telling the life of another, especially one whose world-design is so extravagant. At the end of the account the reader is left with a feeling of the substantial gap between Ellen's experience and Binswanger's interpretation.

To conclude this chapter, it is useful to consider how the centrality of the body in this case illuminates Binswanger's broader view of the self conceived as *Dasein*. Towards the end of his analysis of Ellen's case he states that the body 'signifies the sphere of our existence which is on hand here and now, spatially expanded, present *here*'.[255] What differentiates this statement from the classical philosophical definition of physical bodies extended in space is the notion that the body is 'on hand' and actively presents itself in the world. Because the world of *Dasein* is threefold, the bodily self is properly understood as a phenomenal body (Merleau-Ponty's *corps propre*), by which it is embedded both spatially in the organic and social worlds and temporally through the possibility of self-becoming. By limiting her existence to *Eigenwelt*, Ellen conceptually empties the self of meaningful content and cuts herself from the potentiality of time. Consequently, the manifold 'sphere of existence' is reduced to the image of the 'worm of the earth . . . cut off from the future [which] no longer sees wideness and brightness before her, but now only moves in a dark, tight circle'.[256] Elsewhere, Binswanger uses the same image of the closed circle to symbolize the entrapment cycle of gorging and fasting in which she is caught.[257] Although this image is Binswanger's (rather than Ellen's), the closed circle beautifully

[254]   Ibid., p. 283.
[255]   Ibid., p. 341.
[256]   Ibid., p. 306.
[257]   Ibid., p. 289.

symbolizes an existence which is self-enclosed and cut off from the possibility of temporal change. Instead of Emerson's optimistic image of connecting circles, or James' figure of the ascending spiral, the closed circle becomes a prison which prohibits, to use Binswanger's language, both a communal Being-in-the-world-with-others or an authentic Being-towards-death.

Chapter 4

Erik Erikson:
The Biographical Romantic

The analytic models which Rank and Binswanger offered as alternatives to Freudian orthodoxy were devised as counterpoints to what they conceived to be the constraining presuppositions Freud had uncritically inherited from nineteenth-century natural science. Both Rank and Binswanger addressed existential issues, but only Binswanger self-consciously adopted the language of existentialism fashionable in European cultural circles in the 1940s. Erik Homburger Erikson (1902–94) also developed Freudian insights in an existential direction, but he did so from a more distant historical vantage point than his predecessors. Born into a Danish family and raised in Karlsruhe, Erikson's early education was in the European arts; only later did he consider psychology as a way of bridging the life of the mind with the practical application of knowledge. Unlike the other romantic scientists, Erikson worried less about the problematic dualisms buried within psychoanalysis, wishing to extend its scope to consider historical and socio-cultural forces impacting on an individual life. Trained by Anna Freud in the newly emerging discipline of child psychology in the late 1920s and early 1930s, Erikson applied his abiding interest in aesthetic creation and child's play to a growing awareness of the psychic development and social adaptation of individuals. Like Rank, Erikson became a practising lay-analyst whose work was fundamentally polymathic: combining empirical science, art, religion and anthropology with a rigorous training in both adult and child analysis.

With the rise of European dictatorships in the 1930s, Erikson emigrated to America with his Canadian wife Joan, where his analytic apprenticeship made him eminently employable on his arrival in Boston in 1933. In the late 1930s he moved to work at the University of California and then back East to the Austin Riggs Centre in Stockbridge, Massachusetts and, later, Harvard University. Throughout his career he developed an integrative and hermeneutic approach to analysis which helped redirect the course of American psychology in the 1950s and 1960s, particularly his insights into psychological growth in *Childhood and Society* (1950), the elaboration of the life-cycle in *Identity and the Life Cycle* (1959), his consideration of identity crises in *Identity: Youth and Crisis* (1968), and his development of the 'psychohistory' as an alternative to the analytic case study, most famously, in *Young Man Luther* (1958) and *Gandhi's Truth* (1969). Erikson retained his romantic interest in the creative life of individuals and succeeded in loosening psychoanalytic terminology from its natural-scientific base towards a narrative understanding of psychic life.

Reorienting Psychoanalysis

One of the most significant differences between Erikson and his two European predecessors was that he felt less constrained by the forceful influence of Freud on his work. Consequently, his analytic writings are on the whole less reactive than Rank's and Binswanger's work. Although Freudian orthodoxy was established as the major analytic practice in America at mid-century, the influence of Binswanger's existential psychology upon American popularizers such as Rollo May in the 1950s, the impact of the Frankfurt School thinkers Erich Fromm in *Fear of Freedom* (1942) and Herbert Marcuse in *Eros and Civilization* (1956), together with the Freudian critiques of Philip Rieff and Norman Brown, represented a widespread revaluation of the presuppositions of psychoanalysis and its applicability to mid-twentieth-century life.[1] Erikson also worked in the wake of Freud, but sought to reorient psychoanalysis rather than to attack Freudian orthodoxy directly. Erikson, Fromm and Marcuse all emigrated from Central Europe in the early 1930s and established a fresh direction for European psychoanalysis in the country in which Freud felt so uncomfortable during his only visit in 1909. The hybrid narratives of these transatlantic thinkers provide a very different model of intellectual exchange from Freud's desire to discuss ideas only within his exclusive ring of colleagues. Whereas Jung, Rank and Ferenczi embraced the possibilities that America had to offer to the development of psychoanalysis, Freud viewed it as a threat to his cause. Up until his death in 1939 he tried to control the dissemination of psychoanalytic ideas in Europe, whereas he feared his theories would be bastardized in America and the profession run out of control.[2]

In the second volume of his valuable study of psychoanalysis in America, Nathan Hale Jr charts the rise and crisis of Freudian thought from its close association with psychiatry during the Second World War to its theoretical demise in the mid-1980s. Despite the various attacks on its basic presuppositions, Hale concludes his study by claiming psychoanalysis is compatible with the American frontier mythology in its protection of 'individual autonomy and expressiveness' and reinforcement of 'an older American tradition of individualism'.[3] However, although psychoanalytic training expanded rapidly in the post-war years, paradoxically, at the same time Freud's ideas were institutionalized in America by teaching establishments and disseminated through popular culture, they became contaminated by forms of psychosomatic medicine and alternative psychological

---

[1]   See Norman O. Brown, *Life Against Death* (New York: Vintage, 1959) and Philip Rieff, *Freud: The Mind of the Moralist* (New York: Doubleday, 1961).

[2]   For an interpretation of Freud's misgivings about American psychoanalysis see Martin Halliwell, 'Freud at Coney Island: The European Imagination in America', *Over Here: A European Journal of American Culture*, 17(1) (Summer 1997), 53–66.

[3]   Nathan Hale, *The Rise and Crisis of Psychoanalysis in the United States: Freud and the Americans, 1917–1985* (New York: Oxford U.P., 1995), p. 392.

therapies which offered a broader creative scope than the 'formulaic quality' of orthodox psychoanalysis.[4] Optimized by Marcuse's desire to blend Freudian insights with a Marxist commitment to social analysis, the psychoanalytic climate in the 1950s and 1960s marked a revolt against the individualistic concerns of the early twentieth-century analysts towards a broader conception of the individual situated firmly within an interpersonal and social continuum. For example, in a 1951 review of Erikson's *Childhood and Society*, the critic Walter Bromberg calls for a 'reintegration' of psychoanalysis and the social sciences and a 'recapitulation' of 'analytic findings in the evolution of the psychobiologic-social unit'.[5] However, the development of 'a coordinated social psychiatry' was not straightforward, either in terms of method or ideology.[6] In the standard work on radical psychoanalysis *The Freudian Left* (1969), Paul Robinson discusses the post-war debates over the implications of Freudian thought in America and how different aspects of psychoanalysis were co-opted across the whole ideological spectrum 'from conservatism to radicalism'.[7]

The publication of Marcuse's extremely influential *One Dimensional Man* in 1964 represented a 'radically pessimistic critique of the social and ideological structures inherent in advanced industrial societies', criticized the heavy emphasis on scientific thinking during the Cold War years and asserted that individuality *per se* was under threat from the dehumanizing forces of liberal capitalism.[8] These kinds of critiques were not just reserved for the social sciences, but filtered across the intellectual continuum to writers like Norman Mailer who, in his 1957 essay 'The White Negro' (heavily influenced by Wilhelm Reich), bemoaned that the general 'absence of personality could mean . . . that we might still be doomed to die as a cipher in some vast statistical operation . . . in the midst of civilization'.[9] Mailer's antidote to this erosion of identity was to cultivate the attitude of the hipster or 'white negro' in active defiance to conformist pressures, in much the same way that Rank recommended the adoption of different masks to contend with the paralysing effects of psychic pressures.[10] Despite his public refusal to sign a loaylty oath at the University of California against the pernicious

---

[4]   Ibid.

[5]   Walter Bromberg, Review of *Childhood and Society*, *Mental Hygiene* (October 1951), 642

[6]   Ibid.

[7]   Paul Robinson, *The Freudian Left: Wilhelm Reich, Geza Roheim, Herbert Marcuse* (Ithaca: Cornell U.P., 1990), p. 3.

[8]   Richard H. King, *The Party of Eros: Radical Social Thought and the Realm of Freedom* (Chapel Hill: The University of North Carolina Press, 1972), p. 139.

[9]   Norman Mailer, 'The White Negro', *Dissent* (1957), republished in *Advertisements for Myself* (London: HarperCollins, 1994), pp. 290–91.

[10]   A provocative essay by Susan Reid Al-Dossary, 'Non-violence vs. Violence and the Teaching of Otto Rank' compares Rank's attempt to tap into hidden unconscious energies with Marcuse's work on social revolution; *Journal of the Otto Rank Association*, 7(1) (June 1972), 91–9.

influence of communism (leading to his resignation in 1950) and his later work on militant nonviolence in *Gandhi's Truth*, Erikson was not as radical a cultural figure as either Marcuse or Mailer.[11] Nevertheless, his exploration of identity crises was not born in an analytic laboratory, but responded directly to the post-war American condition, the social and ideological implications of the Cold War and the momentum of the counter-culture and youth movements in the 1960s.[12]

Erikson's European childhood and early adulthood actually contributed significantly to his perspective of the social character and cultural concerns of post-war America. The child psychiatrist Robert Coles argues that Erikson's early aesthetic priorities were influenced by Kierkegaard's existential psychology. Rather than an individual's history being conceived as 'a uniform or predictable series of events', Kierkegaard suggests that identity is formed as 'an endless number of trials between man and himself'.[13] This romantic-existential belief in self-cultivation through tribulation resembles closely the artistic persona which Rank devised in his *Tagebücher* and Binswanger's belief that the self is inseparable from the personal and social environments which give rise to it. The ethical implications of Kierkegaard's thought represent another dimension of romantic science which can be traced from James' conception of selfhood, the interpersonal dynamics of Rank's and Binswanger's therapies, to Erikson's work on ethical commitment in *Gandhi's Truth*. As a Dane, Kierkegaard was influential on Erikson's early life and his decision to adopt the guise of the *Künstler*, to leave school at eighteen and to wander across Europe forging his own identity as a cultural nomad.[14] Erikson later returned to Kierkegaard, granting him a privileged place in an exclusive tradition of 'heroic' individuals alongside Martin Luther, Thomas Jefferson and Mahatma Gandhi.[15] In his essay 'The Ontogeny of Ritualization in Man' (1966), Erikson comments that Kierkegaard's isolated life actually enhanced his 'freedom of communicating with memories and phantasies,

---

[11]   Lawrence Friedman notes that although Erikson refused to sign the 'most invidious form' of the oath he did sign 'the "new form of contract" containing the very loyalty oath language that he had publicly opposed'; Lawrence Friedman, *Identity's Architect: A Biography of Erik H. Erikson* (New York: Scribners, 1999), pp. 251, 250.

[12]   For a transcript of Erikson's resignation statement see Robert Coles, *Erik Erikson: The Growth of His Work* (New York: Da Capo Press, 1970), pp. 156–8.

[13]   Ibid., p. 6.

[14]   Coles notes that Erikson's mother informed her adolescent son that she had read the existential work of Kierkegaard and the romantic writings of Emerson during his childhood; ibid., p. 13. Erikson was later self-deprecating when he admitted that a *Künstler* is 'a European euphemism for a young man with some talent, but nowhere to go'; Erikson, *Insight and Responsibility* (New York: Norton, 1994), p. 20.

[15]   In 1977 Erikson drafted an unpublished study 'Themes from Kierkegaard's Early Life' (held in the Erikson Papers at the Houghton Library, Harvard University), in which he deals with the Danish philosopher's abiding interest in identity formation and existential choice as experiential, rather than rigidly philosophical, conditions.

traditions and rules, aspirations and prophesies'.[16] Although this introspective individualism seems to contradict Erikson's later interest in personal interaction, he continued to stress the individual's ability to reflect upon his existential condition, in much the same way that James recommended intimate acquaintance with the first-person should balance the behaviourist's understanding of the self.

Erikson's mobile lifestyle between Germany and Italy ended in his mid-twenties when he settled in Vienna in 1927 to teach at the Wattmanngasse School, established by Dorothy Burlingham and Anna Freud. Teaching sparked his interest in the creative world of children, later documented in *Childhood and Society* (1950) as a preoccupation with beginnings, identity formation and 'moments of concentrated and representative happening'.[17] In 1927 he also began therapeutic sessions with Freud's daughter Anna, who had already started to develop the implications of her father's work on the psycho-sexual life of children. By observing children at play Erikson could indulge his artist's sensibility in visual images and romantic perceptions, whilst formulating statistical data, to which he returned in *Childhood and Society*. Robert Coles stresses the observational dimension of child analysis and the way in which Erikson quickly transferred his attention from 'the demands of the canvas' to the exigencies of the classroom.[18] Dorothy Burlingham's Montessori school in Vienna actively encouraged improvisation, 'playful new experience, careful experiment, and free discussion' based on the American pragmatist John Dewey's insights into the child's ability to respond to tasks only when fully engaged in them.[19] The tension between Erikson's dual role as teacher, whose 'work involves continuous talking', and analyst, who 'is obliged for the most part to remain a silent observer', marks a site of tension which resurfaces in his writing, but which also characterizes the double perspective adopted by the other romantic scientists and the 'art-and-science' methodology later refined in his psychohistories.[20] Erikson was not ashamed to admit his artistic leanings. For example, in *Life History and the Historical Moment* (1975) he claims:

> I came to psychology from art, which may explain, if not justify, the fact that at times the reader will find me painting contexts and backgrounds where he could rather have me point to facts and contexts.[21]

---

[16]   Erikson, 'The Ontogeny of Ritualization in Man', A *Way of Looking at Things: Selected Papers from 1930 to 1980*, ed. Stephen Schlein (New York: Norton, 1987), p. 593.

[17]   Erikson, *Childhood and Society* (London: Vintage, 1995), pp. 38–9.

[18]   Coles, *Erik Erikson: the Growth of his Work*, op. cit., p. 21.

[19]   Erikson, 'Dorothy Burlingham's School in Vienna' (1980), A *Way of Looking at Things*, op. cit., p. 4.

[20]   Erikson, 'Psychoanalysis and the Future of Education' (1930), A *Way of Looking at Things*, op. cit., p. 14; Erikson, *The Life Cycle Completed* (New York: Norton, 1995), p. 22.

[21]   Erikson, *Life History and the Historical Moment* (New York: Norton, 1975), p. 14. Lawrence Friedman stresses the 'unsystematic' nature of Erikson's thought, claiming it is 'better to score the eclectic, vague and other incompatible qualities' than

What Erikson lacked in scientific rigour he compensated with his painterly sensitivity; it informed his work on dreams and fantasy and his free-association experiments documented in *Childhood and Society*, in which he asked a large sample of children to take their turn to 'construct on the table an exciting scene out of an imaginary moving picture'.[22]

An early essay written in Vienna, 'Children's Picture Books' (1931), is particularly insightful for understanding the way in which Erikson hoped to link artistic insight, close empirical observation and methods of educational reform. Here he describes two kinds of children's books: firstly, the 'ominous' *Struwelpeter* in which creatures are tortured and children punished and, secondly, the 'strange pages that delicately render a pale, sweet, mimosalike world'.[23] The *Struwelpeter* are designed to prevent children acting in an inappropriate manner, 'advise caution' and serve to develop the child's prohibitive superego, whereas the other 'advise capitulating immediately to the threat by behaving like a doll or a delicate mimosa, the kind of child so many adults like to see'.[24] Erikson argues that these two types of books actually create a split in the child's psyche, banishing the child's unruly 'instinctual energy' and replacing it with sadistic impulses which are potentially dangerous to the self and others, or sanitised images of goodness and a 'doll-like infantility that is without punishment only because it has no drives'.[25] He concludes that the removal of these kinds of book at school may not prevent the child encountering such images elsewhere: 'a child who is frightened by a picture book has already been disturbed and has merely been waiting for a chance to express it'.[26] Rather than systematically removing any image that frightens or disturbs, the adult should be sensitive to the particular child and adopt a genuine 'inward' relationship 'to create a truly different environment for a child'.[27] This kind of pedagogy is typical of Erikson's early works and epitomises his interest in the complexion of psychic disturbance rather than focusing on isolated manifestations of anxiety. Echoing Rank (and pre-figuring Sacks), Erikson claims that, unlike picture books, folk-fairy tales which have 'trickled down' across generations can bring a 'reconciliatory strength' to adults as well as children and serve to link isolated conditions to larger cultural issues (as Bruno Bettelheim later developed in his work on fairy tales).[28]

---

criticizing him for a lack of scientific rigour; Lawrence Friedman, 'Erik Homburger Erikson's Critical Themes and Roles: The Task of Synthesis', in eds Robert S. Wallerstein and Leo Goldberger, *Ideas and Identities: The Life and Work of Erik Erikson* (Madison: International Universities Press, 1998), pp. 356, 365.

[22]   Erikson, *Childhood and Society*, op. cit., p. 86.

[23]   Erikson, 'Children's Picture Books', *A Way of Looking at Things*, op. cit., p. 31.

[24]   Ibid., p. 34.

[25]   Ibid.

[26]   Ibid., p. 37.

[27]   Ibid., p. 38.

[28]   Ibid., p. 37. See Bruno Bettelheim, *The Uses of Enchantment* (London: Penguin, 1991).

Even in his early essays Erikson displays an engagement with Freudian language, but, rather than adopting an orthodox perspective, he uses Freudian terminology piecemeal when his work requires psychoanalytic ballast. His early work on children derives from Anna Freud's theoretical applications of her father, but whereas her work tends to deal with 'the defensive aspects of the ego', Erikson explores the more creative channels and the active adaptability of ego-formation.[29] Erikson is sympathetic to the hybrid aspects of psychoanalysis much more than its natural-scientific orientation. He praised the kind of close observation which psychoanalysis encouraged, not only in relation to manifest symptoms but also to the 'hidden' forces of 'transference, resistance, repression and regression' which can only be inferred from behaviour and personal interaction.[30] Erikson's work continued to develop in relation to Freudian orthodoxy, but his address 'The First Psychoanalyst' presented at the University of Frankfurt in 1956 to commemorate the centenary of Freud's birth best characterizes his relationship to the psychoanalytic legacy.

The centenary address begins with the rather ambiguous comment that 'to some of us, the field created by Sigmund Freud has become an absorbing profession, to some an inescapable intellectual challenge, to all the promise (or threat) of an altered image of man'.[31] He characterizes himself as a Freudian thinker and he acknowledges Freud as the 'creator' of this 'absorbing profession' (granting him a higher status than Rank usually did), calling him a 'man of rare dimensions, rare contradictions': on the one hand, the discoverer of 'mankind's daimonic inner world' and, on the other, an 'expert of warped biography' whose later work was characterized by 'grandiose one-sidedness'.[32] He goes on to compare Darwin and Freud as thinkers who possessed the 'superior gift' of a 'selective' mind and both of whom developed a serious 'method': geology for Darwin and physiology for Freud.[33] However, Erikson argues that Freud's commitment to the 'physiological laboratory' was a manifestation of his one-sidedness: 'as an ascetic reaction to romantic indulgence', rather than the more balanced hybrid view which Erikson was developing in line with the tenets of romantic science.[34] Reflecting his interest in identity crises, the substance of the address outlines Freud's 'threefold crisis: a crisis in therapeutic technique; a crisis in the conceptualization of clinical experience; and a personal crisis'.[35]

In Erikson's account, the first crisis Freud faced was the need to replace the neurologist's 'dominance' over the object of study with the analyst's role of

---

[29]   Erikson, *Childhood and Society*, op. cit., p. 175.
[30]   Erikson, 'Studies in the Interpretation of Play' (1940), *A Way of Looking at Things*, op. cit., p. 144.
[31]   Erikson, *Insight and Responsibility*, op. cit., p. 19.
[32]   Ibid., p. 20.
[33]   Ibid., pp. 21, 23, 24.
[34]   Ibid., p. 24.
[35]   Ibid., p. 28.

dividing himself 'into an observer and observed'.[36] Interestingly, here Erikson transforms his psychoanalytic ancestor into a figure moulded in the image of a romantic scientist. His comments from the 1930s suggest that the analyst's observational role should complement the teacher's interaction with children: in other words, the analyst should devise a dual role in which he is simultaneously passive and active. Freud's second crisis emerged from his clinical work, in which he encountered the same difficulty of finding an appropriate language to conceptualize the creative urges with which James had wrestled in his later work. Freud's commitment to the language of physiology and neurology often proved inadequate, forcing him to revert to metaphorical language despite his best intentions in, for example, 'A Note Upon the "Mystic Writing-Pad"' (1924). While Erikson commends Freud for formulating the 'sexual libido' as a means of describing the instinctual 'fuel' of individuality, he criticizes his theory that libidinous drives are connected closely to 'passive sexual experiences in the first years of childhood'.[37] Lastly, Erikson's description of Freud's 'personal' crisis portrays him as a vulnerable individual and transforms him into the kind of emotional pillar which Binswanger recommends the analyst should embody (despite the problems of transference). However, Erikson actually credits Nietzsche with the idea of the analyst as 'life-saver': 'it is in those moments when our divided selves threaten to drag each other down, that a friend, as Nietzsche said, becomes the life-saver which keeps us afloat and together'.[38]

These three crises reveal as much about Erikson's relationship to Freud's legacy than they do about Freud himself. Although the ceremonial occasion dictated that the speech was essentially a tribute to Freud, at times Erikson's comments are barbed and at others he transforms Freud into a figure moulded in his own image: Freud is attributed with 'passionate introspection', of giving a 'new direction to human awareness' and of emphasizing the 'inescapable responsibility' of psychoanalytic methodology.[39] For these reasons the commemorative speech functions more as a clarion call to a new generation of analysts who can learn from Freud's discoveries and mistakes than it is a balanced appraisal of the Viennese thinker. Erikson concludes his speech by stressing the ethical imperative which awaits all individuals (not just analysts), especially the adult's relationship to childhood which can only be repudiated at the risk of 'massive regressions endangering the very safeguards' of the self.[40]

If psychoanalysis represents one dimension of Erikson's thought, then James

---

[36]   Ibid., pp. 30, 29.
[37]   Ibid., p. 32.
[38]   Ibid., p. 36.
[39]   Ibid., p. 42.
[40]   Ibid., p. 45. Despite these criticisms, Freud remained one of the cornerstones of Erikson's thought, as he returned to him at significant moments in his psychohistories on Luther and Gandhi.

strongly influences his work on identity formation.[41] James appealed to Erikson because he bridged the divide between the introspective sick soul and the expansive self-reliant individual who confronts obstacles directly. Rather than just focusing on the psychic conflicts which an individual faces, Erikson, like James, was also interested in the pragmatic skills which can be adopted to contend with exacting pressures. However, the meditative dimension of James' thought approximates to Kierkegaard's claim in *The Sickness Unto Death* (1849) that in all individuals there dwells 'an uneasiness, an unquiet, a discordance, an anxiety in the face of an unknown something'.[42] Erikson observes anxiety in children as well as adults but does not trace its origin to some primal moment of trauma, viewing it as firmly embedded in the human condition (as is made evident in his comments on the picture book). Although Kierkegaard and James both suffered with the 'curse' of 'a precocious and relentless conscience' which prevented them from relaxing or acquiescing to a comfortable mode of being, such subliminal anxiety does not necessarily lead to the paralysis of self: in its most useful form it acts as a constraining force for capricious impulse.[43] On this account, creative exuberance and 'the self-determination of free will' cannot be fully inhibited by anxiety.[44] To substantiate this idea, Erikson quotes from James' correspondence: 'Life shall [be built in] doing and suffering and creating.'[45] Countering Freud's therapeutic emphasis on the curbing of libidinous energy, Erikson claims that the 'restoration of the patient's power of choice' is the chief end of psychoanalysis, an idea most fully expressed in Jamesian pragmatism.

In addition to his general sympathy with James' work and the influence of John Dewey on his early educationalist philosophy, Erikson's description of the 'path' of James' life matches his own transatlantic experience very closely:

> the path led from artistic observation through a naturalistic sense of classification and the physiologist's grasp of organic functioning, to the exile's multilingual perceptiveness, and finally through the sufferer's self-knowledge and empathy to psychology and philosophy.[46]

Whereas James' 'multilingual perceptiveness' is more figurative than literal, Erikson's transatlantic journey in the 1930s gave him an authentic double cultural

---

[41]   Interestingly, Erikson aligns Freud and James as 'bearded and patriarchal founding fathers of the psychologies on which our thinking on identity is based'; Erikson, *Identity: Youth and Crisis* (New York: Norton, 1994, p. 19).

[42]   Søren Kierkegaard, *The Sickness Unto Death*, trans. Alaistair Hannay (London: Penguin, 1989), p. 52.

[43]   Erikson, *Gandhi's Truth: On the Origins of Militant Nonviolence* (New York: Norton, 1993), p. 128.

[44]   Erikson, *Identity: Youth and Crisis*, op. cit., p. 155.

[45]   Ibid., p. 154; quoted from *The Letters of William James*, ed. Henry James (Boston: Atlantic Monthly Press, 1920), p. 148.

[46]   Erikson, *Identity: Youth and Crisis*, op. cit., p. 152.

perspective to supplement his fusion of painterly art and scientific investigation. Moreover, as he outlines in *Childhood and Society*, his wish to link the anxiety of individuals to a study of 'the immediate and extended group' enables him to fulfil one of the directives of romantic science which James could only gesture towards.[47] Just as Binswanger desired to combine the three dimensions of the individual's world (*Umwelt*, *Mitwelt* and *Eigenwelt*) into one field of study, Erikson's integrative approach to human existence links what he calls three 'processes' drawn from different scientific disciplines: firstly, the somatic process deriving from biology which examines the anatomy of the individual; secondly, the societal process drawn from the social sciences which links an historical view of collective behaviour with analysis of social meaning and significance; and, thirdly, the psychological ego process which focuses on personality formation and identity crises. Erikson claims that taken individually each mode of study dissects the 'total living situation in order to make an isolated section of it amenable to a set of instruments or concepts'; consequently, he argues that fundamentally these three processes should be viewed as 'three aspects of one process', by which he means an examination of 'human life'.[48] Just as Rank worried that too much attention to the social construction of identity detracts from the creative personality and too little attention leads to a narcissistic view of the self, so Erikson sought to balance each of these processes against each other. Indeed, this formulation of a three-dimensional approach can be seen to integrate elements of Freudian thought, Jamesian pragmatism and Binswanger's existential analysis in a historically sensitive study of the 'simultaneous changes' involved in identity formation.[49]

Identity and Crisis

One of the most significant aspects of Erikson's thought which emerges from his transatlantic experience is his concern with the 'roots' of identity and the possibilities of adapting the self to new psychic, cultural and social demands. Rather than adopting the negative perception of personality as a 'cipher' as Mailer does in 'The White Negro', Erikson views identity as always in the process of becoming, in accordance with Stanley Cavell's theory of moral perfectionism and Marx's understanding of *Entstehungsakt* ('the act of becoming').[50] However, like Rank, Erikson outlines the inhibiting factors in the development of stable identity, as much as he is interested in the adaptability of self. He argues that identity is formed in the interstices between bodily, psychic

---

[47]   Erikson, *Childhood and Society*, op. cit., p. 30.

[48]   Ibid., p. 31.

[49]   Ibid., p. 38.

[50]   Mailer, *Advertisements for Myself*, op. cit., p. 290; Erikson, *The Life Cycle Completed*, op. cit., p. 95.

and social forces, but imbalances often occur when the impetus of one force skews the influence of the others. On this homeostatic model, only when the forces are symbiotically supportive of, and mutually restrained by, each other will a well-integrated individual emerge who does not acquiesce passively to any one set of forces. Erikson maintains that all selves have the capacity to negotiate a space of individuation between internal and external forces, but the degree to which this is possible depends largely on personal circumstances and social environment. Rather than trying to define 'neurosis' or focusing on the specific role which an individual plays within the family romance, Erikson encourages a more global outlook on identity, within which both the constraints and vicissitudes of selfhood can be positioned.

In his address, 'Identity and Uprootedness in Our Time', presented to the World Federation of Mental Health at the University of Vienna in 1959, Erikson discusses the difficulties facing individuals whose cultural roots have been severed by historical dislocation or migrational change. Erikson sees the condition of identity as one which must be engaged with seriously, but which is also in danger of becoming an unhealthy obsession. He uses the example of an individual whistling a tune, 'first without being especially aware of it, but then with an obsessiveness that can become a mild state of malaise'.[51] He cautions against excessive introspection and the dangers of being seduced by the melody, arguing that only if one attends to the whole range of sound patterns (reminiscent of James' 'fringes' of experience) can identity be located within a broader cultural context. His particular example is Antonin Dvořák's 'New World' Symphony No. 9 (1893), which on one level is an extended 'historical lullaby', but also inscribes the issues of migration and generativity in its complicated semiotics.[52] Erikson's personal experience draws him repeatedly towards the phenomenon of 'transmigration', but his comments also apply to 'in-migration' or any condition in which the individual's locality is disturbed, either by enforced separation from a 'homeland' or by psychic dislocation from familial roots.[53] Although these conditions are sometimes imposed (for example, the experience of European Jews in the early 1930s), migration is often of the individual's own choosing: the choice 'to actively transplant old roots, and . . . to find new roots in Change itself'.[54] The personal mantra of 'Change, Chance, Choice' (associated with the spirit of Roosevelt's New Deal philosophy) provides Erikson with a model of mobile identity, which benefits from not being fully grounded in one locality, but is consequently vulnerable to the unsettling experience of uprootedness.

Erikson echoes James' liberating description of the 'shifting self' in flux, but he is not as optimistic as James about the individual's resilient ability to adapt to

---

[51] Erikson, *Insight and Responsibility* (New York: Norton, 1964), p. 83.
[52] Ibid., p. 84.
[53] Ibid.
[54] Ibid.

a mutable environment, especially when the individual may be estranged from the new locality in terms of culture or ethnicity.[55] In the case of refugees such as many East European Jews in America, Erikson detects an inability or unwillingness to assimilate to the new culture, 'holding on, instead, to the world that had disowned them'.[56] Such an allegiance to the past has important implications for ensuring the continuity of identity, but can often settle into a jaundiced view of the future in which the 'flicker of hope' is in danger of being extinguished by negativity.[57] Here Erikson argues that whatever the initial cause of migration, travellers retain the ability to affirm themselves through the activity and discourse of travel and what Iain Chambers calls 'restless interrogation, undoing its very terms of reference as the point of departure is lost along the way'.[58] In most circumstances mobility is a mark of freedom from external constraints, although the desire to root the self in a particular culture often checks the restless impulse to travel. However, Erikson does not just write in terms of the activity of travel: he maintains that one can 'actively stay put' or 'actively hide'.[59] Only when activity ceases does the actual self lose the protean possibility of becoming a new potential self. For this reason, Erikson agrees with James in his claim that 'patienthood . . . is a condition of inactivation', whereas health is associated with agency: 'an inner sense of being unbroken in initiative and of acting in the service of a cause which sanctions this initiative'.[60] Here Erikson relies on the existentialist vocabulary of commitment to reorient the self in a potentially bewildering environment, a task for which the solitary individual often requires the assistance of a guide to accomplish.

Erikson constructs a three-stage model to differentiate those who are 'driven' by 'unmanageable impulses' from those who are 'compelled' or 'coerced' by inner psychic urges and those who are 'persecuted' by external social forces.[61] He argues that whichever case applies, the stronger the stimulus the more likely the individual will be impelled into colluding with the hostile force, intensifying the feelings of oppression and increasing the individual's 'sense of worthlessness'.[62] For example, Erikson argued that the environment of conspiracy and collusion in American society in the 1950s only succeeds in seducing the disaffected individual into accepting his or her own fate as inevitable.[63] Although writers such

---

[55]   He rarely discusses gender difference in these terms, an issue to which I return in the next section in relation to Carol Gilligan's criticisms of Erikson's life cycle.

[56]   Ibid., p. 85. For a quasi-fictional exploration of the problems of Jewish assimilation see Henry Roth's influential novel *Call It Sleep* (1934).

[57]   Ibid., p. 85.

[58]   Iain Chambers, *Migrancy, Culture, Identity* (London: Routledge, 1994), p. 2.

[59]   Erikson, *Insight and Responsibility*, op. cit., p. 86.

[60]   Ibid., p. 87.

[61]   Ibid., p. 89.

[62]   Ibid.

[63]   This is evident at the end of Saul Bellow's second-world-war novel *Dangling Man* (1944) when the protagonist finally capitulates to the restrictions of the army: 'I am

as William Burroughs came to mistrust all institutions and social figures (for example, physicians who claim to act in the patient's best interest), Erikson concurs with Mailer's claim that without 'cunning and resolution' the fate of the passive patient/victim has already been sealed with persecutory 'delusions' which are in danger of running out of control.[64] The 'overlapping' of 'outer and inner conditions' may serve to collapse the stable subject-object world in the philosophical sense, but creates an experiential environment of uprootedness in which the individual has no secure ground for the self.[65] The proximity of dislocation and distrust in such conditions can only be remedied by a renewed sense of the individual's agency, partly through individual resolve and partly by trusting others, whether friends, physicians or social advocates.

In order to deal with the 'elusive subject of identity' Erikson adopts a double conception of the self.[66] Firstly, he claims that a 'sense' of identity entails 'being at one with oneself', rather than leading a fractured or fragmented existence.[67] He explains that the articulation of 'I' is intrinsically linguistic, but represents 'the verbal assurance according to which I feel that I am the center of awareness in a universe of experience in which I have a coherent identity'.[68] Such 'coherent identity' is the culmination of all the shifting selves which have gone before and is directed by an 'inner agency' (Erikson's version of willing) which guards against the ego dissolving into the flux of experience.[69] Once again deploying Jamesian language, Erikson describes the experience of sentient awareness a 'subjective halo', which, in part, derives from self-knowledge but mainly by acting meaningfully in an environment.[70] This dimension provides the bridge to his second definition of identity which entails having 'a sense of affinity with a community's sense of being at one with its future as well as its history – or mythology'.[71] Although the majority of Erikson's studies focus on European subjects, his attention to individuals whose life is future-oriented derives from a combination of Emersonian idealism and Jamesian pragmatism, the 'newness' of which Erikson acknowledges is essentially American.[72]

---

no longer to be held accountable for myself; I am grateful for that. I am in other hands, relieved of self-determination, freedom cancelled'; Saul Bellow, *Dangling Man* (London: Penguin, 1996), p. 191.

[64] Ibid., p. 89.
[65] Ibid.
[66] Erikson, *Dimensions of a New Identity*, op. cit., p. 27.
[67] Ibid.
[68] Erikson, *Identity: Youth and Crisis*, op. cit., p. 220.
[69] Ibid., p. 218.
[70] Ibid., p. 220.
[71] Erikson, *Dimensions of a New Identity*, pp. 27–8.
[72] Ibid., p. 79. The exceptions to this European bias are his short study of William James, a chapter, 'Reflections on the American Identity', in *Childhood and Society* and his essays on Jefferson in *Dimensions of a New Identity*.

Like James, Erikson focuses on the disjunction between the elusive language of self – 'I' and 'You' – and the realities designated by those words. In autistic children he observes a common inability to 'feel' the words as they are articulated (the internal *ich*, not the external ego) and an associated 'fear that life may run out before such feeling has been experienced'.[73] Although most individuals adopt a stable identity which constitutes a compromise of internal and external pressures, within this 'coherent or 'composite Self' are other 'various selves' which become manifest at times of crisis or shock. For this reason, Erikson wishes to separate an understanding of an 'ideal self' (which an individual desires to become) from the 'ego' which maintains

> our coherent existence by screening and synthesizing . . . all the impressions, emotions, memories, and impulses which try to enter our thought and demand our action, and which would tear us apart if unsorted and unmanaged by a slowly grown and reliably watchful screening system.[74]

Although the ego can be inferred by the existence of a continuous self, under most conditions it cannot be perceived directly; only in those individuals where the 'various selves' are not controlled or balanced is the lack of ego stability evident.[75] Erikson's fundamental concern is to help develop an amenable situation for those individuals for whom the language of self and their inner feelings do not correspond. Although the individual needs to nurture an inner conviction in him or herself, he believes only within a 'psychosocial "territory" of trusted mutualities' can such a transformation be brought about. Indeed, in an environment of mistrust and hostility the sundering of self from expression can be dangerously exacerbated.[76] The danger lies in an individual being coerced into adopting a 'debased self-image' or 'negative identity', rather than choosing a truly oppositional identity (such as Mailer's 'white negro') by refusing to submit to a certain set of pressures.[77]

A literary example of this environment of mistrust and the related formation of negative identity is articulated in the African-American writer Ralph Ellison's

---

[73]   Erikson, *Identity: Youth and Crisis* (New York: Norton, 1968), p. 217. Erikson maintains that such anxiety cannot be expressed in scientific language which is why he claims that existential problems are most often 'left to poetry or metaphysics'; ibid., p. 217.

[74]   Ibid., p. 218.

[75]   Although there are no references to Jacques Lacan in his work, the parallels between Erikson's theorization of identity and Lacan's work on the role of the 'mirror stage' in identity-formation are notable. See Lacan, 'The Mirror Stage as Formative of the Function of the I', *Écrits: A Selection*, trans. Alan Sheridan (New York: Norton, 1977), pp. 1–7.

[76]   Erikson, *Identity: Youth and Crisis*, op. cit., p. 220.

[77]   Erikson, *Insight and Responsibility*, op. cit., p. 97.

existential novel *Invisible Man* (1952), to which Erikson alludes in *Identity: Youth and Crisis*.[78] At the beginning of the novel the nameless protagonist refers to himself as 'invisible . . . simply because people refuse to see me'.[79] In other words, no-one can, or will, see beyond the colour-line which separates the invisible man from those people whose worlds collide with his. His invisibility results from the way in which others refuse to see him or passively construct him as a 'phantom' of their minds. Invisibility does allow him the freedom of subterfuge, but he aches 'with the need to convince' himself that he exists in the 'real world' and although he physically retaliates against his ignorant persecutors even then he remains invisible to them.[80] Despite his uprooted and near abject condition he has found for himself a rudimentary home, even if it is really only a 'hole in the ground' with few recognizable marks of identity.[81] But, even in this culturally and existentially impoverished environment, the protagonist shows his resilience: 'my hole is warm and full of light'; despite his invisibility he claims 'light confirms my reality, gives birth to my form'.[82] *Invisible Man* illustrates Erikson's notion of uprootedness as a form of racial and spatial dislocation in a hostile environment in which the protagonist is forced to adopt a negative identity, but it also suggests that the protagonist has the ability to survive in the face of adversity. Erikson would argue, however, that only with the counterbalance of an intimate and mutually supportive environment to provide 'organic nourishment' can the protagonist's plight be relieved.[83]

One way in which Erikson claims that such affirmation and intimacy can be regained from the 'depth of nothingness' is through sincere face-to-face contact between individuals, when divisive issues such as regional difference, gender or race are (as far as is possible) set aside.[84] This does not mean the face under scrutiny must submit to the controlling gaze and belief system of the observer, nor should differences be erased, but a mutual reciprocity of contact or 'dialogue' can be nurtured between two individuals.[85] Erikson argues that this moment of recognition derives from the 'earliest mutual recognition of and by another face', most often between mother and child.[86] The gaze of the mother in this instance is not a controlling one (although it can warn against dangers), but is itself partly defined by the recognition of the child: 'in the playful encounters, the light of the *eyes*, the features of the *face*, and the sound of the *name* become essential

---

[78]   Erikson, *Identity: Youth and Crisis*, op. cit., p. 25.
[79]   Ralph Ellison, *Invisible Man* (London: Penguin, 1965), p. 7.
[80]   Ibid.
[81]   Ibid., p. 9.
[82]   Ibid., pp. 9, 10.
[83]   Erikson, *Insight and Responsibility*, op. cit., p. 101.
[84]   Erikson, *Youth: Identity and Crisis*, op. cit., p. 25.
[85]   Erikson, *The Life Cycle Completed*, op. cit., p. 40.
[86]   Erikson, *Insight and Responsibility*, op. cit., p. 94.

ingredients of a first recognition of and by the primal other'.[87] The origins of this
psychic need strike at the heart of the 'dreaded estrangement' when nurturing
recognition is removed and the solitary individual suffers from a 'loss of face'.[88]
Here the invisibility of Ellison's protagonist is more a curse than a blessing:
although it may allow him to achieve a set of goals prohibited by the controlling
gaze of social surveillance it also prevents him from establishing a sense of
reciprocity with others. In the broader sense of social recognition, Ellison's
invisible man is banished from the daylight into the anonymity of night: only his
love of light saves him from complete obliteration. Such isolation is be found in
the fugue states of children robbed of parental care and also characterizes the
mood of one of Erikson's own dreams, in which 'a motionless image of a faceless
face' appears in the place of a specular image of the self.[89]

As a development of Erikson's discussion of the 'face' as the zone of
recognition, in *A Thousand Plateaus* (1980) Gilles Deleuze and Félix Guattari
argue that identity is both inscribed by and lodged within the face. They
conceptualize this double movement in terms of the 'white wall' and 'black hole'
of identity in which the self is simultaneously visible and invisible.[90] The face
provides a metaphor for suggesting that 'the dimensionless black hole and
formless white wall are already there to begin with' in an interdependent way,
representing both the form or appearance of the face and the darkness which is
lodged behind its mask.[91] Like Erikson, Deleuze and Guattari do not claim that
the face is ontologically prior to the body. Rather, it constitutes a 'zone of
frequency or probability' which both reveals and conceals identity; it is at once
extremely personal but also a punctured mask: the seeing eyes recognize, but the
'gazeless eyes' stare blindly.[92] As such, the face (and not the head), forms the
white wall/black hole dynamic: an incomplete surface which simultaneously
lends identity to, and robs identity from, the self.

Their main point is that what is the most human of all – the face – is also the
most inhuman, revealing a 'horror story' behind the mask.[93] As such, standing
face-to-face may provide a glimpse of the unrepresentable void of Being. But
rather than forcing the self into negative positions, divided from self-identity and
separated from the social world of others, the bilaterality of the face (both visible
and invisible, human and inhuman) may provide the stimulation to relocate
identity within different spatio-temporal coordinates or within an alternative
mythical pattern. The Jewish philosopher Emmanuel Levinas shares Deleuze's

---

87   Erikson, *The Life Cycle Completed*, op. cit., p. 40.
88   Erikson, *Insight and Responsibility*, op. cit., p. 95.
89   Ibid., p. 62.
90   Gilles Deleuze and Félix Guattari, *A Thousand Plateaus: Capitalism and
Schizophrenia*, trans. Brian Massumi (London: Athlone, 1988), p. 167.
91   Ibid., p. 168.
92   Ibid.
93   Ibid.

and Guattari's emphasis on the 'black hole' of Being as a recognition of finitude, but goes on to outline the primary responsibility of the self to 'face' the Other: 'the proximity of the other is the face's meaning . . . in a way which goes beyond those plastic forms which forever try to cover the face like a mask'.[94] Instead of these masks serving to undermine any attempt to respond to the Other, Levinas argues that 'always the face shows throw the forms' and 'calls for me' in a moment of 'self-expression'.[95]

To relate these notions of faciality to Erikson's work, it is useful to compare the isolation and invisibility of Ellison's protagonist with the intimacy expressed by another African-American writer James Baldwin in a letter to his nephew 'The Dungeon Shook' (1963).[96] In this letter Baldwin bemoans the racial intolerance of 1960s America and warns his nephew not to be 'defeated' like his grandfather (Baldwin's father), who died believing 'what white people said about him'.[97] To counteract this hostility Baldwin stresses the family resemblance by emphasizing the shape and form of his nephew's face: 'I keep seeing your face which is also the face of your father and my brother.'[98] Baldwin does not claim there is an essential self lurking behind the facial mask that can be fully revealed, but realizes that only by fostering an environment of love can any kind of recognition take place: 'other people cannot see what I see whenever I look into your father's face, for behind your father's face as it is today are all those other faces which were his'.[99] For Baldwin, his brother's face is an open book of past experiences, but lodged beneath this face is a pain which grows with feelings of racial persecution, but which has no name nor substance. Although Baldwin echoes Erikson's suggestion that invisibility can only be countered by love, there is something desperate in his brother's condition: 'no one's hand can wipe away those tears he sheds invisibly today'.[100] Here the black hole of identity and the awareness of human anxiety rupture even the most intimate of interpersonal moments.

In 'Identity and Uprootedness in Our Time', Erikson blends Kierkegaardian existentialism and Emersonian self-reliance with an idealistic and quasi-democratic commitment to nurture an environment of trust. Like Ellison and Baldwin, he declares an allegiance to existentialism, but claims that 'no school of thought has any monopoly on it'.[101] He is attracted to Albert Camus' figure of

---

[94]   Emmanuel Levinas, 'Ethics as First Philosophy' (1984), in ed. Seán Hand *The Levinas Reader* (Oxford: Blackwell, 1989), pp. 82–3.

[95]   Ibid., p. 83.

[96]   Erikson discusses Baldwin directly in his essay 'Race and the Wider Identity' (1966), *Identity: Youth and Crisis*, op. cit., p. 297.

[97]   James Baldwin, *The Fire Next Time* (London: Penguin, 1964), p. 13.

[98]   Ibid.

[99]   Ibid., p. 14.

[100]   Ibid.

[101]   Erikson, *Young Man Luther: A Study in Psychoanalysis and History* (New York: Norton, 1993), p. 22.

l'étranger as an example of 'active self-uprooting' (or a 'positive' negative identity), but maintains that the individual must first have an awareness and experience of locality and home: 'in order to lose one's identity, one must first have one; and in order to transcend, one must pass through and not bypass ethical concerns'.[102] Appropriating Camus' claim in *The Rebel* (1951) that there are certain ideological limits beyond which the self should not and will not capitulate and for which '"no" affirms the existence of a borderline', Erikson argues that most individuals (under most conditions) have the resources to resist coercive internal or external pressures, although often these resources remain untapped due to fear and uncertainty.[103] Only by developing a more secure environment (which for him is a societal as well as an interpersonal concern), in which the self can choose either to be mobile or to set down roots, will the climate of mistrust and suspicion be averted.

These ideas epitomize Erikson's ideological perspective as much as Cavell's description of moral perfectionism as an 'evolving configuration' which, in Erikson's words, integrates 'constitutional givens, idiosyncratic libidinal needs, favored capacities, significant identifications, effective defences, successful sublimations, and consistent roles'.[104] In a supportive environment the integration of the various potential selves into a coherent actual self maintains and develops the 'style of one's individuality'.[105] In other words, the creative growth of the individual is partly his or her own responsibility, but also dependent on the contingencies of a supportive environment. However, as the examples from Ellison and Baldwin imply, whatever the circumstance there is something from within that disturbs the development of identity.[106] The 'horror' (or 'black hole') which Erikson detects in the autistic child who struggles to express his or her identity and 'grasp the meaning of saying "I" and "You"' is a general condition ultimately deriving from existential uprootedness.[107] Although Erikson focuses on spatially dislocated individuals and the negative identities with which they are faced, he repeatedly returns to this general condition which shifts its character and name depending on the exigencies of the environment, but disturbs even the most homely and intimate of moments.

---

[102] Erikson, *Insight and Responsibility*, op. cit., pp. 99, 100.
[103] Albert Camus, *The Rebel*, trans. Anthony Bower (London: Penguin, 1971), p. 19.
[104] Erikson, *Identity: Youth and Crisis*, op. cit., p. 163.
[105] Ibid., p. 50.
[106] Although Erikson hints at the darker aspects of psychoanalysis, which link to Freud's uncanny and Rank's *Doppelgänger*, Oliver Sacks more fully expands on their implications in his neurological practice.
[107] Erikson, *Identity: Youth and Crisis*, op. cit., p. 217.

Self-Development and the Life Cycle

Erikson is best known for his formulation of the life cycle in which he charts the development and maintenance of the self through various stages in life. Between *Childhood and Society* (1950) and *The Life Cycle Completed* (1982) he refined an eight-stage developmental model charting the critical phases which define the fluctuating changes of life. Binswanger's world-designs and Erikson's life cycle are two articulations of the parameters in which the shifting self is defined in both personal and social terms. Corresponding to his notion of psychosocial identity, only an alignment of the three processes of selfhood – somatic, social and psychological – enables the individual to maintain a degree of autonomy with the mutual support and symbiotic restriction of these forces; only 'the integration of the whole ensemble', together with the harmony of the individual's size and function, can provide the network of support which enables growth and development.[108] Erikson searches for sensory descriptions of this integration: 'as the eyes begin to focus, to isolate, and to "grasp" objects', the ears 'learn to discern significant sounds, to localize them, and to guide a searching turn toward them', the arms 'reach out aimfully' and the hands 'grasp firmly'.[109] He claims this is not the 'simple causal effect of training on development', but the 'mutual assimilation of somatic, mental, and social processes'.[110] The collision between the personal needs of the individual and the demands of the social world may be potentially destructive but also can be mutually supportive, helping to attune the individual to the environment.

Although Erikson can be accused of over-generalizing in such a schema and not attending closely to individual circumstances, the eight stages which he defines (infancy, early childhood, play age, school age, adolescence, young adulthood, maturity and old age) are purely heuristic in outlining the basic qualities that emerge in different phases of life: for example, he does not give specific ages which define the temporal parameters of a particular stage. Nevertheless, a 'cycle' does seem to imply a teleological progression which only comes to fruition in its final stage (which for Erikson would be the wisdom of old age). However, the schema does not define attainment levels which would leave the individual floundering in a perpetual state of immaturity if he or she were to fall short of the goal. Instead, his model follows an embryonic (or epigenetic) precedent in which traces of residual and emergent phases are present at the same time that a single mode of existence rises to ascendancy. As such, the model defines a process of becoming, rather than a progression to a higher attainment goal.

Erikson's cyclical model is not incompatible with the spiral design which James and Rank deploy: the dialectic oscillation between two terms produces a

---

[108]   Erikson, *The Life Cycle Completed*, op. cit., p. 29.
[109]   Ibid., p. 36.
[110]   Ibid.

third which absorbs as well as develops the previous phase. Each stage is defined by a set of alternative terms between which the individual oscillates and an emerging modality which characterizes the next transition of the self. A brief consideration of the two initial stages of the life cycle illustrate this point. The first stage of infancy is defined by the poles of trust and mistrust. In an initially trustful environment of reciprocity (as outlined in Baldwin's letter to his nephew) the infant often develops an optimistic outlook, whereas an environment of mistrust can often affect infants who become withdrawn and depressive. Similarly, the second stage of early childhood oscillates between the 'battle for autonomy' and wilfulness which defines the child as distinct from the mother and feelings of doubt which emerges from the child's continued dependence on parental care.[111] Throughout the eight stages Erikson implies that the basic existential parameters which define selfhood are always in play, but are expressed in particular ways and through specific mythic patterns.

He is careful to stress that the cycle is never actually completed: in *The Life Cycle Completed* he claims 'it is hoped that this title sounds just ironic enough not to be taken as a promise of an all-inclusive account of a perfect human life'.[112] Although there is general deflecting of attention away from the self defined in personal terms and a concomitant increase in the awareness of the social Other ('the communal sense of "we"' in the broadest terms), this does not imply a hierarchy of moral attainment.[113] Erikson graphically expresses his model as an ascending line, but this diagram does not adequately illustrate his theory that the seeds of an earlier stage are still present in a residual, and often subliminal, form during later phases. Indeed, Erikson goes as far as to reverse the direction of the chart in *The Life Cycle Completed*, suggesting that the notion of unilateral linear progress from one stage to the next provides a bogus model. This notion of reversibility is under-theorized in Erikson's thought, but nevertheless develops the bilateral dynamic implicit in his analytic interest in faciality (as developed by Deleuze and Guattari) and therapeutic emphasis on mutual reciprocity.

Erikson's model concurs with James' and Rank's ladders of social types, in the respect that a failure or refusal to align the ascending potential self with the actual self can lead to debilitating regressions or an inability to cope with changes in mental and physical constitution, especially at early stages of life when the boundaries of the self are very fragile. Erikson associates balance with health and an active willingness to embrace the changes in life patterns, whilst he links the inertia and paralysis of self with a passive submission to a single force. However, the consequence of too much flux and too little stability leads to the disconnected and fragmented existence which Anaïs Nin anxiously encountered in her early meetings with Rank. While Erikson's dynamic of becoming potential or future

---

[111] Ibid., p. 108.
[112] Erikson, *The Life Cycle Completed*, op. cit., p. 9.
[113] Ibid., p. 85.

selves is not prescriptive, it does lay down guidelines and paths along which the self can adapt to internal forces and social expectations. For Erikson, a negative identity is only desirable in situations in which the individual actively opposes intolerant or dehumanizing conditions, rather than acquiescing passively to environmental forces.

Each stage in the life cycle revolves around a particular crisis in which the individual is faced with two or more forces that impinge upon a settled or stable world. The crises do not always entail the 'threat of catastrophe', but a rupture or critical 'turning point, a crucial period of increased vulnerability and heightened potential'.[114] Interestingly, Erikson claims this notion of crisis derives from 'the most radical change of all, from intrauterine to extrauterine life', echoing Rank's theories about the symbolic restaging of birth trauma in later stages of life. Erikson's definition of crisis indicates a double experience of 'vulnerability' and 'heightened potential' which may lead to either the enhancement or diminishment of self. At these critical moments the dyad of health-illness is radically problematized as well as other previously stable loci in the individual's world. Rather than a smooth transition from one stage to the next, crisis implies a rupture and discontinuous process of anxiety which usually manifests itself in particular periods such as puberty and middle age. Crises become manifest in different ways – the frailty of posture and difficulty of walking in early childhood, the confusion of bodily identity in adolescence and psychic fragmentation in old age – but they all entail a re-evaluation of personal identity and a realization of the importance of an 'interpersonal perspective'.[115] The all-too-human possibility of 'negative identity' corresponds to Sartre's definition of bad faith and Erikson's understanding of ideologically suspect 'pseudospecies' (both of which fall short of the 'potential of all-human maturity'), but also provides the base line of potential recovery.[116] While some individuals may shore up their boundaries to preserve the self, the positive orientation which can emerge from these crises is linked to the ability to assimilate past traces of the self with a willingness to open the self (in Heidegger's sense of 'open') to new experiences and modes of being.

Erikson is wary about claiming that his schema is a universal and ahistorical template. Arising from his clinical and classroom observations, he attends closely to the modalities of infancy through to adolescence and less on adulthood, but he does not assert that his analytic methodology provides a blueprint for studying all human activity. He continues to stress the idiosyncratic 'style' of an individual life and the individual's space of self-expression within the broader picture of the life cycle. Similarly, although the chart seems to overlook the manner in which some phases of life are extended or contracted in some historical periods (for example, the elongation of the period of adolescence in the post-war years or the reconception of 'the elderly' in an era of ageing populations), he is keen to stress

---

[114]   Erikson, *Identity: Youth and Crisis*, op. cit., p. 96.
[115]   Ibid.

the issue of 'historical relativity'.[117] As such, he is aware that the pattern needs to be constantly reassessed in the light of further evidence or a revised understanding of psychosocial identity. At root, his model conforms to the pattern of flux and crisis established by James in *Principles* and developed in the existentialist work of Rank and Binswanger. Where Erikson extends such study is his emphasis on narrative patterns in which the unfolding self can be situated in a dynamic and constantly evolving way. This is homologous to his notion of the uprooted or migrant self defined by change rather than stasis: to recall Kierkegaard, a transitional self characterized by repetition of difference rather than slavish continuity.

Erikson provides a number of concrete examples of how the various life crises unfold, but the most detailed application of the life cycle to a particular case is found in his essay 'Reflections on Dr. Borg's Life Cycle' (1976), which deals with the crises facing Isak Borg, the ageing protagonist of the Swedish director Ingmar Bergman's existentialist film *Wild Strawberries* (*Smultronstället*, 1957). Erikson admits his biographical interest in the film (as a child he visited his uncle who lived twenty miles from Lund near the Ore Sund) and is sympathetic to the Scandinavian connections between Kierkegaard's brooding existentialism and the dark moods of Bergman's films. Although Bergman claimed that *Wild Strawberries* was not intended as a psychoanalytic film, it provides Erikson with an opportunity to indicate how the different stages of the life cycle can be applied to the desires and memories of the doctor.[118] The film follows a car journey with his daughter-in-law to Lund, where Borg (played by Swedish director Victor Sjöström) is to receive a Jubilee doctorate in recognition of his distinguished career. But, as Erikson comments, the initial journey 'through familiar territory also becomes a symbolic pilgrimage back into his childhood and deep into his unknown self'.[119] Interestingly, Erikson resists using the Freudian terminology of the unconscious in order to stress dimensions of the self which demand interpretation but lie just out of reach. He is particularly interested in the way in which Bergman shuttles between internal and external worlds, especially the 'intricate composition of facial expressions and postures, of landscapes and seaviews' and the 'tender earthiness' with which he imbues his characters, encouraging an existential analysis 'in the most concrete sense of the world' rather than a reduction of the human drama to Oedipal principles.[120]

In a 1968 interview Bergman emphasized the unpredictable and protean nature of Isak's journey in which the different stages of life are often startlingly juxtaposed with each other: one scene shows him 'walking into his childhood'

---

[116]   Erikson, 'Reflections on Dr. Borg's Life Cycle', *Daedalus*, 105(2) (1976), 16.

[117]   Erikson, *The Life Cycle Completed*, op. cit., p. 9.

[118]   Stig Björkman et al. (eds), *Bergman on Bergman*, trans. Paul Britten Austin (New York: Simon & Schuster, 1973), p. 138.

[119]   Erikson, 'Reflections on Dr. Borg's Life Cycle', op. cit., p. 1.

[120]   Ibid., p. 3.

and the next 'opening another door and walking out into reality . . . then walking round the corner of the street and coming into some other period of his life'.[121] Erikson replicates his reversal of the direction of the life cycle in *The Life Cycle Completed* and echoes Bergman's aesthetic in his argument that Borg's journey 'demonstrates how a significant moment in old age reaches back through a man's unresolved adulthood to the dim beginning of his awareness as a child'.[122] Just as the logic of faciality is reversible (the 'I' who is looking and the 'you' who is looked at are interchangeable), so too is the direction of Borg's psychological journey: as the image of the clock with no hands suggests, looking back and looking forward turn out to be in the same temporal direction.[123] This method of storytelling suggests that the stages of Borg's life are not resolved in a linear direction but spiral back on themselves as further crises emerge. As such, Erikson's attention to problems of interpretation and the need to relate the film 'in [his] own words' are matched by Bergman's own interest in the dynamics of telling a story.

Erikson detects three 'conditions' which collide during the course of Borg's journey, corresponding to the three processes of the self: 'old age' (somatic), personal confrontation (psychological) and ceremony (social) which all serve to dramatize the psychic conflicts in Borg's life. Although the film adopts an analytic study of Borg's life, progressing from an anatomy of his personal crisis (as illustrated by the dream-sequence at the beginning of the film) to a confrontation with other characters and the ceremony at the end, the three processes can be seen to fold into each other rather than being categorically distinct. Erikson displays his fondness for the modernist aesthetic of epiphany in his claim that these processes often collide in intense and fleeting instants of self-awareness. These 'special moments' cannot be explained by religious language (what Erikson calls 'mystical rapture') or rational discourse ('intellectual reconstruction'), but manifest themselves as a form of wordless and 'transcendent simplicity'.[124] This does not mean that epiphanic moments are always positively enlightening: often they are characterized by the awareness of loss or fragmentation such as in the opening dream sequence when Borg 'comes face-to-

---

121   Björkman et al. (eds), *Bergman on Bergman*, op. cit., p. 133.

122   Erikson, 'Reflections on Dr. Borg's Life Cycle' op. cit., p. 1.

123   Bergman recalls his surprise when he realized that Isak Borg and he share the same initials, suggesting that, in some way, they are doubles or mirror-images of each other (the film is full of distorted images). This bleeding of experience into aesthetics is strengthened as the image of the coffin in the opening dream-sequence recurred in Bergman's own dreams; Björkman et al. (eds), *Bergman on Bergman*, op. cit., p. 146.

124   Ibid., p. 3. It is noticeable that Bergman conforms to this modernist aesthetic by focusing on concrete images, such as the wild strawberries, as stimuli for these reflexive moments of insight. As such '*smultronställe* has the figurative connotation of a moment in the past to which someone looks back and which they would like to revisit or recapture'; Philip and Kersti French, *Wild Strawberries* (London: BFI Publishing, 1995), p. 23.

face with his own corpse'.[125] Replicating Rank's analytic interest in *The Student of Prague* as a filmic illustration of the double and self-obsession, *Wild Strawberries* offers Erikson visual and dramatic ways of describing the complexion of Borg's psychosocial identity. As such, the filter of art provides a tool for analysis, enabling the interpreter to focus closely on experience and to 'bridge' the interconnecting dimensions of identity.[126]

The narrative of *Wild Strawberries* is structured as an episodic series of events, involving Isak's encounters with people on his journey to Lund. Following Rank's filmic analysis and Binswanger's dream interpretation, Erikson interprets these events as symbolizing a dramatic restaging of crises which had beset him earlier in life and which continue to impact on the present. The events are interwoven with a number of dreams and flashbacks, suggesting that neither Isak's 'reality' nor his personal identity (as opposed to his public identity) are stable certainties. For example, the narrative begins with a surrealist dream which features a faceless man who collapses when confronted: 'on the sidewalk lay only a heap of clothes with some liquid oozing out of them: the person was gone'.[127] Not only does this moment epitomize Erikson's interest in faciality and the fragility of identity, but the emptiness in Isak's own life is mirrored by the empty features of the faceless man. This dream is augmented by a Proustian memory sequence, stimulated when he stumbles upon a patch of wild strawberries (symbolizing passion and mystery), and a 'humiliating' dream, in which his wife is 'seduced by a disgusting but virile man'.[128] These moments are reminders that other incomplete selves are hiding behind the public persona of the distinguished doctor, inextricably linked to a number of unresolved crises concerning his sense of self and his relation to others: his wife, son, first love and housekeeper.

Erikson argues that Isak's 'murdered' selves need to be resurrected before he can hope to resolve these crises. Despite the solemnity of the Jubilee ceremony, the film ends with Isak's triumphant vision: 'a truly primal scene' in which 'his father waves and his mother nods, both smiling in recognition'.[129] As such, Isak's actual journey is complicated by a retrogressive journey which stimulates a growing awareness of the factors contributing to his self-estrangement. But, rather than struggling to resolve an Oedipal crisis, Erikson sees the engagement with the past as the 'sheet anchor' of hope in the future.[130] Although Isak is in the last stage of his life, only by engaging with his previous crises can he gain the wisdom of old age: Erikson suggests that Isak's second childhood does not represent a demise into senility, but embodies a positive identity which does not deny, nor is in conflict with, his earlier selves. Erikson argues convincingly how

---

125   Ibid., p. 4.
126   Ibid., p. 3.
127   Ibid., p. 4.
128   Ibid., pp. 12, 13.
129   Ibid., pp. 7, 17.
130   Erikson, *Gandhi's Truth*, op. cit., p. 35.

elements of Isak's journey correspond to stages of the life cycle, but he admits 'a good story does not need a chart to come alive'.[131]

As 'Reflections on Dr. Borg's Life Cycle' demonstrates, the life cycle can prove a useful framework in which to examine the shifting or uprooted sense of self, but the criticism remains that Erikson does not attend closely enough to ethnic and gender difference. However, his work with Yurok tribes, his comments on Jewish identity and the inclusion of the case of the African-American child who listens to 'Red Rider' in *Childhood and Society* all indicate that he is much more sensitive to cultural difference than many of his psychoanalytic predecessors. The 'double consciousness' which many black American adolescents felt in the 1950s and 1960s in confronting white cultural identity was often in conflict with their childhood sense of self. The consequence of such conflict may lead to either a fragmented sense of identity which Baldwin fears for his nephew, or the manifestation of 'violent discontinuities' as the individual is caught between identification with the 'minority' group and assimilation into the myths of dominant culture.[132] By adopting the available negative identity (as illustrated in Ellison's *Invisible Man*), many African-American adolescents were complicit in denying themselves the self-determining choice of an identity 'won in action'.[133]

To defend Erikson's work against the charge that he deals only cursorily with gender difference is a more difficult task because, in assigning particular attributes to boys and girls, he can be accused of being a biological essentialist with gender difference based on 'inner design'.[134] In *Childhood and Society* he documents an experiment carried out over eighteen months in which he invited 150 children to play with building blocks. His observations lead him to claim that boys preferred to build high structures like towers or buildings while girls concentrated on the interior of scenes. He concludes that boys tend to make extrusive structures, whereas girls prefer inclusive spaces: in other words, the 'masculine variables' are 'high' and 'low' and the 'feminine modalities' are 'open' and 'closed'.[135] These modalities are directly connected to genital modes which seem to determine the spatial orientation of each sex. However, Erikson indicates that these spatial preferences seem to be connected to gender-specific roles to which the child can aspire, suggesting that parents have an active hand in the early formulations of the child's own sense of him or herself. Rather than pushing the argument for biological essentialism, Erikson indicates that these modalities are intricately connected to those social and historical forces in the West which have driven 'a division of function between the sexes'.[136] Indeed,

---

131   Erikson, 'Reflections on Dr. Borg's Life Cycle', op. cit., p. 27.
132   Erikson, *Childhood and Society*, op. cit., p. 220.
133   Erikson, *Identity: Youth and Crisis*, op. cit., p. 300.
134   Erikson, *Childhood and Society*, op. cit., p. 86.
135   Ibid., p. 91.
136   Ibid., p. 95. In 'Womanhood and the Inner Space' (1964), Erikson argues that, although women are partially determined by their 'anatomy', they retain psychological

Erikson emphasizes the value of play in a child's life and its creative role in widening the experiential and imaginative sphere of activity, rather than the child functioning within narrow stereotypical parameters.

In her influential book *In a Different Voice* (1982), the American feminist and Erikson's former teaching assistant Carol Gilligan argues that his model of the life cycle replicates the mistakes of Freud's theory of sexual development by focusing primarily on the psychic development of boys into young men. Like Erikson, she makes the existential claim that 'to have a voice is to be human', but her book focuses on the psychic and social impediments which prevent women finding their own 'voice'.[137] On one level, her desire to discover an expressive space for women approximates to Rank's psychology of difference: she wishes to posit 'difference' as a positive term and not as a mark of subordination or deviance.[138] But, on another level, she is wary about the female patient's story being told from the perspective of a male analyst. Thus, she focuses on voices that are rarely heard or whose mode of expression is subdued by another dominant discourse and, like the other romantic scientists, describes the relationship between self and Other as a moral as well as an existentialist problem. While Erikson's life cycle accounts for both boys and girls up to adolescence, she claims that the 'celebration of the autonomous, initiating, industrious self' only describes the emergence of a male self, while female identity is often held 'in abeyance'.[139] She interprets Erikson to imply that 'while for men, identity precedes intimacy and generativity' as sequential stages, 'for women these tasks seem instead to be fused'.[140] As such, she is critical of Erikson for not modifying his life cycle to account for these differences: for the French feminist Luce Irigaray, this would amount to a failure to 'leave space' for the possibility of a different 'feminine' language.[141]

Whereas Erikson focuses on Isak Borg's psychodrama in *Wild Strawberries*, Gilligan is keen to resurrect the story of his daughter-in-law Marianne from her passive role as 'the catalyst who precipitates the crisis that leads to change' in the old doctor's life.[142] Rather than the individuation which usually accompanies the

---

creativity and social 'capacity on many levels of existence'; Erikson, *Identity: Youth and Crisis*, op. cit., p. 285. Whilst overgeneralizing at times, he argues that only by adopting 'feminized' qualities in social activity ('that which nurses and nourishes, cares and tolerates, includes and preserves') can the technological forces of 'masculinized' society be checked; ibid., p. 293.

[137] Carol Gilligan, *In a Different Voice: Psychological Theory and Women's Development* (Cambridge, MA: Harvard U.P., 1993), p. xvi.

[138] Ibid., p. xviii.

[139] Ibid., p. 12.

[140] Ibid.

[141] Luce Irigaray, 'The Power of Discourse and the Subordination of the Feminine' (1991), in ed. Margaret Whitford, *The Irigaray Reader* (Oxford: Blackwell, 1991), p. 127.

[142] Gilligan, *In a Different Voice*, op. cit., p. 107.

emergence of the male self, Marianne's commitment to the intimacy of relationships enables Borg to escape from his introspective world and break 'the cycle of repetition' which had led to his loneliness.[143] However, for Gilligan, Marianne's story remains largely 'untold': 'it is never clear how she came to see what she sees or to know what she knows'.[144] One argument which could be mounted against Erikson is that his narrative focus on the dramatic encounters of a single life relegates the significant Others in that life to the status of catalyzers or impediments in the development of the protagonist. While this is certainly an issue which Erikson does not fully address, as Binswanger's two case studies demonstrate the same problem is endemic to the telling of this kind of story whatever the gender focus (for example, Ellen West's husband remains a shadow in the background of her life story). However, while Gilligan criticizes the gender bias of Erikson's work, she does adopt his model of the life cycle as a narrative of 'development through crisis' and his belief that the fullest experience should entail a recognition of the 'communal sense of "we"' which transcends an introspective and potentially solipsistic life.[145] Indeed, Erikson's emphasis on care as a mode of generativity (in all its senses: '"to be careful", "to take care of," and "to care for"') and his interpretation of *Wirklichkeit* as a mode of 'mutual activation' suggests his ideal of selfhood is close to Heidegger's description of *Dasein* and is not incompatible with Gilligan's revisionist agenda.[146]

The Dynamics of Psychohistory

Much of Erikson's work in the late 1950s to the 1970s was dedicated to refining a narrative model which could chart the development of, and crises facing, a number of different individuals, ranging from 'great' historical, political and literary figures such as Martin Luther, Thomas Jefferson, Maxim Gorky, George Bernard Shaw, Adolf Hitler and Mahatma Gandhi to detailed case studies of his patients.[147] The biographical mode he adopted has the advantage of conforming to an established genre of writing, but also presents problems of biographical authenticity, in the sense that the central character must conform to the literary

---

[143]   Ibid.

[144]   Ibid. Marilyn Blackwell's criticism of *Wild Strawberries* is very similar to Gilligan's critique of Erikson. Although the film 'confronts issues of gender in terms of family structure and rigidly proscribed gender roles', the last child-like vision can be interpreted as Isak's 'reintegration' into the patriarchal structure rather than a disruption of it; Marilyn Blackwell, *Gender and Representation in the Films of Ingmar Bergman* (Columbia, SC: Cambden House, 1997), p. 200.

[145]   Erikson, *The Life Cycle Completed*, op. cit., p. 85

[146]   Ibid., pp. 59, 89.

[147]   Most of Erikson's clinical compositions held at the Houghton Library, Harvard University have restricted access, controlled by his son Kai Erikson, Yale University.

conventions of a fictional protagonist with the story structured around a series of dramatic events. Erikson's psychohistories roughly follow a conventional linear pattern in which he charts the psychic development of his case study, interspersed with clinical analysis and analytic interventions. However, he is also drawn to Bergman's use of reversible biography in *Wild Strawberries* to indicate that the various crises associated with the stages of the life cycle are not permanently resolved, but may arise again in later life. Where Erikson's psychohistories differ from Freud's and Binswanger's diagnostic case studies is that he is eager to record the ways in which individuals can adapt to their environment and 'maintain a significant function in the lives of others', as much as he is interested in the crises or turning points in their lives.[148] As such, he fuses ego psychology, an existential focus on the development of self, and psychoanalysis as a tool for probing the causes and manifestations of neurosis. He is particularly interested in the early period of the individual's life which provides a source of hope for later life, but broadens out his study to consider the historical and cultural forces which continue to impact on the formation of selfhood.

Young Man Luther and *Gandhi's Truth* are by far the lengthiest of his psychohistories, but he adopts the same methodology for shorter studies. For example, in his Jefferson Lecture delivered in Washington DC in 1973 and published in *Dimensions of a New Identity* (1974) he is careful to distinguish his own use of psychohistory from other appropriations of this model. He claims that in his sense of the term,

> psychohistory, essentially, is the study of individual and collective life with the combined methods of psychoanalysis and history. In spite of, or because of, the very special and conflicting demands made on the practitioners of these two fields, bridgeheads must be built on each side in order to make a true span possible. But the completed bridge should permit unimpeded two-way traffic.[149]

The development of this dual discourse is central to Erikson's adaptation of romantic science and provides him with a narrative vehicle for considering the intertwined dimensions of an individual's life. Although Freud's biographer Peter Gay claims that 'history has never enjoyed a cordial relationship with psychoanalysis', Erikson emphasizes the discursive 'bridge' must be 'two-way', with neither history nor psychological study dominating or obliterating each

---

[148] Erikson, *Dimensions of a New Identity* (New York: Norton, 1974), p. 13. Erikson inherits a methodological problem from Binswanger's work on Ellen West. Both analysts struggle to balance an objective analysis of a 'finished' life with the intimacy and care which they both recommend. However, whereas Binswanger does not really attend to Ellen's ethnic and cultural backgrounds, Erikson emphasizes the crucial impact of historical forces on an individual life.

[149] Ibid.

other.[150] Reacting against conventional modes of history which rely on strictly empirical evidence (verifiable documentation and reliable witness), Erikson claims that historians must be aware that they have 'always indulged in a covert and circuitous traffic with psychology' in terms of the depth and angle of study adopted by the historian.[151] Similarly, he argues that orthodox case histories rarely address the broader 'historical determinants' which are as important in influencing the configuration of the self as is the individual's role in the family romance. In order to correct these over-confidences, psychohistory locates a psychoanalytic 'method of observation' within an historical 'system of ideas': in other words, it both passively 'takes' and actively 'makes' history.[152]

Erikson's view of great or gifted individuals who have affected (or redirected) the course of history by offering new perspectives or values fits within an American liberal tradition which derives in part from Emerson's proto-psychohistorical study *Representative Men* (1850).[153] Emerson argues that 'great men' are not just interesting from an historical perspective, but reveal something about the present construction of 'greatness'. The influence of Emerson's six figures (all of them European: Plato, Swedenborg, Montaigne, Shakespeare, Napoleon, Goethe) on mid-nineteenth-century American culture is obviously important, but so is the sense that they are representative of their age: 'other men are lenses through which we read our own minds'.[154] He argues that rather than greatness being an eternal value, its applicability depends crucially on the dominant values and beliefs of the present. Moreover, Emerson insists that the 'great man' is not categorically distinct from other individuals; indeed, in his romantic argument, it is the obligation of all individuals to strive for greatness: 'all men are at last of a size; and true art if only possible, on the conviction that every talent has its apotheosis somewhere'.[155] In developing Emerson's ideas, Erikson's psychohistories serve a double purpose: firstly, they are exemplary studies of influential leaders who may inspire the reader to adopt an active and committed life and, secondly, they provide a touchstone for understanding the present by providing an historical bridge to the past. The exceptional 'gifts' which these figures develop in their lives stimulate both Emerson and Erikson to examine 'the whole ecology of greatness' which transcends the case study's narrow focus on 'the inner economy of a person'.[156]

---

[150] Peter Gay, 'Psychoanalysis and the Historical', in ed. Michael Roth, *Freud: Conflict and Culture: Essays on His Life, Work and Legacy* (New York: Knopf, 1998), p. 117; Erikson, *Dimensions of a New Identity*, op. cit., p. 13.

[151] Ibid.

[152] Erikson, *Young Man Luther*, op. cit., p. 17.

[153] Emerson, *Essays and Lectures*, op. cit., p. 615.

[154] Ibid., p. 616.

[155] Ibid., p. 630.

[156] Erikson, *Dimensions of a New Identity*, op. cit., p. 55.

Erikson is particularly interested in influential figures like Luther, Jefferson and Gandhi who lived during crisis moments in history (the European Renaissance, the American Revolution, the dissolution of the British Empire) whose private sense of self conflicts with the received opinion of their public persona. On one level, he treats these figures as patients whose 'life-work' is available for analysis, but, on another level, he is attracted to their ability to contend with exacting personal and public pressures. In the preface to *Young Man Luther* he claims that the similarity between Luther and his clinical patients is not in terms of psychiatric symptoms, 'but is oriented toward those moments when young patients . . . prove resourceful beyond all professional and personal expectation'.[157] These 'powers of recovery' contribute to their greatness, together with their lonely 'willingness to do the dirty work' when others are immersed in conventional or habitual modes of existence.[158] Erikson is very keen to stress that the psychohistorian should declare his or her interest in the subject of study (especially when it involves an historical figure), because the interpretative perspective is crucial for determining the way in which a life-story is told, both in terms of selection of detail and the symbolic trajectory of the narrative. As such, Erikson shares the concerns of New Historicism in giving equal weight to textual discourses and empirical evidence, in addition to self-consciously interrogating the ideological perspective of the historian. Erikson's concern with the constructed nature of history bleeds into his interest in the political and religious dangers which forced Luther and Gandhi to forge distinct identities, 'to mobilize capacities to see and say, to dream and plan, to design and construct, in new ways'.[159] This emphasis upon the possibilities of an actively created existence follows Binswanger's interest in the potentiality of his patients' world-designs and recalls Erikson's earlier work with children who literally construct their mythological play-world out of building blocks.

Most of Erikson's extended psychohistories are characterized by a tone of intimacy which suggests an empathic relationship between interpreter and the subject of interpretation (although of course this is only an imagined intimacy in his historical studies). Taking the example of Luther's famous dialogues with the devil, he claims 'I heard him, ever again, roar in rage, and yet also in laughter' during the composition of *Young Man Luther*.[160] Only by approaching a individual's life in a sensitive and receptive way (like 'facing a face, rather than facing a problem') can Erikson's focus on personal experience complement historical significance and social impact.[161] This does not always mean that the 'personal' and 'public' are always in tension, but indicates that there is a hidden

---

[157] Erikson, *Young Man Luther*, op. cit., p. 8.
[158] Ibid., pp. 8, 9.
[159] Ibid., p. 15.
[160] Ibid., p. 29.
[161] Ibid., p. 17.

dimension inaccessible to all but an analytic method. His empathic technique enables him to adopt 'a position on the borderline of what is demonstrably true and of what demonstrably *feels* true'.[162] However, this raises questions about historical accuracy and problematizes dimensions of the figure's biography, upon which Erikson imaginatively expands without substantial evidence.[163] Thus, his psychohistories approximate more closely to mythmaking than historically verifiable accounts (although, to his credit, he cites regularly his Latin and German sources). By calling attention to the role of the interpreter, Erikson does not tell a seamless story but one which hovers between the discourses of history, biography, analysis and portraiture, calling attention to its own constructed nature. He insists that this orientation is intrinsically 'ethical', counteracting the fatalism and moral inertia he detects around him by alerting the reader to 'those early energies which man, in the very service of his higher values, is apt to suppress, exploit, or waste'.[164]

The analytic framework of *Young Man Luther* fits within Erikson's life-cycle model and is typical of his other psychohistories in exploring the effects of temporal and environmental change upon identity formation. He is particularly interested in the formative part of Luther's life leading up to his public renunciation of the Catholic Church and the nailing of the ninety-five theses on the church door in Wittenberg in 1517. After a short introduction, the study begins with a dramatic episode which parallels the fictional dramatization of James' breakdown in *Varieties*. The second chapter 'The Fit in the Choir' explores a significant incident in Luther's early twenties, in which he reputedly collapsed in Erfurt monastery, '"raved" like one possessed, and roared with the voice of a bull: "*Ich bin's nit! Ich bin's nit!*" or "*Non sum! Non sum!*"'.[165] Erikson admits that the validity of this mythological moment is debatable, but remains important as 'half-legend' or 'half-history' because it fascinates even those critics 'who would do away with it'.[166] Most commentators interpret this, almost dehumanized, moment either as an expression of Luther's doubts about his vocation as priest or as a moment of demonic possession. Erikson discusses at length the various interpretations of the incident – a theologian claims Luther was inspired by a divine agency, a psychiatrist argues that he suffered from

[162]   Ibid., p. 21.
[163]   Harold McCurdy from the University of North Carolina offers a hostile review of *Young Man Luther*, claiming that Erikson and Luther inhabit 'entirely different' worlds. As such, McCurdy argues Erikson's psychohistory is far from authentic, 'where Luther himself can be mistily perceived as an ectoplasm undergoing various interesting transformations, now as Freud, now as Hitler, and now again as a Latin-American revolutionary clutching an anti-capitalistic sword instead of a Bible in his big fist'; Harold McCurdy, 'Luther: Psychoanalyst', *Contemporary Psychology*, 4(7) (1959), 202.
[164]   Erikson,*Young Man Luther*, op. cit., p. 19.
[165]   Ibid, p. 23.
[166]   Ibid, pp. 25, 37.

psychopathology and a psychoanalyst traces the incident to a traumatic early childhood – but he concludes that all these commentators are partial in their judgement as they 'slice' him 'in different ways'.[167] Only by historicizing the incident within the 'total existence' of Luther's life, 'when his future was as yet in an embryonic darkness', can its real significance be gauged.[168]

Erikson views this incident from historical and biographical perspectives as a manifestation of uprootedness: an equivocal 'borderline' episode which can only be positioned between the discourses of psychiatry and religion.[169] Luther's cry 'I am *not*' is a repudiation of his expected identity as a meditative monk and a moment of profound self-doubt when 'he half-realizes that he is fatally overcommitted to what he is not'.[170] But, Erikson argues, from this moment of 'two-facedness' derived the future-oriented impetus of his life, enabling him 'to break through to what he was or was to be'.[171] This incident has obvious biographical resonances, but also profound historical implications: Erikson interprets Luther's break from medieval monastic obedience and his proclamation of justification by faith in 'Concerning Christian Liberty' (1520) as 'decisive' steps 'in human awareness and responsibility' that characterize the philosophy of Renaissance humanism.[172] Erikson, like James and Binswanger, focuses on the mood of religious revelations and concludes that 'excessive sadness' or '*tristitia*' characterizes both the monastic order and Luther's bilious early life.[173] The violent fit broke through this 'veil of sadness' and stimulated his counter-life of religious revolt, later marked by 'occasional violent mood swings between depression and elation, between self-accusation and the abuse of others'.[174] This transition conforms to James' distinction between the healthy-minded, whose lives are marked by clarity and uniformity, and sick souls, whose existence is more complex and, consequently, more creative. Rather than his body melting away in a moment of mystical rapture (as the late nineteenth-century mind-curists would argue), Luther's corporeality – his constipation, urinary retention and bodily discomfort – constitutes the 'dirt-ground' of his Being 'where one meets with the devil, just as one meets with God'.[175] As Erikson

---

[167] Ibid., pp. 214, 35

[168] Ibid., p. 24.

[169] Ibid, p. 38.

[170] Ibid., pp. 36, 43.

[171] Ibid, pp. 36, 38.

[172] Ibid., p. 39. Erikson argues that Luther was most definitely a Renaissance figure even though he largely 'ignored' the Renaissance flowering in painting and architecture during his trip to Italy; ibid., p. 175.

[173] Ibid, p. 39.

[174] Ibid., p. 40.

[175] Ibid, p. 206. In *Life Against Death* (1959), Norman O. Brown also analyses Luther's life from a psychoanalytic perspective, with a particular interest in Luther's conception of the Devil as a mediating term between anality and Protestantism. Although he makes no reference to Erikson, Brown is critical of Erich Fromm for transforming

emphasizes throughout his later work, human beings are fundamentally both biological and spiritual beings and neither dimension can be denied without risking the integrity of the self.

Although Luther himself never referred to the fit in the choir (Erikson surmises he suffered from partial amnesia), the episode coincides with the reorientation of his life in a therapeutic, as well as a theological direction, leading to his later religious revelations. What emerged from the incident was a much more chaotic and tormented mode of existence, marked by Protestant self-reliance, theological questioning and individualistic self-expression. As such, Erikson argues that Luther's own language was full of extravagant, 'blustering, and often unreliable words'.[176] The reliability of a stable belief-system and medieval world order was thus epitomized by the absolutism of the Roman Catholic Church, whereas Luther's revolutionary counter-life is characterized by contradiction and an introspective interrogation of self in his problematic relation to religious tradition and divinity. Luther replaced 'the religious style of life prepared by tradition' with his own innovative style of worship and theological interpretation.[177] This idea of 'style' suggests a wilful existence of self-fashioning, but one held in check by unconscious manifestations of another self (or other selves) which manifest themselves at moments of crisis and cannot be contained within a coherent life-design.

To this end, Erikson explores the psychodynamics of the mythological figure of Proteus who can 'make many things of oneself'.[178] A protean life is not a rigid template, but may correspond to either a virtuous 'man of many gifts, competent in each; a man of many appearances, yet centred in a true identity' or a more elusive 'man of many disguises; a man of chameleonlike adaptation to passing scenes'.[179] At its most extreme, the protean mode of existence seems to be formless, but, so long as it is guided by the active agency of the self 'into patterns of action, into character, into style' (homologous to Erikson's earlier configurational study of child's play), it enables individuals to overcome the demands of habit and external pressures.[180] Although Erikson detects that the protean model is most closely applicable to an appraisal of American national identity (epitomized by Jefferson's 'intellectual and esthetic style'), it is also

---

Freud's biological emphasis on anality into an understanding of 'character structure' which is 'autonomous from the body'; Norman O. Brown, *Life Against Death: The Psychoanalytical Meaning of History*, 2nd edn (Middletown: Wesleyan U.P., 1985), p. 204. More recently, Dean Simonton has criticized the psychobiographical method for its evasion of the 'biological roots' of trauma, however this is countered in Erikson's consideration of Luther's corporeality; Dean Simonton, *Greatness: Who Makes History and Why* (New York: Guilford Press, 1994), p. 47.

176   Ibid., p. 50.
177   Ibid, p. 254.
178   Erikson, *Dimensions of a New Identity*, op. cit., p. 52.
179   Ibid., p. 51.
180   Erikson, *Young Man Luther*, op. cit., p. 254.

central to his work on Luther and Gandhi. In *The Protean Self* (1993), the American psychologist and close friend Robert Jay Lifton commends Erikson's 'psychoanalytic openness' in refining a mutable conception of identity.[181] Lifton shares Erikson's enthusiasm for psychohistory and develops his own work on proteanism in the late 1960s by applying it to the lives of 'ordinary' patients.[182] Although the impact of most single individuals on history is negligible, Lifton maintains that many of his patients display the same 'capacity for flexible imagination and action' as Erikson's exceptional psychohistories, in future-oriented lives which refuse to be determined by traditional roles or delimited by childhood trauma.[183]

If Erikson's interest in the condition of uprootedness derives from his own transatlantic experience in the 1930s, Lifton argues that proteanism represents a dominant 'modus vivendi' in the later part of the twentieth century (by his account, not just in the West).[184] By articulating a mode of existence which galvanizes the possibility of alternative selves, Erikson and Lifton outline a performative model of selfhood as an antidote to the threat of losing a rooted or grounded self.[185] However, rather than conforming to the kind of postmodern relaxation identified by Jean-François Lyotard, Lifton characterizes proteanism as 'a quest for authenticity and meaning, a form-seeking assertion of self', suggesting an intermediary experiential state somewhere between modernist and postmodernist aesthetics.[186] While Lifton agrees that men have a greater chance of nurturing proteanism than women because of social differentiation, he argues that 'protean juggling' and 'bold protean forays' for women are becoming increasingly possible in the late twentieth century.[187] Indeed, although the philosopher Margaret Mead remained sceptical about the flexibility of gender roles and the 'style of relations' between sexes, Luce Irigaray defines *écriture feminine* as having a protean and 'fluid' style which 'resists·and explodes every firmly established form, figure, idea or concept'.[188] Irigaray's comments parallel Lifton's suggestion that proteanism is not prescriptive, but defines 'an ideal model . . . for what one constantly seeks', and also correspond to James' active pragmatism, Rank's description of self-becoming and Erikson's focus on

---

[181]   Robert Jay Lifton, *The Protean Self* (New York: Basic Books, 1993), p. 26. See also Lifton's earlier essay 'Protean Man', *Archives of General Psychiatry*, 24 (1971), 298–304.

[182]   Ibid., p. 26.

[183]   Ibid, p. 120.

[184]   Ibid., p. 3.

[185]   Ibid., p. 9.

[186]   See Jean-François Lyotard, *The Postmodern Condition*, trans. Geoff Bennington and Brian Massumi (Manchester: Manchester U.P., 1984), pp. 71–82; Lifton, *The Protean Self*, op. cit., p. 9.

[187]   Ibid.

[188]   Margaret Mead, 'The Life Cycle and Its Variations: The Division of Roles', *Daedalus*, 96(3) (Summer 1967), 87; Irigaray, 'The Power of Discourse', op. cit., p. 126.

exceptional individuals who are more often than not 'beyond' their allotted place in history: what Erikson calls being 'ahead in the future'.[189]

Mythmaking and Revolutionary Ethics

If Luther epitomizes Erikson's belief in the possibility of experimenting with a therapeutic mode of existence instead of conforming to social expectations, then *Gandhi's Truth* demonstrates the triumphs as well as the pitfalls of living an ethical life in the most exacting of circumstances. *Gandhi's Truth* is in many ways the pinnacle of Erikson's achievement in applying his model of the life-cycle to 'great' historical figures. Although he had access to the authorized translation of Gandhi's autobiography, *An Autobiography, or The Story of My Experiments with Truth* (1927) and the galleys of the *Collected Works of Mahatma Gandhi* (1958), as well as having the benefit of extensive testimonies from the Indian leader's close associates, *Gandhi's Truth* is an exploratory work which investigates what Iain Chambers calls a 'drama of the stranger'.[190] Whereas the study of Luther raised methodological difficulties in terms of historical accuracy, in his work on Gandhi he encounters a number of cultural and linguistic barriers standing between him and his subject of study. However, both studies are closely related on three key issues: the political dimensions of the two lives; their successful attempts to transcend religious and social constraints; and their ability to reconcile the conflicting demands of their public and private lives.

Initially stimulated by his interest in Gandhi as an 'alienated youth in Europe' in the 1920s, Erikson began detailed biographical research during a series of lecture tours to India in the 1960s, but began to realize that collecting 'truthful' accounts of the leader's life would prove no easy task.[191] He soon learnt to treat the reliability of the witnesses he interviewed with some scepticism, not because they sought deliberately to deceive him, but because the dynamics of resistance and transference are embedded in every conversation he had, especially as they concerned a figure who had grown to mythical proportions in Indian culture. Whereas the romantic scientists before him focused primarily on the difficulties of approaching the 'psychic Other', *Gandhi's Truth* enabled Erikson to expand on his earlier field-work with Yurok Indians and provided him with a vehicle to consider the 'cultural Other', whose historical proximity belies a strange and mysterious story behind the public mask. His major methodological difficulties lay not only in endeavouring to tell Gandhi's story in the fullest sense possible, but attempting to make the account more than 'a distorted mirror' of the personal

---

[189]   Ibid., p. 9; Erikson, *Dimensions of a New Identity*, op. cit., p. 36.

[190]   Chambers, *Migrancy, Culture, Identity*, op. cit., p. 4.

[191]   Erikson, *Gandhi's Truth*, op. cit., pp. 9–10.

testimonies given to him or of his own analytic interests.[192] In order to resist the biographer's sleight of hand in producing a seamless narrative, he firmly embeds these predelictions in his account of Gandhi's life-journey and his own personal encounters with the interwoven myths and realities which surround the figure. The themes of travel and cultural movement stretching throughout Erikson's work reappear in the metaphor of the journey, both in terms of Gandhi's life (he spent significant periods in England and South Africa) and in guiding the reader through a vivid account of his own lecture tours of India in the 1960s. As such, he positions himself as a philosophical ethnologist and cultural critic who highlights his own ideological baggage in order to immerse himself freely in this 'foreign' experience, even if this means that initially he confronts India from a romantic perspective 'as one dreams it to be'.[193]

As with Luther, the private sphere of Gandhi's life proves the most difficult to unravel for Erikson, but he realizes that despite (or because of) the proliferation of stories presented to him the Indian leader's social persona is equally mysterious. This understanding is evident when he records his confusion over the coexistence of 'the myth affirming and myth-destroying propensities of [this] post-charismatic period'.[194] The 'relativistic' method he adopts derives from his understanding of the inseparability of myth and history:

> If history is a collection of events which come to life for us because of what some actors did, some recorders recorded, and some reviewers decided to retell, a clinician attempting to interpret an historical event must first of all get the facts straight. But he must apply to this task what he has learned, namely, to see in all factuality some relativities which arise from the actors', the recorders', and the reviewers' motivations.[195]

Just as Gandhi refined an ethical code in the face of external and internal pressures in his own life, so Erikson wishes to develop an ethical mode of relativisitc enquiry based on Gandhi's conception of 'truth method', which balances myth and truth without allowing one to dominate or destabilize the other.[196] However, both Gandhi and Erikson were aware that biography and history cannot be conflated: literary reconstruction of a life cannot fully account for the forces impacting on it. For example, Erikson highlights a section in the middle of the *Autobiography* when Gandhi proclaims:

> I understand more clearly today what I read long ago about the inadequacy of all autobiography as history. I know that I do not set down in this story

---

[192] Ibid., p. 13.
[193] Ibid., p. 19.
[194] Ibid., p. 66.
[195] Ibid., p. 55.

all I remember. Who can say how much I must give and how much omit in the interests of truth?[197]

Despite the methodological risks entailed by such an 'in-between' approach entails, whereas his previous psychohistories only received critical attention from American and European critics, *Gandhi's Truth* attracted glistening reviews from Indian psychologists such as Sudhir Kakar, who praised the book for matching the revolutionary content of Gandhi's life with 'profound and brilliant' innovations in psychohistorical methodology.[198] Because *Gandhi's Truth* represents a culmination of the major themes introduced in the 1960s and developed in essays published in *Life History and the Historical Moment* in the 1970s, it is Erikson's methodological approach to the subject of study which provides the chief focus for the concluding part of this chapter.[199]

The structure of *Gandhi's Truth* is carefully conceived to highlight the multiple dimensions of the leader's life and what Kakar calls the 'four-fold complementarity of the individual's developmental history'.[200] The prologue offers a leisurely account of Erikson's lecture tour of India, in which he mixes personal experience with an account of the historical, religious and cultural forces which impinge on the future of the nation. As with his earlier studies, he maintains that only by interweaving these four trajectories can the psychodynamics of Gandhi's existence be identified. In the first section Erikson highlights the significance of 'The Event' (a strike against mill-owners in Ahmedabad in 1918) for understanding Gandhi's development as leader. Taking an incident comparable to Luther's fit in the choir as the stimulus for his rejection of the Catholic church, Erikson argues that in this 'elusive' and what is usually seen as a 'minor event' can be detected the seeds which were later to grow into Gandhi's philosophy of militant non-violence.[201] This section is followed, secondly, by an account of Gandhi's early life and the conflicting pressures between the social expectations of his caste and the pressures of his own hybrid identity (after education at the English bar) and, thirdly, by a detailed account of 'The Event' in the development of leader and nation. Erikson is eager to show how, like Luther, Gandhi managed to overcome the constraints of the existing religious and political institutions by improvising an experimental code of ethics.

---

[196] Ibid., p. 245.

[197] Ibid., p. 230.

[198] Erikson, *Life History and the Historical Moment*, op. cit., p. 114; Sudhir Kakar, 'The Logic of Psychohistory', *Journal of Interdisciplinary History*, 1(1) (Autumn 1970), 194. The Japanese version of *Gandhi's Truth* was also well-received.

[199] In the introduction to *The Gandhi Reader*, Homer Jack discusses the 'elusive quality' of greatness, but asserts that Gandhi is one of the truly great creative personalities of our age'; Homer A. Jack (ed.), *The Gandhi Reader* (New York: Grove Press, 1956), p. v.

[200] Kakar, 'The Logic of Psychohistory', op. cit., p. 188.

[201] Erikson, *Gandhi's Truth*, op. cit., pp. 12, 45.

In this third section Erikson switches from a balanced third-person account to a first-person testimony, in which he confronts Gandhi and analyses his view of spiritual truth in a very direct and, at times, confrontational way. Not only does this alert the reader to the interpersonal dynamics of the study, but indicates Erikson's own motivational interest in his subject-matter. The fourth section, 'The Leverage of Truth', reverts back to a biographical mode (mixed with analytic commentary) in an examination of Gandhi's later life as 'father of his country', in which his earlier developmental stages culminate in the maturity of an ethically responsible public role.[202] Lastly, an epilogue with the symbolic title 'March to the Sea' recounts the end of the leader's life and reconsiders the existing relationship between Europe and Asia. This multi-layered structure, moving between the travelogue, cultural criticism, personal intervention and analytic investigation, enables Erikson to retain a shifting focus on the intersecting dimensions of Gandhi's story, while allowing him to emphasize the internal structure and historical implications of Gandhi's life.

The clash between the beliefs and mythologies of East and West is a major concern of the study, conveyed through Erikson's personal experiences in the opening section and then later in his analysis of Gandhi's psychic and ethical development as someone caught between cultures. Throughout the study he compares his model of the life cycle with Hindu scriptural traditions, indicating homologous points and others where they prove incompatible. Lawrence Friedman notes that Erikson adds two additional elements to his life-cycle model in the mid-1960s which enable him better to account for the public and private dimensions of Gandhi's life: the social rituals of Hinduism and the 'schedule of virtues' practised in accordance with his spiritual and political ideals.[203] Erikson argues that rather than Eastern and Western mythic patterns being comparable 'point for point', it is their epigenetic dynamic which is most closely comparable, 'according to which in each stage of life a given strength is added to a widening ensemble and reintegrated at each later stage'.[204] Even the points of divergence do not devalue the analytic worth of this psychohistorical model, as Erikson's commentary continually alerts the reader to the manifold problems of how to engage in a cultural system which is variably similar and distinct from the observer's own. The experience of uprootedness (which he theorizes elsewhere) is conveyed here in an experiential sense, especially at times when he admits that his impressions become 'too varied and too crowded' to be fully assimilated.[205] Erikson does not draw definitive conclusions from such experiences, but acts as

---

[202] Ibid., p. 395.

[203] Wallenstein and Goldberger (eds), *Ideas and Identities*, op. cit., p. 361. Erikson acknowledges that the private and public split is perhaps misleading in his claim that 'it is always difficult to say where, exactly, obsessive symptomatology ends and creative ritualization begins'; Erikson, *Gandhi's Truth*, op. cit., p. 157.

[204] Ibid., p. 38.

[205] Ibid., p. 33.

the cultural critic who changes his tone, at times focused and at times hesitant, to indicate the difficulty and magnitude of the material with which he is dealing.

In outlining his approach to Gandhi's life, Erikson emphasizes the narrative dimension of his study over his analytic goals: he acknowledges that he is neither a 'historian' nor 'an expert on India', but someone who is very interested 'in the telling of its story'.[206] In the opening section of the study he declares that the story stems from his interest in Gandhi's 'existential experiments' (as expressed in the subtitle of his autobiography, *My Experiments With Truth*), but had no real direction to begin with: its coherence emerged from his initial 'strong esthetic impressions' of India to an understanding of 'what survives' throughout culture 'as ethically urgent'.[207] The embodiment of ethical urgency in Gandhi's life proves inspirational to Erikson in his desire to develop his analytic commitment to understanding the hidden trajectories of the self and complementing Gandhi's 'spiritual truth' with his own 'psychological truth'.[208] Rather than accepting the received version of Gandhi's life and his interventionist role in India's colonial history, Erikson becomes increasingly aware that there is a secret dimension to Gandhi's life existing behind the image of the public saint. He is particularly interested in the unsaintly aspects of Gandhi's character: the 'cruel love' which he displays to his wife and the uncertainty he faces about his familial role indicate a profound level of personal confusion behind his firm commitment to political, religious and ethical goals.[209] Such evidence of the coexistence of responsibility and bewilderment in Gandhi's life substantiates Erikson's claim that 'the dynamics of moral and ethical conflict are built into the values themselves'.[210]

What emerges from Erikson's narrative is a simultaneous empathy and respect for Gandhi's public role and a growing scepticism about the implications of the Indian leader's definition of 'Truth'. Initially Erikson is drawn to Gandhi's belief in *Satyagraga* as the dynamic 'truth-force' which provides the basis for non-violent opposition to coercion and tyranny.[211] This combination of freedom to experiment and firmness to believe (comparable to James' combination of yielding and willing), corresponds to what Erikson describes as 'detachment and commitment . . . an almost mystical conflux of inner voice and historical actuality' and echoes Luther's paradoxical claim that a Christian is 'the most free lord of all, and subject to none' and also 'the most dutiful servant of all, and subject to everyone'.[212] As such, the 'meaningful flux' of truth for Gandhi blends

---

[206] Ibid., pp. 10–11.

[207] Ibid., p. 12.

[208] Ibid., p. 231.

[209] Ibid., p. 233.

[210] Erikson, *Life History and the Historical Moment*, op. cit., p. 263.

[211] *Satyagraha* can be defined variously as 'Truth-force of soul-force (*sat*, truth; *agraha*, firmness); non-violent direct action; passive resistance; civil disobedience; non-violent-cooperation'; Jack (ed.) *The Gandhi Reader*, op. cit., p. xix.

[212] Erikson, *Gandhi's Truth*, op. cit., p. 411; Martin Luther, *Basic Luther* (Springfield, IL: Templegate, 1954), p. 115. Although, on this view, self-fashioning seems

critical 'appraisal' with responsible 'action' until they become inseparable.[213] However, although Gandhi adopted a number of different roles as a young man (the 'delinquent' and the 'dandy' among others) and later mixed generative caring with strident belief, instead of successfully balancing proteanism and commitment, Erikson considers his conception of 'Truth' to be at times too inflexible and even perverted: 'I seemed to sense the presence of a kind of untruth in the very protestation of truth; of something unclean when all the words spelled out an unreal purity.'[214] This outburst may be the result of negative transference on behalf of Erikson, but does indicate the frailties of even the most committed individuals. Erikson argues that instead of truth remaining a metaphysical concept or an inviolable inner belief, it is his ability to skilfully manipulate the 'lever of truth' in social situations which defines the core of Gandhi's greatness and enables him to build spiritual bridges between individuals, 'the ethics of family life, communities and nations'.[215] On this view, in accordance with the fusion of idealism and pragmatism at the heart of romance science, truth is both an ideal and a pragmatic 'tool' to facilitate face-to-face contact with others on their own terms and in an environment of 'heightened mutuality'.[216]

One of Gandhi's primary influences in developing his belief in *Satyagraga* was Thoreau's 'Essay on Civil Disobedience' (1849), which he had read while in jail in South Africa in 1893 (although Erikson argues that he is likely to have read Thoreau in 1891 during his involvement in the vegetarian movement in London). Thoreau's promethean resistance to paying his poll-tax and his 'experiment' with life in *Walden* (1854) imbue Gandhi's belief in *Satyagraga* with an almost Western romantic view of individual liberty, but tempered by a commitment to improve the social and spiritual welfare of others. This ethical dimension distinguishes Gandhi's experimental life from the reckless individualism of, for example, Mailer's 'white negro', who resists conformity as an act of freedom from dehumanizing social pressures but at the risk of self-integrity. In a thoughtful essay 'Psychoanalysis: Adjustment or Freedom' (1974), Erikson argues that this example of negative freedom (in classic liberal thought, 'freedom from' constraints) cannot be truly ethical, just as he asserts that 'Truth is more

---

to be fully within the individual's own control, Gandhi comments that the 'inner voice' sometimes 'would speak unexpectedly' to him, which suggests that there are various fragments of the self competing to find an active outlet; Erikson, *Gandhi's Truth*, op. cit., p. 412. Although Gandhi and Nietzsche seem to be philosophical opposites, Erikson detects that they agree on this point.

[213]   Ibid., pp. 410, 44.

[214]   Ibid., pp. 140, 231.

[215]   Ibid., pp. 198, 413. Erikson admits the metaphor of the 'lever' is a 'hopelessly primitive analogy in an electronic age', but provides an image to embody the antimodernist tendencies of romantic science and, in the context of *Gandhi's Truth*, keeps alive the possibility of a revolutionary ethics 'even against the cold and mechanized gadgetry of the modern state'; ibid., p. 198.

[216]   Ibid., pp. 198, 413.

than not lying' and 'Courage is more than not to be cowardly, Faith much more than the absence of existential doubt'.[217] Only the kind of positive freedom embodied by Gandhi ('freedom to' act with, or in spite of, constraints) can succeed in harnessing therapeutic creativity with a willingness to work with and for others, or what the philosopher Charles Taylor calls 'an excellence of moral development'.[218] Although this distinction between negative and positive freedom may be only a matter of degree, the existentialist and ethical currents of Erikson's work fuse with his claim that the 'obligation' of clinicians and analysts should be

> to heal by gaining and giving insight into the unconscious sources of irrational anxiety in order to free the energies and the shared pleasures of a sensual existence. Only such mastery of the irrational frees us for the capacity to fear discerningly those factual dangers which demand competent action. Beyond anxiety and fear we may face the existential dread which awakens universal resources of faith and fellowship.[219]

He argues that psychic energies need to be freed from constraints, but only 'competent action' (as epitomized by Luther and Gandhi) can bring about the 'self-responsible life choice' characteristic of positive freedom.[220] These ideas are developed at length throughout Erikson's clinical and cultural writings, but the precarious combination of idealistic beliefs (freeing energies and awakening 'universal resources'), pragmatic activity ('competent action') and existentialist awareness ('anxiety', 'fear' and 'dread') are clearly evident in this quotation and neatly summarize Erikson's distinctive contribution to the theoretical trajectory and practical application of romantic science.

---

[217] Erikson, *Life History and the Historical Moment*, op. cit., p. 263.
[218] Charles Taylor, 'Liberal Politics and the Public Sphere', *Philosophical Arguments* (Cambridge, MA: Harvard U.P., 1995), p. 258.
[219] Erikson, *Life History and the Historical Moment*; op. cit., p. 264.
[220] Taylor, *Philosophical Arguments*, op. cit., p. 258.

Chapter 5

Oliver Sacks:
The Storytelling Romantic

The life and work of Oliver Sacks (1933–) is full of doublings: he is a serious neurologist and a popular psychologist; he currently practises neurology on a private basis and is a public figure dedicated to institutional reform; he is committed to empirical scientific research and writes of his love for literature, music and visual art; and he is a Briton from a Jewish family who has lived in New York City since the early 1960s. Currently professor of neurology at the Albert Einstein College of Medicine in the Bronx, Sacks' work with post-encephalitic patients at Mount Carmel chronic hospital in the late 1960s, recounted in *Awakenings* (1973) and the Yorkshire TV documentary (1974), brought him widespread critical attention. When Sacks appropriated Alexander Luria's phrase 'romantic science' in the 1980s to define his overall project (which has dealt with neurological illnesses as diverse as migraine, encephalitis, Parkinson's disease, Tourette's syndrome, body image disorders and colour blindness), he chose a term which crystallizes all the doublings that exist uneasily, but often creatively, in his work.

This chapter discusses the question of selfhood in Sacks' work as a development of the tradition of romantic science represented by James, Rank, Binswanger and Erikson. In his seven books, he has questioned accepted modes of inquiry into neurological illness, self-consciously reacting against scientific epistemology which aims to classify disease objectively, but which tends to overlook the subjective experience of illness. He claims classical neurology and cognitive science view brain disorder as objectively analysable, whereas he is also interested in illness as it is embodied in, and experienced by, the patient. In his work, he emphasizes the subjective experience of illness: how it feels for patients to be ill and how they narrate their condition in an attempt to accommodate debilitating illness. But, because this type of inquiry often runs tangential to the hard science of neurology, critics argue that he is merely a scientific popularizer, whose publications replace the serious pursuit of science with literary pseudo-science.[1] Sacks insists, however, that his is a complementary approach to classical neurology, not a substitute for it: his focus on the phenomenological aspects of illness is rarely at the expense of neurological integrity.

---

[1] For example, Ella Kusnetz's hostile opinion is evident in 'The Soul of Oliver Sacks', *Massachusetts Review*, 33(2), 175–98 (discussed in the fourth section of this chapter).

Awakening the Self

Sacks' main criticism of classical neurology is that it is too narrowly interested in the physical aspect of disease, usually at the expense of illness as experienced by the patient. Two traditional neurological approaches particularly trouble Sacks: the holist and the mosaic views. The former approach conceives of the 'total energy' of the brain as 'uniform, undifferentiated, and quantifiable' and the latter describes a myriad of energy-centres, breaking down the total brain-energy into smaller 'sub-systems'.[2] While each approach has certain strengths, he claims both are inadequate when pursued in isolation because 'they are completely alien to the experiences of the patient'.[3] Instead, he argues that the physician should be less the disinterested inquirer and more the intimate collaborator who is able to 'feel with' the patient and who is receptive to the patient's own expressions.[4] Consequently, his interest is not only in the physiological aspects of neurology, but also the experience of the patient suffering from the disorder. As such, Sacks' attempt to facilitate an expressive space, where the patient can figuratively describe the nature and scope of his or her illness, may provide the physician with a set of techniques to challenge these two traditional neurological approaches.

Although Sacks often calls for institutional reform, he understands the patient's expressive space to be crucially determined by the severity of the illness and whether the patient has the neuropsychological capacity to collaborate in the production of such a space. For Sacks, the physician's responsibility is to encourage therapeutic discourse, even if this means challenging the presuppositions of existing analytic models. Following Binswanger's and Erikson's emphasis on care and love, this recommendation for a more humanistic approach is central to his work, but raises a number of theoretical implications. In order to give some substance to his retreat from classical neurology, Sacks has appropriated Alexander Luria's conception of romantic science and has also implicitly associated himself with the strain of scientific counter-culture of Anglo-American anti-psychiatry trends in the 1960s, epitomized by the work of Ronald Laing and Thomas Szasz.[5]

The argument that Sacks is a romantic reactionary against a hegemonic neurology which still conforms to the Enlightenment project for knowledge, can be defended by considering his attempt to reconnect what Habermas calls the

---

[2]   Oliver Sacks, *Awakenings* (London: Picador, 1991), pp. 239–40.

[3]   Ibid., p. 240

[4]   This ideal of empathic understanding does not mean the physician should identify completely with the patient's position, but should be able to shift his or her subject-position as the different modes of his inquiry demand: sometimes the sympathetic friend, sometimes the collaborative partner and sometimes the authoritative physician.

[5]   Luria's romantic science is most evident in his last two books, *Mind of a Mnemonist* (1968) (Cambridge, MA: Harvard U.P., 1987) and *The Man with a Shattered World*, trans. Lynn Solotaroff (1972) (Harmondsworth: Penguin, 1975).

'separation of the spheres' (as discussed in the Introduction).[6] If the now autonomous spheres of science, morality and art could be reunited then the 'hermeneutics of everyday communication' may be released from the knot of specialization.[7] Sacks does not want to debunk the whole neurological programme; rather, he wishes to broaden its foundations and, following the other romantic scientists, to phenomenologize neurology by bridging the human and natural sciences. Because he highlights a central place for an aesthetics of expression, his writing can be positioned at the interface between diverse modes of inquiry and a variety of written forms, some clinical and some literary. To introduce Sacks' particular vision of romantic science it is useful to turn to two essays, 'Neurology and the Soul' (1990) and 'Luria and "Romantic Science"' (1990), to examine the impact of his version of romantic science on traditional neurological inquiry.

Sacks' professional work as a physician is not only to study neurological dysfunction for its own sake, but to help suffering patients regain their ability to live in the most expansive manner possible. Here the archaic meaning of physician as 'healer' is a legacy which the institutionalization of medicine in the nineteenth century did much to erase, but indicates the physician's primary task is to help restore patients to health. Biological investigations and technological development has turned the body into a medical arena where reparation is increasingly possible, but, despite the intense interest in neuroscience in the 1990s, the complexities of brain activity remain beyond current understanding. The double science of neuropsychology approaches the brain less as a physical entity and more as a centre for processing 'higher cortical functions'.[8] Sacks notes the manner in which Luria, in his last two books, began to fuse the two 'modes of anatomy and art': firstly, by the conceiving of the 'individual as a *being*', 'a living being, containing (but transcending) organic functions and drives, a being rooted in the depths of biology, but historically, culturally, biographically unique' and, secondly, by constructing a 'biological biography' to reconstruct the life-world of the patient. As such, Sacks follows Luria's (and Erikson's) attempts to restore the patient from the 'statistical entity' of medical records to a narrative account of an experiencing being.[9]

In conceiving of the self in this manner, Sacks does not rely on Cartesian dualism, but, like James, deploys 'mental' and 'physical' as two different descriptive categories. In 'Neurology and the Soul', he conceives of the body and mind as two levels of meaning which require two quite different modes of description, but both underpinned by biological monism. As the American

---

[6]   Habermas, 'Modernity – an Incomplete Project', op. cit., p. 131.

[7]   Ibid., p. 132.

[8]   Sacks, 'Luria and "Romantic Science"', in Elkonon Goldberg, ed., *Contemporary Neuropsychology and the Legacy of Luria* (Hillsdale, NJ: Erlbaum, 1990), p. 189.

[9]   Sacks, *The Man Who Mistook His Wife For A Hat* (London: Picador, 1986), p. x (hereafter *The Man Who Mistook*).

philosopher John Searle claims, consciousness is not connected to the brain by a quasi-mental, quasi-divine umbilical cord, but should be seen as a metastructural sector which emerges from the organic brain and enables the assignation of meanings to different perceptions.[10] While ultimately dependent on the unimpaired functioning of the organic brain, this sector is overdetermined and cannot be adequately explained on the level of synapses or neurone-firings. Rather, like Binswanger's world-designs, consciousness demands a mode of description on the level of semantics: the organization of meanings and the assignation of values are seen as essentially cultural rather than physical.

Gerald Edelman, another multiple scientist working on the boundaries of biology, neuropsychology and philosophy, asserts that the 'mind-brain' can only fully be understood through an interactive study of morphological and evolutionary processes, both biological and cultural.[11] Like Erikson's psychobiographies, only by situating the patient within a narrative framework can the physician unearth facets of the self which lie in the intersecting spheres of neurobiology and social psychology. As such, Sacks posits a self 'rising from experience, continually growing and revised' as 'a confederation, an organic unity, of innumerable categorizations'.[12] The 'organic and historical unity' of the self results from the complex interaction between primary and higher-order consciousness: the former providing a sentient awareness of the world and the latter enabling the individual to reflect upon conscious awareness.[13] Sacks does not view the self as a seamless metaphysical whole, but as a series of evolving constructs deriving from higher-order consciousness, and, when positioned within a symbolic structure, generating a sense of self. Much of Sacks' work deals with patients who have lost a significant function of their primary consciousness (usually eliciting feelings of wholeness, solidity and depth), creating a hole or lacuna in their higher-order consciousness. By stimulating other areas of consciousness, these patients are encouraged to devise ways of compensating for neuropsychological loss.

Most contemporary practitioners of neuroscience have attempted to cultivate a humanistic approach to deal with patients as living people with particular needs, but also maintain a behavioural perspective in order to map neurone circuits in the brain and nervous system.[14] In this manner, neurological illness is conceived

10   Searle, *The Rediscovery of Mind*, op. cit., p. 14.
11   Edelman outlines his theory of neural Darwinism in *Bright Air, Brilliant Fire: On the Matter of the Mind* (New York: Basic Books, 1992).
12   Sacks, 'Neurology and the Soul', op. cit., p. 50.
13   Humans and some other primates are unique in the sophistication of their higher-order consciousness, granting them 'the powers of language, conception and thought' and the ability to describe their condition; Sacks, *A Leg To Stand On* (London: Picador, 1991), p. 186.
14   For a discussion of the project to humanize medical protocol see Michael Schwartz and Osborne Wiggins, 'Science, Humanism and the Nature of Medical Practice: A Phenomenological View', *Perspectives in Biology and Medicine*, 28(3) (1985), 331–61.

as the disorder of brain functions stemming from damaged or altered neural groupings. While patients are treated as holistic organisms with their own phenomenal life-world, the primary aim is to approach illness at the level of brain mechanisms and processes. By understanding the hard-wiring of the brain it may be possible to reduce the intensity of the 'dis-ease' for the suffering patient and those who may develop similar conditions; if the pain can be isolated and materially rooted it may be possible to treat it chemically or physically. Sacks agrees with this agenda, insofar as he conceives neurological illness to be grounded in an understanding of bodily disorder, but he subscribes to a more sophisticated understanding of illness, conceived as both fact *and* value: on one level, a factual description of bodily impairment and, on another, a qualitative evaluation of the individual's 'ability'. Although he endorses the importance of test-based approach for mapping neural complexities, he claims the mysteries of brain functions are such that sometimes the best possible route to understanding illness is through close interaction with the suffering patient, who should not be conceived as a problem to solve, but as a unified person who is undergoing a variety of debilitating effects: physical, mental, moral and spiritual.

Unlike scientific approaches which claim to be value-neutral, this medical development of romantic science demonstrates the impossibility of separating the object of study from the method, because the former is always already construed through the conceptual presuppositions of the latter. As Sacks comments in 'The Great Awakening' (1972), the terms which he deploys to understand the conditions of his patients 'are ontological definitions, and as such cross the boundaries of the usual neurological or psychiatric "diagnoses"'.[15] In addition to being at root a disorder of function, neurological illness can be conceived as having ontological and existential proportions. Sacks' attempt to glean knowledge of these supra-physical dimensions is a development of Heidegger's conception of *Dasein* as a radical questioning of the parameters of selfhood. By carefully interpreting his patients' conditions (through a mixture of behavioural observation and their own expressions of illness), he hopes to facilitate an interpretative space which broadens the conception of the health-illness complex. In this way, romantic science maintains the balance between the art of medicine (figuring and imagining) and the science of medicine (inquiring and examining), in which experimentation is balanced by 'poetic vision'.[16]

In considering the role of illness for understanding the romantic scientist's conception of self, a useful starting point is Sack's first major book, *Awakenings* (discussed more fully in the third section). He recounts the spread of the *encephalitis lethargica* pandemic ('sleepy sickness') from Vienna in the winter of 1916–17, resulting in world-wide death over ten years. Of those that survived the

---

[15]   Sacks, 'The Great Awakening', *The Listener* (26 October 1972), p. 522.

[16]   Quoted from a letter of comment to 'The Great Awakening', by R.M. Allott in *The Listener* (30 November 1972), p. 756.

coma, some recovered to the extent that they returned to working life, but many of the sufferers later developed neurological disorders. Such disorders were not easy to categorize, as can be gauged by the five hundred 'distinct symptoms and signs' documented by the Austrian pathologist Constantin von Economo in 1929 and the 'twenty thousand clearly different disorders of function' recorded by Sacks in the 1960s.[17] Mount Carmel Hospital in New York was set up after the First World War for victims suffering from profound damage to their nervous system, but later admitted some of the most chronic post-encephalitic patients. In the mid-1960s Sacks was commissioned to treat these patients, but such was the strange nature of the disorder that he was uncertain about how to implement rehabilitative treatment. *Awakenings* comprises the close study of twenty of these post-encephalitic patients with the primary aim of understanding the illness, or cluster of illnesses, in order that he might treat them.

Despite the range of symptoms displayed by the patients, Sacks distinguishes two moods of illness: 'negative' and 'positive' disorders. Negative disorders are displayed by patients who recovered from the coma but 'failed to recover their original aliveness': they were 'conscious and awake – yet not fully awake . . . They were as insubstantial as ghosts, and as passive as zombies . . . ontologically dead, or suspended, or "asleep".'[18] Other patients displayed more active symptoms, 'over-animations and excesses and perversions of behaviour' closely connected to Parkinsonism.[19] Those displaying negative behaviour appeared to have entirely lost their selfhood, whereas the bodily 'coercions' exhibited by the other patients seemed 'as if their very selves were being clenched or tensed or twisted or torn'.[20] Sacks is concerned not only with the mechanics of dysfunctional illness but how, in an ontological sense, the disorder undermines a sense of selfhood. Although the 'positive' patients retain an ability to express themselves and remain ontologically alive, he believes that even the 'negative'or catatonic patients continue to exist in a minimal sense.

Much of the philosophical content of *Awakenings* stems from Sacks as physician standing in relation to the catatonic patients as individuals, rather than bodies emptied of mental life. The exact impairments of their brains are a mystery to the methods of classical neurology; not until Sacks approaches the problem through a series of intuitive guesses, do the enigmas of the illness begin to unfold. When periodically an 'oculogyric crisis' (a sudden spasmodic activity characteristic of Parkinsonism) breaks through the catatonia, Sacks soon realizes there are as many different crises as there are patients and concludes that only by positioning the patients within their individual life-stories can he unravel the intricacies of the disorder:

---

[17]   Sacks, 'The Great Awakening', op. cit., p. 521.
[18]   Ibid., p. 522.
[19]   Ibid.
[20]   As can be gleaned from von Economo's description of the patients as 'extinct volcanoes'; Sacks, *Awakenings*, op cit., p. 14; Sacks, 'The Great Awakening', p. 522.

what seemed an impersonal or even depersonalizing disease had, in fact, a strong quality of the personal, and could not be understood without reference to the personal. It was not merely humanly, or ethically, necessary to see these patients as individuals, it was scientifically necessary to do so as well.[21]

Accordingly, the central section of the book is a collection of individualized cases which record the responses of the patients to the drug L-DOPA, administered to them during the summer of 1969.[22] Although some do not respond to the drug and others react badly, the remarkable awakening displayed by a number of catatonic patients enables them to regain many of the lost facets of selfhood. Sacks places particular emphasis on two primary abilities corresponding to the central aspects of romantic science: firstly, the ability to give expression to the personal qualities of their illness and, secondly, the capacity to act practically. He claims that L-DOPA presents them with a chemical window, providing their mental life with a physical outlet. While the patients' physiological responses to L-DOPA represent the empirical data for Sacks' technical commentary, the cases are presented in a manner closer to symbolic narrative (through which Sacks hopes to convey the drama of recovery) than to clinical record.[23] A cursory comparison between the huge number of impersonal entries to his earlier book *Migraine* (1970) and the extended narratives of *Awakenings* indicates this shift in emphasis.

Awakenings is a clinical study which is simultaneously scientific and existential, but, in Sacks' attempt to fuse diverse perspectives, is often in danger of being essentially neither. The condition of the patients at Mount Carmel is primarily neurological, but the poetic triumph of the book stems from the possibility that they might rediscover an active and expressive self: the title suggests the tearing away of a veil to reveal a more authentic self behind the catatonic mask of illness. In conveying the general exuberance which the awakenings release, Sacks often treads a precarious line between his tribute to the patients and the preservation of his professional bearing as an inquisitive, but impartial, scientist. Consequently the book reads as a hotchpotch of ideas and observations which derive from clinical experience, but also reveal his faith in a repository of health which survives despite illness and can transcend, if only momentarily, the severity of bodily constraints. As such, he maintains an optimistic and often sentimental vision which frequently jars with the bodily disorders he depicts. But, as he claims in his preface to the first edition:

[21]   Sacks, *Awakenings*, op. cit., p. 65.

[22]   L-DOPA was administered in order to compensate for a defective level of the nerve-transmitter dopamine, but, as Sacks frequently indicates, the long duration of the illness (both in its latent and manifest stages) meant that, for the patients of *Awakenings*, much of the neurological damage was of a structural nature and could not simply be 'topped up'; ibid., p. 30.

[23]   This is seen most clearly in the Yorkshire TV documentary 'Awakenings' in the *Discovery* series (1974).

> The general style of the book – with its alteration of narrative and reflection, its proliferation of images and metaphors, its remarks, repetitions, asides and footnotes – is one which I have been impelled towards by the very nature of the subject-matter. My aim is not to make a system, or to see patients as systems, but to picture a world, a variety of worlds – the landscapes of being in which these patients reside.[24]

The painterly verbs suggest a figurative quality which blurs the boundaries between orthodox case study, symbolic narrative and philosophical contemplation. This is a deliberate strategy, since Sacks is not a systematic thinker who wishes to categorically define selfhood: he relies heavily on metaphors by which he hopes to gesture towards a Being that cannot be adequately described. His later editions of the text (in 1982 and 1991) read as a series of revisions and reflections, constantly widening the focus of his study: where the fragments of the text intersect, meanings proliferate in centrifugal rings without leading back to a centre. If selfhood can be said to exist for the patients it is always fleeting and fugitive, inscribed within discourse but unable to be fully revealed by description. This is reflected in the difficulties which Sacks encounters in writing the book: just as the patients struggle to discover a voice from within illness, so Sacks struggles to produce a study which is both descriptive and explanatory, romantic and classical.

His chief aim is to combine an understanding of his patients' phenomenological condition with an interrogation into behavioural patterns resulting from neurological disturbance. He attempts to retrieve 'a combined vision of body and soul' ('It' and 'I'),[25] which he believes has been lost in the professional bifurcation of neurology and psychology early in the twentieth century. He wishes to supplement the study of 'It' (the object of neurological study) with an investigation into 'the terms in which we experience health and disease', terms which 'neither require nor admit definition; they are understood at once, but defy explanation; they are at once exact, intuitive, obvious, mysterious, irreducible and indefinable'.[26] Closely resembling Binswanger's and Erikson's recommendations, Sacks' notion of giving the self over to the experience of dialogic relation correlates with a belief that the physician should cultivate a human relationship in an 'existential encounter' to complement a technical understanding of symptoms and syndromes.[27] His ideal dialogue would acknowledge institutional constraints and therapeutic limitations, but move beyond them into a realm of 'direct and human confrontation, an "I-Thou"

---

24   Sacks, *Awakenings*, op. cit., p. xviii.
25   Sacks, *The Man Who Mistook*, op. cit., p. 88.
26   Sacks, *Awakenings*, op. cit., p. 224.
27   Ibid., p. 226.

relation, between the discoursing worlds of physicians and patients'.[28] Such an intimate dialogue does not entail a complete effacement of the physician's role, nor complete intimacy with the patient: Sacks argues that the physician 'must inhabit, simultaneously, two frames of reference, and make it possible for the patient to do likewise'.[29] In this ideal model of 'real communication', the physician becomes 'a fellow traveller, a fellow explorer, continually moving *with* his patients, discovering with them a vivid, exact, and figurative language which will reach out towards the incommunicable'.[30] But in deploying the existential language of 'I and Thou', Sacks sets for himself a project which is always in danger of falling short of the mark or breaking into monologue.[31]

Clinical Tales

Sacks' interest in the form of medical writing is closely linked to his attempts to develop traditional empirical study in order to do justice to the idiosyncratic psychic worlds of his patients, by juxtaposing and assimilating different types of written discourse. Romantic science offers a distinctive way of seeing illness and identity, facilitated through close interpretation and to the form of what Sacks calls his 'clinical tales', which challenge the way in which medical knowledge is conventionally documented.[32] By relying on symbolic language, Sacks seems to

---

[28] Ibid., p. 255. The major influence on this aspect of Sacks' thought is the religious existentialist Martin Buber. Sacks shares with Buber (and Erikson) an optimism that, while an interpersonal intimacy is rarely, if ever, achieved, the possibility of dialogic understanding should be a continual motivation. In *I and Thou* (1923) Buber claims 'primary words are not isolated words, but combined words' which indicate 'relations'; Martin Buber, *I and Thou*, trans. R.G. Smith (Edinburgh: T. and T. Clark, 1987), p. 15. He describes two 'primary words' – 'I-Thou' and 'I-It' – which hold the 'I' in relation with another, and by which the signification of 'I' is determined. His work reveals a faith that others can be met through interpersonal relation (elevating the Other to a Thou rather than obliterating the Other through the projections of selfhood or reducing 'I' to 'It') and indicates that only the 'I-Thou' relation is authentic, for it 'can only be spoken with the whole being' whereas the 'I-It' relation 'can never be spoken with the whole being'; ibid., pp. 15–16.

[29] Sacks, *Awakenings*, op. cit., p. 226.

[30] Ibid., pp. 225-26.

[31] Although Sacks' desire to render the feelings of the post-encephalitic patients is well-meaning, at times he seems to do little more than project his own preoccupations onto them. In a review of *The Man Who Mistook*, J.K. Wing comments that 'one is . . . uneasily aware . . . that Sacks' patients all talk as he does. He cannot quite give them their own accents and idiom.'; J.K. Wing, 'Distorting Mirrors', *Times Literary Supplement* (7 February 1986), p. 146. One indication that this occurs in *Awakenings* is when Sacks relates that Leonard L. 'using his shrunken, dystrophic index-fingers . . . typed out an autobiography 50,000 words in length', but he does little more than comment on it in a short footnote; Sacks, *Awakenings*, op. cit., p. 212.

[32] Sacks, 'Clinical Tales', *Literature and Medicine*, 5 (1986), 16–23.

endorse two anti-rationalist directives: firstly, he suggests that Cartesian lucidity does not always lead to a clear and distinct view of truth; and, secondly, following Binswanger, he acknowledges that Being may only be 'unconcealed' through a non-intellectualized beholding of it. But, far from being an irrationalist, he shifts his stance between an explanation of mental states with reference to material causes and an expression of his own enchantment with ineffable phenomena. His aim to romanticize science is not merely a self-conscious echo of the romantic emphasis on expressive language: he is primarily a practitioner, and secondarily a theorist. Accordingly, he uses his tales to redescribe the conditions of health and illness, rather than indulging in theoretical speculation for its own sake.

Although his clinical tales fuse both theory and practice, the terms in which he couches his project have often been understood by critics to heavily outweigh his empirical accomplishments. It can be claimed that Sacks uses the literary accoutrements of romanticism to hide essentially institutional polemics. As such, the form of his tales may be seen as secondary and peripheral to his role as a clinical neurologist. However, it is more appropriate to argue that the clinical and literary aspects of his work cannot be easily separated without doing an injustice to both. Sacks' riposte to the dominant practices of neurology relies as much on his consideration of literary form as it does upon his theoretical arguments as he seeks to refigure disability in terms of aesthetic expression. His aim to find an appropriate genre of study mirrors what he believes to be his patients' primary need to rediscover their expressive voice to affirm a sense of selfhood. He argues that if one is to experience an 'awakening' (of whatever kind) it is essential to develop both aesthetic and neurological awareness. Rather than adopting the role of a mind-curist, who would argue that once the body is cured the mind will be freed, Sacks senses that cure must be of the whole Being.

The form which Sacks adopts for his clinical tales can be understood by comparing them to standard medical formats. His tales are not simply introduced into the medical arena from the outside, but represent hybrid developments from accepted forms; nor do they conform to a generically distinguishable model. As such, it is important to consider the medical forms from which Sacks' work emerges, before returning to a close reading of *Awakenings* in the next section. One of the crudest forms of clinical documentation, the medical chart, provides an insight into the development of his clinical tales.

In the collaborative essay 'Charting the Chart', published in *Literature and Medicine* (1992), Suzanne Poirier and the Chicago Narrative Study Group assess the limitations of the medical chart.[33] The chart documents a patient, Mrs R, who had been admitted to a rehabilitation centre after initial hospital treatment for a stroke. Her two-month stay was recorded by a team of two physicians, a nurse, two therapists, a chaplain, a psychologist and a social worker. The chart is

---

[33]   The contributors are thought to 'form a shadow team, a chorus of voices, to comment on the work of the health care-team'; *Literature and Medicine*, 11(1) (1992), ix.

structured around a series of brief journal entries designed to facilitate communication between the consultants. First devised by the American physician Lawrence Weed in the late 1960s and known by the acronym 'SOAP', the protocol aims at producing a 'problem-orientated record' of a patient in care.[34] SOAP entails a four-point method for documenting aspects of the patient's condition: their *subjective* response; an *objective* description of the condition written by the health official; followed by an interpretative *assessment*; and a *plan* outlining subsequent action. Weed wanted to present medical history as a 'scientific document' by devising SOAP as a version of scientific medicine 'based on the Newtonian model', in reaction to the impressionistic case history which 'often failed to take account of physiological fact'.[35]

The SOAP format fits well with the 'computer-assisted, decision-making field of medical informatics' which Kathryn Montgomery Hunter argues was prevalent in American medicine during the 1980s. She claims that the desire to base scientific knowledge on the model of mathematics is 'primarily rhetorical, a way of addressing the problem of uncertainty' and contends that Weed designed SOAP to control 'the observer's subjectivity in the face of an unattainable scientific objectivity'.[36] The objectivity of scientific description implies the medical official has attained a position of unquestionable authority. But Hunter suggests that 'such an assumption is part of the problem of unacknowledged epistemological uncertainty', an uncertainty which Sacks claims is exacerbated when encountering complex neurological disorders. Influenced by the writings of Sacks in the neurological sciences, Roy Schafer in psychoanalysis and Paul Ricoeur in literary theory, the contributors to *Literature and Medicine* (launched in 1982) have begun to assess the importance of narrative techniques for preserving a humanistic approach towards the patient. Rather than advocating a return to the impressionistic case history denounced by Weed, the essays in *Literature and Medicine* rigorously investigate the institutional and textual politics implied by different medical forms. This type of study not only opens up the medical arena to a variety of theoretical influences, but also indicates that medical writing should be seen as fundamentally part of cultural production.

The co-authors of the essay each respond to particular aspects of the chart – whether they thought the patient, Mrs R, received 'good care' and whether hers was a 'good story'.[37] Although the description of a 'good story' may appear to be an unfair criterion with which to judge a medical record, it serves to provoke responses concerning 'how "reality" is constructed' by particular institutions and

---

[34]   Suzanne Poirier et al., 'Charting the Chart – An Exercise in Interpretation(s)', *Literature and Medicine*, 11(1), 3. See Lawrence Weed, 'Medical Records That Guide and Teach', *New England Journal of Medicine*, 278 (14 and 21 March 1968), 593–600 and 652–7.

[35]   Ibid., p. 166.

[36]   Ibid., pp. 167, 177.

[37]   Poirier, 'Charting the Chart', op. cit., p. 2.

to what extent the reader is able to recoup the patient's personality from a representation on the chart.[38] Some critics express positive responses, but most are wary of such a rigid format. On the positive side, the chart enables 'health professionals to work from a fuller data base and provide more coordinated care', providing a vital means of communication which allows consultants to briefly scan over particular comments and diagnoses to inform their own more fully.[39] However, on the critical side, the information conveyed by the chart is limited: one critic comments 'this format suits small, well-defined problems, such as surgical wounds' but is inadequate for more general or intricate complaints.[40]

The contributors collectively indicate three broad types of problem implicit in the SOAP protocol and the chart itself. Firstly, attention is directed towards the material conditions under which the document is completed: it is 'regarded by many as an onerous task or an administrative or legal exercise'.[41] The brevity of the notes results from the time constraints and pressured circumstances of their composition: the actual conditions under which writing occurs may mean that the report is often cursory and fails to do justice to the patients themselves or their condition. Moreover, the 'charts do not invite interaction and feedback', tending 'to be more a collection of monologues than a true dialogue or colloquy'.[42] This criticism implies that charts are sometimes little more than procedural exercises which (often because of illegibility or indecipherable shorthand) rarely fulfil their potential.

The second type of criticism focuses on the subjective entry of the chart, usually a direct quotation from the patient. Although the patient's contribution is given some priority, the quotation tends to be delimited by the length of the space on the chart as much as the attitude of the health professional to the patient. This 'representative anecdote' is clinically selected from a variety of experiences articulated by the patient and is more likely to be guided by the subsequent categories of description and assessment than by comments the patient feels are important.[43] The patient's personal context (the tone of voice, whether they are cooperative or reluctant to impart information, or whether they experience difficulty in describing certain bodily experiences) are given the barest attention. In short, the SOAP chart limits the contribution of the patient to a selected anecdote, often without any contextualizing remarks.

Thirdly, and most importantly, the critics raise questions concerning the demand that health officials record objective comments. Pure objectivity ignores the observer's perspective, the emotions which he or she may experience and the

---

[38]   Ibid., p. 19.
[39]   Ibid., p. 7.
[40]   Ibid., p. 10.
[41]   Ibid., p. 7.
[42]   Ibid.
[43]   Kenneth Burke, *A Grammar of Motives* (Berkeley: University of California Press, 1969), p. 59.

epistemological uncertainty which Hunter claims challenges the scientific status of the chart. Extensive work with a patient is likely to provoke a range of emotions from the medical team, especially the type of interaction in which nurses and social workers involve themselves. Focusing her comments on the nurse's entry, Lioness Ayres writes:

> SOAP charting excludes from the clinical record any evidence of the feelings or reactions of the nurse, and focuses on the objective (and by implication unbiased) facts. Opinions, impressions, and hunches may form the keystone of expert practice, but they lack objectivity and thus resist clinical documentation.[44]

Although the nurse's feelings may inform many of the objectified entries in the document, the SOAP format does not permit any emotive description and prohibits the patient's story from being told in anything other than flat clinical terms. The critics do not wish to replace the SOAP format with a lengthy series of emotional responses by the consultants; rather, they prefer a form which would enable the reader (both clinical and lay readers) to glean a sense of the patient's story in its personal and institutional dimensions. Another critic, Lorie Rosenblum, sums up this third criticism by arguing that procedures such as SOAP may ultimately effect 'a deeper dismemberment of the patient's story' which may not be recoupable by the reader.[45]

Poirier claims 'Charting the Chart' was written in order to evaluate how far 'perceptions, misperceptions, hasty judgments, professional interest, or prejudice carelessly enter the chart in a way that contributes to a "characterisation" of the patient that may be false or a caricature of a complex human being'.[46] As Erikson's psychobiographies show, characterization is inevitable if the patient is to be placed within a symbolic narrative. But if the patient's story is to be rescued from the isolated comments of the chart then an alternative format may be more representative of the patient and also of the conditions of writing the document. Having asserted this, Poirier comments that nevertheless the attentive reader is able to rescue fragments of the story: the patient's isolation and fear 'cannot be repressed despite the structure of the chart or the concerns of its constructors'.[47] But the chart format exacerbates the reader's difficulty in gaining an insight into the psychology of either the patient or the health official. The impersonal tone in which the entries are expressed ('Pt. found crying in room'; 'Pt. unable to wash'; 'Mrs R has been very tearful throughout') does little more than hint at the fear, despair and inadequacy which Mrs R is sensed to have felt.[48]

---

[44]   Poirier, 'Charting the Chart', op. cit., p. 9.
[45]   Ibid., p. 15.
[46]   Ibid., p. 19.
[47]   Ibid., p. 20.
[48]   Ibid., pp. 3, 5.

In a discussion of modernist writing, the literary theorist Jonathan Culler suggests that 'impersonality depends not on what is said but on the fact that no identifiable narrator speaks'.[49] As the third criticism makes clear, the flat and impersonal voices demanded by the chart imply 'the desire to prevent the text from being recuperated as the speech of an identifiable narrator, to prevent it . . . from being read as the vision of someone who becomes an object that the reader can judge . . . a personality whom we feel we know'.[50] However, unlike the modernist writer, the impersonal clinical author does not refine him or herself out of existence in order to attend more closely to language and form. The scientific narrator stands at the inaccessible locus of authority, increasing the difficulty for the reader of the text to judge how the patient responds to, and is treated by, health officials. As a result, the reader is left with only a vague sense of the subject of the story and its storyteller(s).

Sacks argues in 'Clinical Tales' (1986) that it may be possible to construct a narrative in a manner which highlights the voice of the physician as a writer, as well as the patient's personality. By constructing case histories as 'tales', the reader is made aware that the document is the physician's aesthetic response to the patient and not a value-neutral clinical biography. Thus, Sacks asserts, 'if I write "Clinical Tales" it is because I am *forced* to . . . they do not seem to me a gratuitous or arbitrary compound of two forms, but an elemental form which is indispensable for medical understanding, practice, and communication'.[51] A tale does not merely construct a narrative of character and event, but provides a symbolic medium through which to express the subjective aspects of the illness. The 'elemental' quality of the tale derives from Sacks' claim that 'it is the form *patients* adopt' when they explain themselves to the doctor.[52] The patient's story is not usually presented as an anecdote or a set of symptoms, but a temporal narrative in which the illness experience is integrally influenced by the personality and perspective of the storyteller.[53] Although the SOAP format does not entirely exclude the patient's story (it remains present in 'the margins of the text'), it tends to obliterate the personality of the storyteller beneath the institutional weight of the chart.[54] Sacks counteracts this form of case study, by envisioning a written medium which delineates 'a history of illness from its first

---

[49]   Jonathan Culler, *The Uses of Uncertainty* (London: Elek, 1984), p. 110.

[50]   Ibid., p. 78.

[51]   Sacks, 'Clinical Tales', op. cit., p. 16.

[52]   Ibid.

[53]   Often a patient's response is provoked or prompted by particular questions posed by the health official. This technique can be used to substantiate the idea that the making of the story is in fact a collaborative practice. As such, Sacks argues that it is essential to attend closely to the form and content of the questions as much as the patient's response.

[54]   Poirier, 'Charting the Chart', op. cit., p. 21.

intimations, through all its subsequent effects and evolutions, to (perhaps) its final crisis and resolution'.[55]

He does not ignore the importance of breaking down stories into sequential stages in order to understand the structure of the patient's narrative, nor does he dismiss the importance of using categories and typologies for describing forms of illness. Rather, he emphasizes that the story should be viewed as ontologically prior to the abstraction of stages and components, just as disease cannot be easily separated from how it is manifested in, and embodied by, the patient.[56] Instead of conceptualizing disease and patient as separate entities, he makes the romantic claim that they must be conceived as an ontological whole. This Nietzschean conception of the 'diseased individual' leads Sacks towards a concentration on 'the *subject*, at the deeper, the "existential", level of an identity struggle'.[57] At this level, he figures the individual as undergoing a symbolic battle to recover a sense of agency, to regain 'one's "world" (the integration of one's nervous system, one's mind, one's *self*)'.[58] By reconstructing the life-world of the patient within the textual world of the tale, it is possible to contextualize illness in order to understand how the disease is experienced. Even if the organic disease is incurable, it may be possible for the patient to accommodate their illness through collaborative storytelling. Sacks argues that the patient's world is not to be regained passively, but must be actively remade from within the hostile environment of illness. He is sceptical about approaching this position from a wholly objectivist viewpoint or making wild imaginative leaps into the mind of the sufferer: imagination and observation should be deployed together with extensive research into the patient's personal and family history.[59] The process of constructing the tale may often turn out to be of greater value to the physician and patient than the completed text.[60] Indeed, as the cases in *Awakenings* indicate, the construction of such tales is always in process and under revision.

Although Sacks maintains this position throughout his work, he often uses terminology loosely or ambiguously. The extraordinary and fabulous dimensions of the tales in *The Man Who Mistook* and *An Anthropologist on Mars* (1995) signify a fictive realm which is some distance from the material realities of neurological disorder. Similarly, his formulation of the generic character of his

[55] Sacks, 'Clinical Tales', op. cit., p. 17.

[56] This argument is also expounded by Larry Churchill, 'The Human Experience of Dying: The Moral Primacy of Stories over Stages', *Soundings*, 62 (1979), 24–37.

[57] Sacks, 'Clinical Tales', op. cit., p. 17.

[58] Ibid., p. 18.

[59] Arguably, this approach is evident in Tony Harrison's treatment of Alzheimer's disease in *Black Daisies For The Bride* (1993). The question of imaginative authenticity is raised by Michael Ignatieff in a BBC *Late Show Special* interview with Harrison (June 1993).

[60] Sacks is always aware of the 'other' reader of his tales. Even if the tale offers no tangible assistance for the severely ill patient, it may provide insight into the experience of illness, or provide solace to other isolated or despairing sufferers.

tales often appears confused. For example, in the preface to *Awakenings* he states: 'what I needed to convey, was neither purely classical nor purely romantic, but seemed to move into the profound realm of allegory or myth. Even my title, *Awakenings*, had a double meaning, partly literal, partly in the mode of metaphor or myth.'[61] Here he seems to use 'allegory', 'myth' and 'metaphor' as synonyms without acknowledging their crucial differences. By equating these literary terms he creates a mood, rather than formulating a precise mode of writing. However, to his credit, he shares with theorists of romanticism such as J. Hillis Miller and Paul de Man the tendency to use allegory in a particular way. Usually allegory is understood as a 'twice-told tale' or *allegoria* which relies on the reader's knowledge of a known precedent. Without reference to the archetypal myth on which it is based, allegory loses much of its impact. As such, the world of allegory is often experienced as a 'combination of elusiveness and familiarity'.[62] Thrown into a strange fictional world the reader must interpret images and events with reference to an established mythic framework. As in Spenser's *The Faerie Queene* (1590–96) and Bunyan's *Pilgrim's Progress* (1678), the reader follows the journey of a central figure whose task is to interpret the signs of the world which he or she inhabits. The author must include enough clues to enable the reader to unravel the hidden meaning which is, to a great extent, fixed and limited. However, another interpretation suggests there are a number of possible ways of reading allegory as *diversium*. Even if the author's intention to convey a primary story is evident, there are other configurations which can be elicited from the text. This second reading moves away from the traditional conception of allegory, towards a mode of symbolism popular with romantic and post-romantic writers. Unlike fixed images, symbols connote a multitude of possible meanings. There is no one allegorical reading upon which the understanding of the text depends. Thus, although allegory and symbolism are often understood as contrasting literary tropes, they are compatible if allegory is deemed to be polysemous. Not only can this kind of symbolic allegory surpass the limits of didactic instruction, it has the potential to disrupt fixed patterns.

Contributing to the debate on the reinterpretation of allegory, the literary critic Stephen Greenblatt comments in *Allegory and Representation* (1981), that the 'deeper purpose' of allegory 'is to acknowledge the darkness, the arbitrariness, and the void that underlie, all representation of realms of light, order, and presence'.[63] Rather than uncomplicated mimetic representation, Greenblatt argues that allegory can be developed to avoid the limitations and dangers of portraying

---

[61]   Sacks, *Awakenings*, op. cit., p. xxxvii. Similarly, in 'Clinical Tales', he suggests that the drama of illness can be represented by situating the battling self within 'a sort of allegory or epic'; Sacks, 'Clinical Tales', op. cit., p. 18.

[62]   Gay Clifford, *The Transformation of Allegory* (London: Routledge & Kegan Paul, 1974), p. 2.

[63]   Stephen Greenblatt (ed.), *Allegory and Representation* (Baltimore: Johns Hopkins U.P., 1984), p. vii.

a 'stable, objective reality'.[64] This follows de Man's thesis in *Blindness and Insight* (1974) that in understanding the 'hidden' meaning the reader does not unravel the text in its entirety, for at the moment of insight other, unseen, possibilities remain latent: the act of reading suppresses some potential meanings and activates others. By highlighting the impossibility of understanding 'order and presence', allegorical understanding actually reveals the uncertainty of interpretation.

Far from misusing the term then, Sacks can be understood to deploy allegory in two complementary ways. In the broadest sense, an allegorical tale is a metaphorical and descriptive tale which discloses its own figurative status. Unlike the SOAP format which, although rudimentary, describes a relatively unproblematic social reality, an allegorical tale can highlight its own constructed nature and its indeterminacies. However, at the moment of disruption, allegory helps to preserve and connect: the disruptive potential of symbolic allegory is checked by the conserving structure of the traditional narrative allegory. As Greenblatt comments, 'allegory arises in periods of loss . . . from the painful absence of that which it claims to recover'.[65] By presenting a mythical story the allegorical mode reveals wider cultural archetypes through which the illness experience can be figured.[66] If allegory positions an individual experience within a larger cultural narrative, then such a reclamative story may serve to empower the suffering patient. Sacks does not pretend to fully represent the illness experience in his clinical tales, nor does he claim to possess messianic powers, but, as the patients actively tells their own stories, he allows the possibility that one may glimpse the seeds of hope for a recovery from, or an accommodation of, illness.

Constructing Narrative Voices

The patients documented in *Awakenings* each react to L-DOPA in unique ways, but their narratives share an underlying structure which provides an insight into the nature of encephalitic illness and Sacks' role as romantic scientist. While the cases contain an extra-narrative dimension of commentary and footnoting, Sacks consistently foregrounds the patient's story. This format is distinct from the brevity of the medical chart and the cases presented in *Migraine*, where typology is highlighted over and above the emplotment of story. However, the two books are alike in following a pattern of description, explanation and speculation which, as the critic William Howarth discerns, 'follows classic therapy (observe,

---

[64]   Ibid., p. viii.
[65]   Ibid.
[66]   Three of these archetypal experiences, the quest, the transformation and the fall (archetypes explored, although far from exclusively, by romantic writers) are discussed more thoroughly with reference to *A Leg To Stand On*.

describe, analyse, prescribe) but toward a romantic end': the 'continuous double vision' that sees migraine as both a structure and strategy, reflecting 'the absolute continuity of mind and body'.[67] By switching his focus between neurological 'structure' and emotional 'strategy', Sacks figures different illnesses as both 'physical and symbolic events'.[68] The numerous editions of *Awakenings* update the case studies and commentaries which, in the words of Howarth, are 'unfinished and continuing, very like a series of medical charts waiting for new entries'.[69] But unlike the medical chart, the symbolic and narrative dimensions of the clinical tales locate *Awakenings* outside the scope of classical scientific texts. In addition to a description of neurological details, the cases are situated within a web of interconnecting narratives, which are not only told but *re*-told as the reader is invited to shift across different planes of textual engagement. The 'mystical thread' of symbolic stories is intertwined with the primary narrative relating the long-term effects of encephalitis, the life of the patients in Mount Carmel before and after the administration of L-DOPA.[70]

The story of the patients' illness, moving through, but irreducible to, a series of distinct but continuous stages, is not a stable narrative, but constantly under revision. The shifting relationship between narrator (alternating between physician and patient) and narrated (variously, physician, patient and illness) highlights questions of authority and authorship: who is the storyteller? whose story is being told? and what is the epistemological status of (re)telling? Moving between these unstable textual spaces is the figure of Sacks. By presenting himself variously in the roles of reader, listener and narrator he exemplifies what Jean-François Lyotard has termed the 'pragmatics' of narration: 'all the complicated relations that exist . . . between the story-teller and his listener, and between the listener and the story told by the story-teller'.[71] Sacks highlights the symbolic dimension of storytelling in his belief that scientific explanation should never be totalizing: in splicing explanation with allusion, suggestion and comment, he undermines a closed explanatory structure. Narrative openness provides the retelling of patients' stories with two therapeutic dimensions: firstly, the emotional appeal of the story provides a way in which the patient may be able to connect with his or her lost past and, secondly, because any single account is

---

[67]   William Howarth, 'Oliver Sacks: The Ecology of Writing Science', *Modern Language Studies*, 20(4) (1990), 107.

[68]   Oliver Sacks, *Migraine: The Evolution of a Common Disorder* (London: Faber & Faber, 1970), p. 14.

[69]   Howarth, 'Oliver Sacks', op. cit., p. 107.

[70]   *Awakenings*, op. cit., p. 32. Developing Rank's interest in mythic patterns, the stories in *Awakenings* tend to conform to one of two kinds: stories of strangeness (Lewis Carroll's Alice and Gogol's tales) or stories of rebirth (Lazarus, Oliver Twist, Sleeping Beauty). Both kinds serve to disrupt a continuous narrative by moving through a series of transformations.

[71]   Jean-François Lyotard, 'Lessons in Paganism', in ed. Andrew Benjamin, *The Lyotard Reader* (Oxford: Blackwell, 1989), p. 125.

embedded in other stories, the narratives are always open to revision and extension. In (re)telling stories and taking 'other stories as its reference', Sacks claims it is possible to alter the meaning of the patient's experience of illness in a pragmatic direction.[72]

Arguing from a similar position, the American psychoanalyst Roy Schafer suggests this type of destabilizing narrative can actually be reconstitutive. Instead of the Freudian analyst 'reconstituting a single past' and analysis 'held to be in itself a final criterion of insight', he argues that 'the analyst goes on learning about the analytic encounter as he or she goes on developing . . . life-histories'.[73] Whereas Binswanger narrates the 'finished' stories of Lola Voss and Ellen West, both Schafer and Sacks argue that there is never a safe point outside the narrative from which to tell the story. For Schafer,

> the analyst uses the analytically defined elements of narrative incoherence to begin to *retell* the analysand's presentations and to bring the analysand into the process of retelling . . . the analyst and analysand together construct *a* history [which] will be . . . both more confident and more provisional than those they have replaced.[74]

In order to help the patient find his or her own therapeutic solution, the (re)telling of narrative should be conceived both as reclamative and open-ended. Heuristic narratives inscribe the past with meaningful patterns which may enable the patient to understand and learn to accommodate illness, while the multiple telling of unfinished tales opens up the possibility of future recovery.

Just as this type of flexible story can open up the fixity of a patient's long-term illness to the possibility of change, so, too, the physician has a stake in the revision of cases. David Flood and Rhonda Soricelli have suggested that the construction of the discursive medical study (as opposed to the concise format of the chart) is dependent on 'finding an appropriate narrative voice', which would enable the physician to shift between the positions of medical authority and collaborative researcher of illness.[75] In oscillating between the 'diagnostic utility'

---

[72]   Ibid., p. 126.

[73]   Roy Schafer, *The Analytic Attitude* (New York: Basic Books, 1983), p. 208.

[74]   Roy Schafer, *Retelling a Life: Narration and Dialogue in Psychoanalysis* (New York: Basic Books, 1992), p. 10.

[75]   David Flood and Rhonda L. Soricelli, 'Development of the Physician's Narrative Voice in the Medical Case History', *Literature and Medicine*, 11(1) (1992), 67. In *Speech and Phenomena*, Derrida insists that voice, as a metaphor for expressing the intimate relationship between speech, presence and sense, has traditionally been suggestive of self-evident or unmediated truth, in contrast to the dissemination of meanings indicative of writing. However, by deploying metaphors of speech, Sacks suggests that 'voice' is not confined to intentional utterances: voices can also be unintentional or somatic. As such, he balances the romantic belief in an expressive and authentic 'inner' voice and the post-structuralist notion that there is no one transparent route to truth, only a number of textual detours or possible stories to be told.

of documenting medical complaints and the 'personal encounter' with the patient (drawing from a 'premedical understanding of people'), the physician must tackle the difficulties of constructing a 'narrative voice'.[76] Because the physician should be flexible enough to account for the particular characteristics of illness and the needs of the patient, the voice must be adaptable but retain a sense of continuous personality. In short, rather than refining 'the author out of existence', the case study should 'concentrate on *finding* the author'.[77] Flood and Soricelli suggest that, whilst attending to the patient's voice, the physician must also be able to listen to his or her own 'narrative voice to hear what it reveals about the author's professionalized self'.[78] In order to consider theses issues and the wider framework of Sacks' practice, instead of considering the whole of *Awakenings*, of the twenty patients featured, this section analyses just two, Frances D. and Leonard L., which are representative of his clinical tales, if only because they contain many of Sacks' ideas and revisions that are germane to the other entries.

Frances D.

The narrative of Frances D. is the first and one of the longest presented in the first edition of *Awakenings* (1973). Typically, it is split into five sections. The story moves from an initial medical commentary of Frances' experiences, to her admission at Mount Carmel hospital, her initial reactions to L-DOPA and her later attempts to accommodate her condition. However, the closure of the narrative is deferred, not least because Sacks adds an epilogue in 1982, in which he summarizes the changes occurring in the following decade and her eventual death. But her death is situated outside the frame of *this* story (although it always looms close by) as a future draft or another text embedded in the larger narrative of *Awakenings*. As such, the individual stories are not told in isolation; they are inextricably intertwined with the narratives of the other patients. This larger story frames *'fifty* "awakenings" . . . fifty individuals emerging from the decades-long isolation their illness had imposed on them'.[79] Here, Sacks stresses the importance of the community, or 'body', of sufferers, in whom the contagious laughter of 'camaraderie' and 'conviviality' serve to replace the muted voices of illness.[80] There is a curious tension between the isolation of the ageing and enfeebled bodies and the renewed vitality of a hitherto arrested youthfulness: on one level, hinting to an allegorical passage to a less corporeal form of existence and, on another, conveying the possibilities of recovery among the distressing realities of bodily decrepitude. Although Sacks insists that the primary trajectory

---

[76]   Ibid., p. 71; p. 67.
[77]   Ibid., p. 80.
[78]   Ibid.
[79]   Sacks, *Awakenings*, op. cit., p. 65.
[80]   Sacks goes on to explore the notion of a community of sufferers in *Island of the Colour-blind* (London: Picador, 1996).

of *Awakenings* conveys the personalities of the patients, the stories read as discontinuous biographies in which the narrator pauses to address questions of illness and suggest alternative modes of therapy. However, the patient's story is rarely overshadowed by excursus: in Frances' case, L-DOPA provides a chemical window which enables her to speak of her own illness experience.

While most of the first section is presented in a factual manner by a third-person narrator, the narrative fuses description of the physiological symptoms with philosophical and symbolic subtexts. The opening sentence 'Miss D. was born in New York in 1904, the youngest and brightest of four children' introduces an allegorical strain which runs throughout the narrative.[81] As Bruno Bettelheim indicates in *The Uses of Enchantment* (1976) the hero of the *Märchen* is often the youngest of the family (usually the youngest of three) and tends to be the most spirited and the quick-witted of the siblings. Here, the mythic resonance carries the impact of the tragic illness which strips the youth from this fifteen-year-old girl. This 'cut' severs the orderliness of her life and arrests the promise of her childhood, but, at this stage, does not represent a complete severing from the balance of childhood; she is able to regain her 'previously well-integrated personality and harmonious family life'.[82]

The hyper-kinetic form of *encephalitis lethargica* from which Frances suffers is marked by 'insomnia', 'restlessness' and 'impulsiveness', but within a year it subsides sufficiently for her to reintegrate with a healthy body and a relatively unimpaired school life.[83] Sacks stresses the richness and fullness of her life, particularly her creative talents ('she was fond of theatre, an avid reader, a collector of old china') and her interaction with others: she was 'an active committee-woman' and 'frequently entertained friends'.[84] Although residues of her illness, including panting attacks and oculogyric crises, remain with her until 1949, Frances' story shifts from the vitality of childhood to debilitation and back to a healthy social and emotional life.[85] However, health is more than just a balanced body characterized by homeostasis, it is also a performative background ability which enables her to execute coordinated actions.[86]

A marked change in Frances' condition occurs in the 1950s when she begins to display the contrary tendencies to freeze and hurry in her speech and movements. At this point Frances speaks in her own voice to indicate the

[81]   Sacks, *Awakenings*, op. cit., p. 39.

[82]   Ibid.

[83]   Ibid.

[84]   Ibid., p. 40.

[85]   Oculogyric crises are 'attacks of forced deviation of gaze, often associated with a surge of Parkinsonism, catatonia, tics, obsessiveness, suggestibility . . .'; ibid., p. 392.

[86]   The history of homeostasis and bodily equilibrium encompasses pre-modern, conceptions of health, including Heraclitus' correspondence theory of harmony in the macro- and microcosms (a version of which is to be found in Emerson), the medieval humours (*umor*), and traditional Chinese and Indian notions of balance.

'*essential* symptom' of her experience, in contrast to the 'banal symptoms' Sacks detects ('essential' suggesting the most profound, or intensely felt, of the experiences).[87] By privileging Frances' voice, Sacks indicates the importance of the patient's own description of illness. Even though an observer is able to detect the starting and stopping of her bodily movements, the loss of 'in-between states' felt by Frances can only be spoken in the first-person: she claims to experience these 'urges' passively, as if she has lost regulative control over her previously balanced body.[88] Her knowledge that the attacks usually occur at regular intervals ('like clockwork') enables her mentally to prepare for the routine, even though she is unable to do much about the crises.[89]

Frances' body cannot be explained merely as being out of control, or asserting a will of its own, as she simultaneously displays both a withdrawal into herself (a fixed emotion and expression) and an exaggeration of activity (deviation of gaze and flaying of limbs). There appear to be two competing conceptions of personal identity at work here: one founded on the Cartesian split between body and mind and the other which views the Parkinsonism as affecting the whole and indivisible self. While it may seem naive to conceive of the 'real' Frances as being the sentient prisoner of a wildly malfunctioning body, the images of 'Rip van Winkles or Sleeping Beauties' are potent symbols for conveying her potential recovery from torpor.[90] In contrast, supporters of the second view, such as Gerald Edelman, would argue that, although it may *appear* that the motor part of the brain has lost its regulative capacity, while the cognitive part (or higher-cons-ciousness) reflects on her condition, the two continue to interact in a complex way. The tension between these two rival theories of the self is evident when Sacks implies the opposing Parkinsonian forces of pushing and pulling compete for possession of her whole person; as her staring attacks become more dominant it is useful to describe the 'real' self as locked inside the body. Such crises are figured as anti-epiphanies, where corporeal sensation suppresses a freezing of the expressive voice, resulting in a purely internal moment without transcendence.[91]

By the time Frances is admitted to Mount Carmel in 1969, her condition had worsened to the extent that she is unable to walk or hold her body upright. Her bodily pulsion decreases in inverse proportion to her tendency to freeze; only

---

87   Ibid., p. 40.

88   Ibid., p. 42. Of course, the accuracy of the portrayals has, to a large extent, to be taken on trust. The narratives can be read as purely fictional stories, but the evidence of the 'Awakenings' documentary (1974) and the textual details (both biographical and neurological) substantiate Sacks' claims for realism and accuracy.

89   Ibid., p. 40.

90   Ibid., p. 65.

91   In *Poetics of Epiphany*, Ashton Nichols states the recurrent characteristics of the literary epiphany are its brevity and its personal significance. In this instance, both of these aspects are reversed: Frances' anti-epiphany is marked by temporal extension and repetition.

when external stimuli distract her fixed attention can she continue her movements. Her body continues to show signs of movement and rapid action, but prohibits her from acting in any meaningful sense: the 'flapping' of her right hand, the 'grinding' of her teeth and the 'puckering' of her mouth display a blocked channel of energy. Sacks indicates that Frances' 'restlessness' can only be quelled by occupying her hands, but the most marked changes appear in the loss of her 'personal' qualities, characterized by a masked face and a monotonous voice, varying only in the speed of utterance.[92] Her handwriting also displays these opposing tendencies of pulsion and festination, losing its distinctive quality (normally 'large, effortless and rapid') and resulting in chaotic and illegible scrawl or gradually becoming 'smaller and slower and stickier until it became a motionless point'.[93]

In order to balance these distressing signs of bodily deterioration, Sacks focuses briefly on the maintenance of Frances' personality. She remains alert in observation; she retains a quickness of thought; she is 'punctual, and methodical in all her activities' without being obsessional or phobic; and she is able to recognize similar maladies amongst the other patients in her ward. Frances appears able to retain the qualities of selfhood and to maintain what Sacks perceives as 'a healthy self-respect' for her (virtually) uncontrollable body and the demoralizing effect of institutionalization.[94] Throughout all the narratives Sacks is at pains to indicate the limited methods which patients adopt in order to remain 'alive and human in a Total Institution'.[95] For Frances, it seems that only through such means can she compensate for a loss of bodily mastery and motor control by asserting herself as a meaningful actor in the face of adversity.

Sacks structures the central section of the narrative, 'Course on L-DOPA', in the form of successive journal entries, focusing on the body as a site of display and recording cognitive changes with the aid of Frances' comments. He closely documents the dates and precise quantities of L-DOPA administered, in order to

---

[92]   Ibid., p. 42.

[93]   Ibid., p. 43.

[94]   Ibid., p. 44. Howard Brody suggests that the patient's self-respect is a crucial sign that he or she still possesses the desire to live and is willing to take measures to combat the illness. Similarly, Michael Ignatieff focuses on the significance of personal items ('a comb, a mirror and make-up') which, however inadequately, enable the patient to tend to their bodies: 'the self . . . manoeuvres if it can towards some tiny affirming gesture'; Michael Ignatieff, 'Life at Degree Zero, *New Society* (20 January 1983), 96, 97.

[95]   Sacks, *Awakenings*, op. cit., p. 62. Sacks appropriates the term 'Total Institution' from Erving Goffman's influential study on *Asylums* (1961). In many ways Sacks echoes Goffman's stress on the negative experiences of institutionalized life, but, as a practising physician, he needs to retain a positive and life-affirming outlook both for himself and his patients. Goffman claims that meaningful activity is too dignified a term to describe, what he calls, 'removal activities': 'voluntary unserious pursuits' which are encouraged by the hospital staff in an attempt to 'lift the participant out of himself'; Erving Goffman, *Asylums* (London: Penguin, 1970), p. 67. He argues, that instead of granting the patient any sense of agency, they merely serve to 'kill' time; ibid.

chart the 'complex mixture of desirable and adverse effects' of the drug.[96] The initial dosage of L-DOPA serves to increase Frances' 'drive' to act in any manner that she found possible – crochet, wash, write – because she is 'unable to tolerate inactivity'.[97] An increased dosage appears to stimulate a feeling of 'well-being and abounding energy' in her, together with the ability to channel the energy into the shape of words or pedal movement.[98] Adversely, her ability to breathe is affected even though her urge to respire becomes more intense; an urge which, after modification of dosage, manifests itself in frequent respiratory crises.[99]

Such is the violence and persistence of Frances' attacks that Sacks expresses an unwillingness to continue with the drug, but, on her insistence, he continues with half the dosage. However, although she 'was looking, feeling and moving far better than she had done in twenty years', the crises continue and Sacks is led to a 'therapeutic dilemma': he ceases his hitherto primary role as recorder of physiological detail and now admits his predilections in the narrative.[100] Consequently, like Erikson's 'confrontation' with Gandhi, he shifts his stance from third-person narrator to a central figure in the story. He shares Frances' desire to rediscover the lost equilibrium of health, by finding an amenable drug dosage. At this stage he speaks in the plural ('could we find a happy medium'), but when describing the 'emotional needs and contexts' of Frances' illness, he reverts to the autobiographical first-person: 'I was slow to realize, while noting the causes of Miss D.'s crises, that the most potent "trigger" of all was me, myself . . . When I asked Miss D. if this could be the case, she indignantly denied the very possibility, but blushed an affirmative crimson.'[101] The introduction of the first-person voice can be seen from two contrasting points of view. Firstly, by positioning his persona in the story Sacks adds to its sense of historical authenticity and conveys a sense of intimacy with his patient. Currents of transference and counter-transference indicate that the relationship between patient and physician is an analytic one: the use of drugs to regulate the illness is seen as an addition to, rather than a replacement of, a personal encounter. However, a second viewpoint, expressed by Ella Kusnetz in her essay 'The Soul of Oliver Sacks' (1992), treats such comments with suspicion. Sacks begins to recognize signs of Frances' psychic transference, 'only when an observant nurse giggled and remarked to me, "Dr Sacks, *you* are the object of Miss D.'s crises!",

---

[96]   Sacks, *Awakenings*, op. cit., p. 45.

[97]   Ibid.

[98]   Ibid.

[99]   In a long footnote, Sacks comments that these attacks display a mixture of the 'idiosyncratic' and the communal: they represent a 'remarkable 'fossil' symptom' left over from the onset of Frances' encephalitis in 1919, but also share a characteristic in common with the other Parkinsonian patients; ibid., p. 46.

[100] Ibid., p. 48.

[101] Ibid., p. 49.

that I belatedly tumbled to the truth'.[102] The tone here implies (unlike his more open contact with another patient, Rose R.) that the encounter is one of sublimated desire, revealed through this 'belated' disclosure and the repetition and assertion of 'me, myself'. Moreover, as Kusnetz indicates, when 'the therapy goes wrong' the personal pronoun is replaced by 'we', implying the culpability of the whole medical team.[103] Kusnetz claims that 'he is eager to divest himself of this identity', a criticism substantiated later in the narrative when Sacks records the adverse reactions to L-DOPA: 'it produced more and more "side-effects" which I – *we, her doctors* – seemed powerless to prevent'.[104]

Although the emotional dimension is an important determining factor, the respiratory crises have many causes: organic, psychic and social. Another important 'psychic cause' which Sacks claims he 'could not have known of had Miss D. not mentioned it', is revealed when Frances comments: 'as soon as *I think of getting a crisis . . .* I am apt to get one. And if I try to think about not thinking about my crises, I get one. Do you suppose they are becoming an obsession?'[105] Here, the idea that the body entirely extricates itself from the motor control of the brain is again challenged, as Frances' emotions of 'passion, intransigence, obstinacy and obsession' influence her respiratory attacks.[106] The anxiety accompanying the onset of each attack is described as 'a special, strange sort of fear', in clinical terms diagnosed as pathology and in existential terms a subliminal dread. The drug-as-regulator has been replaced by the drug-as-lifeline and the discontinuation of the drug would be to her 'like a death penalty'.[107] However, both patient and physician note, when she is transfixed and 'rooted to the spot' her body appears to be autonomous and to have a will of its own.[108]

---

[102] Ibid.

[103] Kusnetz, 'The Soul of Oliver Sacks', op. cit., pp. 181–2.

[104] Sacks, *Awakenings*, op. cit., p. 54.

[105] Ibid., p. 49.

[106] Ibid., p. 50. There is a psychoanalytic subtext which Sacks goes some way to avoiding in his emphasis upon the neurological, rather than the psychological, causes of the illness. The references to obsessive somatic compulsions are similar to those discussed by Freud and Breuer in *Studies on Hysteria*, in which the (female) patient is often deemed to be at fault in contracting the illness. Although Sacks shares Freud's perspectives on many issues (especially the existence of psychic barriers) there are clear theoretical differences. Importantly, Sacks (like Binswanger and Schafer) argues that there is no one event, or occurrence (for Freud having libidinal origins) which can be cathartically brought to consciousness during analysis.

[107] Ibid.

[108] This rooting to the spot maybe an unwillingness, as well as an inability, to move from a zone of safety; faced with a traumatic situation, the need to find roots often proves stronger than the need for movement. Luce Irigaray has argued 'hysteria', in this case marked by somatic symptoms and immobility, may be a manner in which to avoid the hegemony of patriarchal speech (here manifested by the controlling voice of the physician): 'a privileged place for preserving – but 'in latency', 'in sufferance' – that which does not speak'; Irigaray, 'Questions', *The Irigaray Reader*, op. cit., p. 138.

Sacks infers that beneath, or bypassing, the level of coordinated neurological control is 'a mass of strange and almost sub-human compulsions'.[109] Here, 'sub-human' has two levels of implied meaning: affirming that humans are fundamentally biological animals (Frances claims: 'I bite and chew like a ravenous animal, and there's nothing I can do about it') and indicating that such dark compulsions cannot always be understood through the rational grammar of thought. The only way in which the physician can glean a sense of these reflexive processes is through close participation with his patient and by encouraging her self-expression. Again, Sacks stresses the mythic elements of the narrative: both patient and physician are figured as ill-equipped travellers in the strange world of neurological disorder. However, unlike classical allegories, the signs do not suggest a single level of meaning; instead, many layers of possible meaning intersect and often contradict each other.

This symbolic pattern is evident in the description of the conflicting symptoms warring on the site of Frances' body. Similarly, the stark contrast of Frances' 'good day' (in which she spends 'a perfect day – so peaceful' in 'the country') is thrown into relief by the following one which 'saw the onset of the worst and most protracted crisis of Miss D.'s entire life', convincing Sacks to discontinue his administration of L-DOPA.[110] The violence of this sixty-hour crisis can be sensed in the verbs 'jam' and 'catapult', which replace the less virulent verbs 'freeze' and 'urge', as she moves from an increase in bodily distraction and palilalic speech to wild movement of the arms and 'a shrill piercing scream'.[111] Waves of excitement accompany a denial of selfhood ('It's not me, not me, not me at all'), as if the only way in which to contend with the anguish is for Frances to extricate her life-world from the outward contortions of her body.[112] Sacks conveys this kind of disjunction with the broken prose of his quotation and the anecdote in which, after the crisis, Frances wryly christens the drug 'Hell-DOPA!' Her extreme states are couched in the symbolic opposition between the heaven of the pastoral day and the hellish distress of these climactic moments of self-negation. The exteriority of the countryside (a recurrent romantic motif for Sacks) is figured outside the tyranny of the body and the hospital, and is presented in stark contrast to the interiority of this crisis: a turning inwards almost to the point of self-annihilation. Sacks notes that for Frances each crisis is inexorably linked to the quality of her voice and an emotional or spiritual need: 'her voice, normally low-pitched and soft, rose to a shrill and piercing scream'.[113] The movement upwards and outwards ('each wave rising higher and higher') suggests a macabre anti-epiphany where the physicality of the body prevents any transcendental experience. In a discussion on asthma, Marsha Rowe has pointed

---

109   Sacks, *Awakenings*, op. cit., p. 51.
110   Ibid.
111   Ibid., p. 52.
112   Ibid.
113   Ibid., pp. 51–2.

out that the verb 'breathe' and the noun 'breath' have two etymologies: the first deriving from the Latin *animalus* – an animal, or being, that breathes – and the second from *spiritus*.[114] In both cases the connection between breath, breathing and soul is evident. Thus, the most severe of Frances' respiratory crises is cast among the most profound of symbolic proportions: her struggle to cling on to a physical and spiritual life seems wholly dependent on her ability to breathe.

The next two sections focus on the changes when Frances stops taking the drug and are the most reflective of the five. One reason for this is that Frances enters a more stable 'subterranean state', which the speech-therapist, Miss Kohl (a source of information whilst Sacks is absent from the hospital), compares to the withdrawn fugue state of shellshock. It is as if her 'appetite for living' disappears for two important reasons: firstly, the drug, which seems to offer life, is now viewed as a false god and, secondly, in Sacks' absence, the other sources of life, the relationships with the physician and fellow sufferers, 'had been disbanded' due to administrative changes.[115] Sacks feels that only if he can re-establish these relationships will Frances recover her motivation to live.

The commentary accompanying this section is not reserved solely for the physician's thoughts; Sacks stresses that he wishes to 'interrupt her "story" for her analysis of the situation': the extremes of illness, the rise and fall of bodily states, the emotions of fear, rage and betrayal are left for Frances to recount.[116] Here, Sacks acts as scribe for the patient and his theoretical comments are reserved mainly for the footnotes (which, perhaps to his discredit, sometimes spill into the body of the text). One significant example of this occurs when he stresses that Frances' body and mind cannot be split:

> the 'things' which gripped her under the influence of L-DOPA . . . could not be dismissed by her as 'purely physical' or completely 'alien' to her 'real self', but, on the contrary, were felt to be in some sense *releases* or *exposures* or *disclosures* or *confessions* of very deep and ancient parts of herself, monstrous creatures from her unconscious and from unimaginable physiological depths below the unconscious . . . and she could not look upon these suddenly exposed parts of herself with detachment; they called to her with siren voices, they enticed her, they thrilled her, . . . they possessed her with the consuming, ravishing power of nightmare.[117]

In passages such as this Sacks loses the precision of his earlier documentation in wildly suggestive rhetoric, where ideas are stressed at the expense of losing the presence of the patient's voice. The passage is beautifully poised between the registers of psychoanalysis and myth, but, as Sacks says of certain of his patients,

---

[114] In a discussion chaired by Roy Porter at the Institute of Contemporary Arts (7 December 1993).

[115] Sacks, *Awakenings*, op. cit., pp. 52–3.

[116] Ibid., p. 54.

[117] Ibid., p. 55.

the language seems over-ebullient, leaving an investigation of the patient's condition behind for a moment of poetic excess. However, such a passage does convey the almost gothic drama of visual scenes (such as Frances' grotesque crisis) in a manner lacking in more clinical prose.[118]

Sacks, as physician, moves from a dispassionate recorder of detail to a central actor, intricately bound up in this emotional drama. Given the patient's dependency on the physician for her physical and emotional needs, such a meeting is inevitably an analytic one, involving transference and resistance. The physician is now seen as the sole arbiter of life and death. For Frances, the physician-cum-alchemist, caught between the images of Redeemer and Devil, or good and wicked doctors, experiments with a drug which promises salvation only to offer her living death. But, here, unlike the coy amorosity of the previous passages, Sacks is able to discuss calmly the transference of Frances' feelings: 'thus Miss D. found herself entangled in the labyrinth of a torturing transference-neurosis, a labyrinth from which there seemed no exit'.[119] In order to bear the mantle of an analyst, as well as physician, he realizes that Frances must work through her anxieties with him. However, the presence of her severe bodily illness complicates matters: the sense of bodily fragmentation is likely to lead to the need to secure relationships (in this case with the physician and with other patients) which the third stage of psychic fragmentation (when the patient must understand the analyst is not an ego-ideal) tends to undermine. In this case, the sense of double fragmentation may be so painful that a resolution is never adequately established.

The image of the closed labyrinth conjures up this sense of blockage, but it is one that Sacks, because he is not a professional analyst, seems unable to resolve. In referring to Freud's paper 'Analysis Terminable and Interminable' (1937), Ella Kusnetz points out, 'if, as Freud discovered . . . the sticky matter of psychic resistance makes a simple transference cure an inadequate model for the physician's treatment of neurotic illness, it is even more inadequate for the physician's management of such intractable medical illness'.[120] Kusnetz argues that unless Sacks is to burden himself with the permanent mantle of analyst, a position which is not amenable to the variety of roles he seeks to adopt, he needs to abandon the 'old transferential figure of "magic" and "authority"', and also his reluctant role as administrator of drugs, and relate to his patients as '"co-equals", "co-explorers" and co-actors in the drama of their illness'.[121] Although many of

---

[118] The visual impact of the Parkinsonian body is, perhaps, inevitably lost in a written account. Consequently, Sacks commends the documentary film as a companion to the text, in which the patients' bodily contortions are visually presented.

[119] Ibid., p. 57.

[120] Kusnetz, 'The Soul of Oliver Sacks', op. cit., p. 181.

[121] Ibid., p. 185. It is noticeable that increasingly in *Awakenings* and, especially, *An Anthropologist on Mars*, Sacks stresses his role as fellow traveller and befriender (using the language of the social-worker) rather than as analyst and problem-solver.

Kusnetz's comments are well targeted, she fails to point out that in this instance Sacks does trace the transference-neurosis to its resolution: over a year later Frances manages to find a 'release from the labyrinth', to turn '*away* from her phantasies, and *towards* her reality'.[122] But, his later description of the way in which the 'impostures of the L-DOPA "situation" fell away like a carapace revealing . . . the real self . . . underneath' seems too neatly resolved on both textual and therapeutic levels.[123]

In the last section, 'Summer 1972', Sacks continues to relate Frances' story, but now from a position of temporal and theoretical distance. His personal involvement in the story recedes to a point of professional concern as he reviews the methods of therapy employed over the three subsequent years. This section updates the story to 1973 and represents a tribute to Frances' courage and 'mysterious reserves of health and sanity'.[124] Sacks believes that the extreme tensions of 1969 had served to 'forge' and 'temper' Frances, rather than break her, enabling her to reconstitute a sense of selfhood. He administers another low dosage of L-DOPA which causes a 'milder' reaction of 'improvements-exacerbation-withdrawal symptoms'.[125] In contrast to the uncontrollable extravagance of the earlier respiratory crises, an awareness of this cycle of responses enables Frances to contend with the approaching stages.

The narrative ends with commentary on other therapeutic techniques deployed in order to help Frances regain a sense of autonomy and selfhood. Of the techniques, the most personal and intimate (laughter, touch and music) contain the greatest capacity to 'awaken' Frances, to 'defreeze' her from the constrained monotony of her torpor and to provide a means through which her attention can be channelled away from her physical condition.[126] For example, Sacks finds, in order to rouse Frances, it is necessary that music appeals to her individual sensitivity. Resembling Binswanger's recognition of the particularlity of a patient's world-design, Sacks realizes that she required music to be 'firm but "shapely"' and '*legato*', because 'raw or overpowering rhythm . . . causes pathological jerking; it coerces instead of freeing the patient, and thus has an anti-musical effect. Shapeless crooning . . . without sufficient rhythmic/motor power, fails to move her – either emotionally or motorically – at all.'[127] Music does not appeal primarily to cognition, but on different levels of reception: physically, to the rhythm of the body; chemically, to 'the "go" parts of the brain'; and, metaphysically, serving to 'move' her soul.[128] Sacks suggests that, by

[122] Ibid., p. 62.

[123] Ibid..

[124] Sacks, *Awakenings*, op. cit., p. 58.

[125] Ibid., p. 59.

[126] Ibid., p. 63. Sacks' conception of an 'expressive' music is thoroughly romantic: the patient's life-world is not revealed through speech, but through the rhythm, or voice, of the body.

[127] Ibid., p. 61.

[128] Ibid.

synchronizing bodily movement with a particular tempo and rhythm, the Parkinsonian patient is able to discover an ability to channel the displays of excessive energy, to coordinate the movements of limbs and so to give a shape and structure to the violent tics. Accordingly, he comments that her bodily movement 'was simultaneously emotional and motoric, and essentially autonomous'.[129] This mode of therapy affirms Frances' individuality and sense of agency and also gestures towards an elusive return to health.

Leonard L.

Forming the backbone of the plot for Penny Marshall's film, *Awakenings* (1990), the narrative of Leonard L. powerfully portrays the creative dimensions of recuperation in contrast to the mental poverty of catatonia. However, as Sacks comments in an appendix to the 1991 edition of *Awakenings*, the film is both 'very different in structure, concentrating on a single patient' and adds dramatic details, such as the romantic involvement with Paula, the 'violent psychiatric ward' and the stereotyped responses of 'repressive' doctors.[130] Consequently, it is important to distinguish between Sacks' neurological case and Robert de Niro's portrayal of Leonard in the film. Having asserted this, Sacks expresses that the film and his text both share, what he calls, the 'emotional truth of the portrayals': that is, 'the imagination and depiction of the inner lives of the characters'.[131] Sacks claims that Steve Zaillian's screenplay shares his attempt to blend emotional sincerity with neurological accuracy (de Niro convincingly mimics Parkinsonian symptoms) and centrally focuses on the 'human' dimension of the narrative.[132] Unlike Frances' narrative, Leonard's story does not begin with a biographical sketch: he is introduced as a 'mummified' patient.[133] Rather than the physician composing a resumé of his life, Leonard is left to describe the significant details and Sacks is only 'able to form any adequate picture of his state of mind and being' over a number of years of contact with him and his mother.[134] Although Leonard is described as 'a man of most unusual intelligence . . . with an introspective and investigative passion', by the time Sacks meets Leonard in 1966, his only means of communication is a letter-board on which he painfully taps out phrases in 'an abbreviated, telegraphic, and sometimes cryptic form'.[135]

---

[129]   Ibid., p. 62.

[130]   Ibid., p. 374.

[131]   Ibid.

[132]   A conflicting view is to be found in an article by Paul Taylor in which he questions Sacks' 'criterion' of 'emotional truth' and he criticizes de Niro's acting as 'both technically dazzling and morally quite unacceptable'; Paul Taylor, 'A Study in the Right Frame of Mind', *The Independent* (28 April 1994), II, p. 23.

[133]   Sacks, *Awakenings*, op. cit., p. 204

[134]   Ibid, p. 206. The film sensitively dramatizes the important and mutually dependent relationship between Leonard and his mother.

[135]   Ibid., pp. 204, 207.

The role of Sacks as reader of the codes of illness is established at this early stage. Leonard's unique characteristics of introspection and enforced introversion stimulate Sacks to comment that 'this combination of the profoundest disease with the acutest investigative intelligence made Mr L. an "ideal" patient' who teaches him a great deal about Parkinsonian illness.[136] In contrast to Sacks' semi-amorous relationship with Frances (diffused in the film through the love interest with the nurse, Eleanor Costello), Sacks displays a mixture of identification with, and compassion for, Leonard which establishes a bond of companionship. The relationship is based upon mutual need: Leonard needs someone to interpret the signs of his illness, whereas, if he is to fully understand Parkinsonism, Sacks requires someone to articulate the illness experience. Although Sacks retains the privileged role of the framing author, more than anywhere else in *Awakenings*, the reader has the sense that this narrative is jointly told. Indeed, when Sacks admits that, although he feels that Leonard 'deserves a book to himself', he has chosen to 'confine' himself to 'the barest and most inadequate outline of his state', he admits that this study is inadequate for fully representing Leonard's illness experience.[137] Sacks acknowledges Leonard's role as co-writer, but this claim is somewhat weakened by the fact that traces of Leonard's written autobiography are only marginally included in the text.[138]

Leonard is described as having rigid and dystrophic hands, facial masking and a toneless voice, symptoms which imply that the neurological damage is mainly in the right hemisphere and lower parts of the brain. Although Leonard describes feelings of mosaic vision similar to those described in *Migraine*, Sacks infers from the signs of 'rapid and sure' eye movements that his cognitive left hemisphere is virtually unimpaired, hinting towards an 'alert and attentive intelligence imprisoned within his motionless body'.[139] Here, the tension between the two conceptions of illness (either affecting the whole self or imprisoning the reflective self inside the body) is developed. The evocative image of the caged animal, taken from Rainer Maria Rilke's poem 'The Panther' (*c*.1903), serves suggestively to convey Leonard's feeling of entrapment in a bodily prison 'with windows but no doors' and what he calls the simultaneous existence of an 'awful presence . . . and an awful absence'.[140] Leonard contrasts the presence of a 'nagging and pushing and pressure' with a feeling of being 'constrained and stopped'. The alternative to this, 'a terrible isolation and coldness and shrinking',

---

[136] Ibid., p. 204.

[137] Ibid.

[138] One example of the argument that Sacks has constructed the character of Leonard as a vehicle solely for probing the autobiographical nature of illness can be discerned in the marked similarity between Leonard's love of the post-romantic/modernist poetry of Rilke and T.S. Eliot and Sacks' own mystical quotations from Eliot and Herman Hesse later in the text; ibid, p. 275.

[139] Ibid., p. 204.

[140] Ibid., pp. 207, 205.

is described as being less torturous, but is figured as a consequence of 'castration': a dumbness which is 'something like death'.[141] This severance echoes the cut which slices across Frances' life, but is couched in terms of male sexuality.

Sacks quotes the first stanza of 'The Panther', followed by a prose translation (he does not cite the translator), while the film presents the whole poem as a central metaphor of the post-encephalitic experience.[142] The withdrawal of the panther's 'gaze' into the body ('a thousand bars') is seen as a retreat from the monotony and the confines of illness and institutionalisation ('no more world').[143] The motivating force of 'some mighty will' lies paralysed at a centre around which the panther's body erratically moves. The 'supply powerful paddings' convey the sense of a padded and locked cell and of a constrained, prowling energy which must find an outlet any way it can. The third stanza offers some relief from the monotony: 'just now and then the pupils' noiseless shutter/is lifted'. Despite the utter interiority of the scene, an engagement with the outside world is retained ('an image will indart'), implying that vision and cognition remain functional. In the last two lines of this final stanza the poem recited in the film differs radically from the standard translation which suggests that this 'image' finds its way to the intact 'heart' of the body (the centres of will and emotion) and there 'end its being'. In contrast, the film version traces the 'shape' of the image 'through the silence of the shoulders', as it 'reaches the heart, and dies'. In this version, the 'heart' has been so long affected by suffering and torment that it can no longer cherish this image of release, as it 'dies' within bodily interiority. The literary epiphany described by Ashton Nichols, 'in which an object . . . or experience reveals itself' and coalesces 'into a sudden disclosure of meaning', is undermined in this translation.[144] As Frances found, there is no epiphanic uplifting as personal significance of the moment is consumed by the disease. This rendering (presumably one endorsed by Sacks, who was consulted in the film's making) concentrates on the rigorous effects of the disease on the whole self (body, mind, spirit and perception), rather than the retention of a stable and vital inner core.

Despite feelings of entrapment and alienation, Leonard elsewhere finds the capacity to acknowledge a kind of grotesque charm in his 'disease and deformity'. He says that 'they are beautiful in a way like a dwarf or a toad'.[145] Sacks sees this as a kind of 'Rabelaisian relish of the world', which, in the

---

[141]   Ibid., p. 205.

[142]   Notably, in the film the poem is read as the Sacks figure sits alone in his place of pastoral meditation, hinting that it is as much his touchstone as a manner in which Leonard understand his illness.

[143]   All references to Rainer Maria Rilke, 'The Panther', *Selected Poems*, trans. J.B. Leishman (London: Penguin, 1964), p. 33.

[144]   Nichols, *The Poetics of Epiphany*, op. cit., p. 10.

[145]   Ibid., pp. 207–8.

language of Mikhail Bakhtin, can be interpreted as a celebration of the festive and carnivalesque disorder of the body (in this instance, the individual body, but, also, the collective body represented by the community of sufferers).[146] Peter Stallybrass and Allon White argue that, rather than the controlling and ordering narrative of classical psychoanalysis, this celebration of the 'grotesque' body signifies a liberation of the 'second narrative fragmented and marginalized', a narrative which 'witnesses a complex interconnection between hints and scraps of parodic festive form and the body of the hysteric'.[147] Here, Leonard finds the voice to speak of his condition in terms other than those of distress and decrepitude, finding 'delight' in release 'from entombment' and intoxicated 'with the sense and beauty of everything around him'.[148] Leonard's desire to visit the hospital garden, where he not only relishes his release from his prison-house, but reveals a pantheistic impulse to touch and kiss the flowers, reminds Sacks of Coleridge's Mariner who finds release from his life-in-death by embracing the sea creatures. Leonard figures this liberation in terms of love, rapture and a reading of Dante's *Paradiso* (1321), but Sacks detects signs of plenitude and fullness (soon to become an over-abundance), which 'started to assume an extravagant, maniacal and grandiose form'.[149]

Frances and Leonard react to the drug in similar but clearly distinct ways, with their own personalities reflected in their symptoms. Frances describes her experiences as follows: 'I'd done a vertical take-off . . . I'd gone higher and higher on L-DOPA – to an impossible height. I felt I was on a pinnacle a million miles high . . . And then . . . I *crashed* . . . I was buried a million miles deep in the ground.'[150] In contrast, Leonard experiences a 'pathological driving and fragmentation' which leads, as he adopts the mantle of spokesman for the other patients, to delusions of messianic grandeur: a releasing of 'a Dionysiac god packed with virility and power', as the entrapped self is converted to a 'wanton' outpouring, which spills over into a promiscuous sexual drive of 'libidinous and aggressive feelings'.[151] His desire to kiss the garden flowers soon turns into a need to kiss and harass the nurses and masturbate 'fiercely, freely, and with little concealment'.[152]

In contrast to the balance and 'suppleness' of Frances' childhood, Leonard's psychic and physical energy at this stage lacks any kind of structure. Sacks records that 'all his reactions had become all-or-none': the ' "middle-ground" of

---

[146] Ibid., p. 207.

[147] Peter Stallybrass and Allon White, 'Bourgeois Hysteria and the Carnivalesque', in Simon During, ed., *The Cultural Studies Reader* (London: Routledge, 1993), p. 286.

[148] Sacks, *Awakenings*, op. cit., p. 208

[149] Ibid., p. 209.

[150] Ibid., p. 54.

[151] Ibid., pp. 210, 211.

[152] Ibid., p. 212.

health, temper, harmony, moderation' dissolves and Leonard becomes 'completely "decomposed" into pathological immoderations of every sort'.[153] Accompanying his awakening is a hyper-awareness of the prosaic reality of the strictures and routines of hospitalized life and the increase of 'general' and 'specific excitements' – 'urge, push, repetition, compulsion, suggestion and perseveration'.[154] Only his painstaking autobiography (typed with 'his shrunken, dystrophic index-fingers') seems to give a coherence and pattern to his excessive bodily and psychic energy. Sacks comments that during his obsessive task of writing, Leonard 'found himself free . . . from the pressures which were driving and shivering his being', the form(lessness) of 'tics and distractions' which return when he finishes typing.[155] Just as music grants Frances a moment of coordination, so writing presents Leonard with a creative channel for expressing his 'floods of sexual phantasy, jokes, [and] pseudo-reminiscences'.[156] Unlike the sedate movements of Frances, these outlets often take the form of 'carnivorous and cannibalistic fantasies', indicating that Leonard's display of autonomy does not match Sacks' romantic outlook. Indeed, Leonard's 'frenzied' movements are so violent and his 'voluptuous and demoniac visions' so vivid (in which, like Ellen West, he schizophrenically creates *and* denies reality), that Sacks is forced to resort to using 'physiological safeguards' and a 'punishment cell', precautions which he admits 'in themselves were highly distressing or disabling'.[157]

At this stage, Leonard's reactions appear entirely pathological and, at the request of his patient, Sacks relates his decision to discontinue the drug. Afterwards Leonard returns to a 'cool' state of 'composure' and an '"elegaic" detachment of mind', in which he appears to accept and accommodate the 'violent feeling of promise and regret'.[158] Like Frances, he overcomes his violent psychic reactions to the drug (figured both as an 'Elixir of Life' and 'a deathly poison'), and, although he admits that the situation is profoundly 'sad', the narrative ends in triumphant self-affirmation: 'I've broken through barriers which I had all my life. And now, I'll stay myself, and you can keep your L-DOPA.'[159]

The narratives in *Awakenings* have three distinct roles: they serve to document the patient's illness; form a channel for neurological and theoretical speculation; and provide a commentary on different modes of therapy. However,

---

153   Ibid., p. 216.

154   Ibid., p. 212.

155   Ibid., p. 213.

156   Ibid.

157   Ibid, pp. 216, 214. Sacks does not usually condone such restraints, but the instruments employed, in Goffman's phrase, serve to 'mortify' or 'curtail' Leonard's grandiose sense of selfhood; Goffman, *Asylums*, op. cit., p. 50. Such 'assaults' or 'encroachments', although they serve to typify the negative vision of 'Total Institutions' and their staff, are sometimes necessary and may, as Goffman concedes, bring about 'psychological relief' for the patient; ibid., pp. 24, 60.

158   Ibid., pp. 307, 218.

159   Ibid., p. 219.

rather than remaining on these planes of generalization, Sacks argues that, in order to do justice to the individual patients, each narrative should be situated in its personal and, in the case of severe long-term illness, institutional contexts. There are also hints, particularly evident in the elegiac letters between Sacks and Mrs L. in the 1982 epilogue, that the narratives are, at least partly, constructed as tributes to the patients' ordeals. Sacks attempts to expand upon the traditional forms of pathography and case history, less for pedagogic reasons and more in an attempt to crystallize these intentions. Even if such attempts fall short of his optimistic goal, his development of the medical form displays an ethos of anti-institutionalism, which, by refusing to categorize patients according to disease type, may help to catalyze the possibility of change and recovery. At times the narratives display perceptible inconsistencies: for instance, when Sacks takes on roles and voices which, elsewhere, he expresses his wish to discard. But, in spite of these dynamic problems, as Flood and Soricelli suggest, the development of the physician's voice is as crucial and as unfinished as the writing of narrative.

A Text to Stand On

Writing in one of his more expressively romantic moods towards the end of *Awakenings*, Sacks considers the possibilities for his post-encephalitic patients 'to re-feel the grounds of their being, to re-root themselves in the ground of reality, to return to the first-ground, the earth-ground, the home-ground, from which, in their sickness, they had so long departed'.[160] The mythical motif of the journey, charting an arduous passage through illness before a circuitous return to health, is a well-worn classical, biblical and romantic allegory of death and rebirth, which in *Awakenings* serves to express a radical accommodation by the patient ('a slower, deeper, imaginative awakening'), in contrast to the briefer and less substantial awakenings induced by L-DOPA. The term 're-root' suggests either an originary self rediscovered in a moment of enlightened introspection or that the roots of the self are laid down elsewhere. On this second reading, '*the sense* that something is going wrong' can be incorporated into the psychological landscape of the patient in an accommodation of the illness experience.[161] This would initiate a cure which does not overcome bodily dysfunction, but associates the execution of certain significant activities with a return to health. In a constantly worsening illness, such redescription may be precarious, but necessary in order to cope with prolonged disability or the threat of death.

Although this second reading may be tenable, the rhetoric of 'the first-ground, the earth-ground, the home-ground' implies a kind of transcendental consciousness which has the power to recover a mystical integration with itself:

---

[160]   Ibid., p. 275.
[161]   Ibid., p. 249.

'a quality of nostalgia, recollection, returning to what one had somehow, always felt and known'.[162] By finding a 'voice' with which to express the experience of illness, the patient is able to reassimilate their altered perception of Being-in-the-world into a fresh pattern of significance. Implied here is a description of an 'inner monologue', or what Derrida calls a 'phenomenological voice'.[163] Derrida criticizes Husserl for privileging expressive language which 'lies outside the empirical sphere' and for bracketing off the cultural network of associations which would disrupt a pure sphere of ideality.[164] On this model, an understanding of health may be conceived as a purely expressive sign by which the self in a triumphant moment of transcendence recovers the sense of itself, whereas illness would be defined as an inability to express this notion of self-referentiality (the self grounded in it*self*) or to transcend the confusion of an inexplicable bodily condition. But Derrida argues that Husserl's theory of time is self-defeating: it 'dictates against any "punctually isolated" moment, for time is a "phasing", a continual movement of protentional and retentional traces'.[165] Husserl privileges the present moment as a still point in which memories and past experiences can be organized, whereas Derrida claims the present is mutable and perpetually in flux: any single attempt to understand the self as an abstraction should be understood in reference to other attempts to frame it.

The expression of Husserl's phenomenological self is evident in Sacks' romantic tendency to refer to an idealized conception of Nature. However, rather than sinking into romantic reverie, Sacks can be understood to mean that the patient is able to imaginatively reconceive him/herself as 'real and alive', like the *Umwelt* of which they are a part.[166] One of his central commitments is that by becoming aware of the richness and multivalency of the human condition (or *Dasein*) the patient may be able to take the first steps away from the mortification of illness towards, as Leonard L. briefly expresses, a celebration of self-existence (*Eigenwelt*). In line with the other romantic scientists, only through a constant revisional *process* (that is, revision without progression towards a full, or self-present, sign) along a series of narrative planes can temporal flux adequately balance the attempt to 'keep the person's behaviour coherent and intelligible'.[167]

Echoing Erikson's work on uprooted identity, Sacks suggests the journey of the patient is 'endless', a quest with an always unreachable goal, the pursuit of which is figuratively rendered as a 'homecoming'.[168] The quotations from Hesse's

---

[162] Sacks, *A Leg to Stand On* (London: Penguin, 1991), p. 168.

[163] Jacques Derrida, *Speech and Phenomena and Other Essays on Husserl's Theory of Signs*, trans. David B. Allison (Evanston: Northwestern U.P., 1973), p. xxxix.

[164] Ibid.

[165] Ibid., p. xxxvii.

[166] Sacks, *Awakenings*, op. cit., p. 274.

[167] Jonathan Glover, *I: The Philosophy and Psychology of Personal Identity* (London: Penguin, 1991), p. 81.

[168] Sacks, *Awakenings*, op. cit., p. 168.

The Glass Bead Game (1943) and Eliot's 'Little Gidding' (1942) in *Awakenings* imply, like Odysseus' circuitous voyage, that 'home' (Ithaca) can only be recognized and understood in the leaving of it: that is, in realizing what it is like to be away from home, whether it is the hostility of foreign lands for Odysseus, or, as Sacks would describe them, the strange distortions of neurological illness. As such, 'home' can only be perceived from a vantage point of its opposite, just as 'health' can only be understood from a knowledge of illness and dysfunction. The symbols of 'health' and 'home' can be productively elided to encourage the patient to reactivate, or to newly learn, abilities, but it is the values attached to the accomplishment of these activities that enables the patient to initiate a renewed sense of selfhood. In these terms, the self is not to be conceived as a metaphysical entity, but an enabling device through which hostile bodily conditions can be dealt with. Thus, Sacks finds it more productive to refer to ability and disability rather than health conceived as an abstract ideal.

He makes the point of stressing that feelings of 'homeliness' correspond to 'knowing in the depths of one's being that one has a real place in the home of the world'.[169] This home is conceived as being both corporeal (an awareness of a perishable body) and topographical (for the patients of *Awakenings*, Mount Carmel is the only home they can remember). Such an awareness of home-in-the-world allows a place which is not-home (Mount Carmel) to *act* 'as a *home*', an understanding which, as Sacks suggests, is 'deeply therapeutic to all of its patients'.[170] But, to the extent that a chronic hospital is an institution, forcing the patients to follow the rules and restrictions which define their role as *patients*, 'it deprives them of their sense of reality and home, and forces them into false homes and compensations of regression and sickness'.[171] Sacks argues that only by nurturing emotional values associated with home (familiarity, friendliness, compassion, comfort and care) can the notion of this figurative return to health be edifying or conciliatory. By stressing the word 'settle', Sacks is able to suggest the simultaneity of pacifying (to settle down), active decision-making (to settle) and a notion of dwelling in the world (a settlement). As this section demonstrates with reference to Sacks' autobiographical book *A Leg To Stand On* (1984), the therapeutic potential of symbol and myth is not miraculous and should be mirrored by amenable conditions of living for the patient.

Many critics have approached *A Leg To Stand On* as a text which, in the corpus of Sacks' writing, requires special consideration. Although *Awakenings* disrupts many of the conventions of classical science, it retains a recognizable structure in its assemblage of case histories. Conversely, the discursive narration of *A Leg To Stand On* places this neurological novel firmly outside the generic realm of scientific writing. However, this does not render it anti-scientific or anti-

---

[169]   Ibid., p. 272.

[170]   Ibid.

[171]   Ibid.

philosophical. In many ways it represents an amalgam of the *Bildungsroman* and the philosophical novel, characterized by the high modernist writings of Thomas Mann and Hermann Hesse. Because the Germanic feel of the novel aligns it closely with the tradition of romantic science, it is not surprising to find Sacks evoking Mendelssohn and Kant at crucial moments in the text. The published version of the book is a severely curtailed edit of an earlier sprawling work which would have been more discursive than narrational. As it exists, the narrative dimension is clearly the focus and provocation for ideas and reflections.

The story represents an autobiographical account of a profound injury to Sacks' leg sustained in the mid-1970s while mountaineering in Norway. Instead of merely documenting the medical complications which arise when he loses feeling and tone in his leg (sustained after an operation to his quadriceps), he sees his 'own story' has the 'shape' of a 'universal existential experience, the journey of a soul into the underworld and back, a spiritual drama – on a neurological basis'.[172] As Teresa de Lauretis comments in an essay on Sylvia Plath's *The Bell Jar* (1963), the claim of a 'universal existential' drama must be established by:

> filtering a unique emotional and experiential content through the sieve of symbolic discourse . . . its success and forcefulness are due in large part to the author's ability to integrate the historical, diachronic self . . . with a synchronic, timeless mythic structure, the descent-ascent pattern . . . or from one state of being to another.[173]

This 'theme of rebirth' presents a similar mythic pattern to Rank's therapeutic narratives in *Awakenings* and can be understood as an attempt by Sacks, as physician, to complement the empathy he shows for his patients. Here, he figures himself as a patient forced to contend with fear, pain, confusion and the slow process of rehabilitation which grants him insights into the experiences of other patients. William Howarth claims that this autobiographical mode 'is a double form of retrospection in which the author plays both character and narrator, one to whom events happened and one who now relates the history'.[174] Developing Erikson's shifting voice in *Gandhi's Truth*, this double perspective allows Sacks to shift between positioning himself as a protagonist (in many ways, an anti-protagonist to whom events happen) in a symbolic drama and also the commentator and committed, but detached, philosopher of illness. The projection of two roles, 'two selves, patient and physician' indicates that the self, rather than something originary, is a textual construction which enables modes of (always incomplete) understanding to cluster around different subject positions.[175] The dual commitment of protagonist and engaged thinker indicates that this novel,

---

[172]   Ibid., p. 146.
[173]   Teresa de Lauretis, 'Rebirth in The Bell Jar', *Women's Studies*, 3(2) (1976), 173.
[174]   Howarth, 'The Ecology of Oliver Sacks', op. cit., p. 110.
[175]   Ibid., p. 111.

despite posing many conceptual problems, is a project of 'good faith' and not the 'false book' which Ella Kusnetz believes it to be.[176]

Although she is generally hostile to Sacks' writing for patronizing his patients or overshadowing them with his own voice, Kusnetz criticizes *A Leg To Stand On* for two reasons. Firstly, she claims he is caught in a conceptual trap:

> although the very science he appeals to subverts any notion of a 'self' or a 'soul' granted by God or metaphysics, he is inclined to override his own logic with emotion. His books, in fact, talk about disease as though it were a falling-away from some pre-wired wholeness, a disintegration longing for reintegration.[177]

Her second criticism stems from what she sees as Sacks' close, but troubled, allegiance with Freud. She claims Sacks' scientific project, to 'describe mental phenomena physiologically', closely resembles Freud's aim, as does his role as 'neurologist qua analyst, the analyst qua guru'.[178] She views this identification to be misplaced, because cures for 'intractable' diseases such as 'organic brain disorders' are very rare.[179] Kusnetz suggests that the reason 'these stories appeal to readers so powerfully' is that 'underlying them all, more basic to the narratives than even the structure of analytic cure or spiritual conversion, is a much more irresistible, primitive fantasy of primal narcissism, of wholeness regained, of return to undifferentiated unity'.[180] She views the text as a strategy on behalf of Sacks to gain his patient's confidence and, thereby, to fulfil a narcissistic wish for himself and the reader: 'Sacks needs to lose the leg . . . in order to regain it.'[181]

While Kusnetz's reaction to Sacks' tendency to 'go mystical' at times of romantic rapture is fair, by interpreting Sacks' project to be a misplaced psychoanalytic one, she reduces a complex body of writing to a predefined genre.[182] Instead of being little more than confidence tricks, his writings may better be appreciated as artful attempts to work in the margins of limited discourses in the Jamesian attempt to account for more, rather than less, experiences. Although Sacks periodically lapses into romantic reverie, this does not render his project defunct. By harnessing his understanding that regular encounters with his patients are inevitably analytic to some degree, together with his piecemeal deployment of Freudian terminology, he is able to appropriate psychoanalytic ideas without them becoming the dominant explanatory discourse. Even if no cure is effected, the telling of stories can often be a lifeline when others have dissolved. As Martin Swales comments in his work on the

---

176 Kusnetz, 'The Soul of Oliver Sacks', op. cit., p. 192.
177 Ibid., p. 178.
178 Ibid., pp. 177, 178.
179 Ibid., pp. 179, 197.
180 Ibid., p. 194.
181 Ibid.
182 Ibid., p. 177.

German *Bildungsroman*: 'the story, then, becomes the guarantor that one is living. Obliteration of the story may seem to promise the realisation of human wholeness, but ultimately it is a wholeness brought at the unacceptable price of stasis, bloodlessness, death.'[183] In other words, the telling of, and participation in, narrative may make bodily existence acceptable. Moreover, Sacks does intimate that there is, in the words of Kusnetz, 'no paradise regained, just as there are no transcendent consolations'.[184] The telling of narratives which lead towards, but never finally reach, a state of idealized health, is a way of organizing sensory data as it is experienced to find patterns of significance in illness. Rather than equating wholeness solely with the fulfilment of a narcissistic wish, stories serve to inform 'the very flux of a character's life and experience'.[185] This is not to deny that stories of recovery can appeal to a primeval wish to recover a lost wholeness, but it may be possible to incorporate this trajectory as part of the texture of narrative without it paralysing the other symbolic dimensions.

However, there is an additional problem in Sacks' narrative which goes some way to vindicating Kusnetz's claim that the text is a kind of confidence trick to gain the reader's (or the patient's) trust. The injury occurred in the early 1970s, but Sacks writes his account nearly a decade later, with the fear and uncertainty which accompanied the experience receding to a point of temporal and emotional distance by the time of writing. Although the meaning of certain experiences are constantly qualified (for example, the shifting metaphors he uses to describe the feeling, or lack of feeling, in his damaged leg), the story cannot be interpreted solely as a mode of self-therapy. If it is to be representative of such, the narrative should have been drafted during rehabilitation. In many ways, Sacks is able to 'stand' outside the frame of the narrative and distance himself from the kind of engaged autobiography that would represent a lifeline for the patient. But, in his defence, Sacks claims that the narrative should not be read wholly in this manner: the text investigates wider neurological issues concerning 'singular disorders of body-image and body-ego', together with an ongoing 'critique of current neurological medicine' and the concepts and taxonomies currently in use.[186] As such, Sacks' journey is twofold: the recovery from illness ('the abyss of bizarre, and even terrifying effects') and the search for a scientific (but not systematic) understanding of neurological damage.[187] Only from the 'double-perspective' of Sacks' existential formulation of a 'clinical ontology' or 'existential neurology' can these issues be explored together.[188]

---

[183]   Martin Swales, *The German Bildungsroman Wieland to Hesse* (Princeton, NJ: Princeton U.P., 1978), p. 33.

[184]   Kusnetz, 'The Soul of Oliver Sacks', op. cit., p. 195.

[185]   Swales, *The German Bildungsroman*, op. cit., p. 36.

[186]   Sacks, *A Leg to Stand On*, op. cit., pp. vii, viii.

[187]   Ibid., p. vii.

[188]   Ibid., p. 166.

Another pejorative charge which Kusnetz levels at Sacks, is that he poeticizes the experience of illness by projecting himself as the hero, or quester, of his texts. But, unlike the criticisms levelled at *Awakenings* and *The Man Who Mistook*, in which Sacks' rhetoric often obscures the voice of his patient, in his autobiographical novel he is, by definition, the protagonist. His quest is both positive and negative: his goal is to redescribe health so he can begin to *think* of himself as an active being again, but, because his is not a journey to discover a graspable object, his grail also consists of the passive process of recovery. For these reasons, Sacks envisages his illness and recovery as a 'pilgrimage' which 'fortune, or my injury, had forced upon me'.[189] He reacts to his doctor's description of the case as an 'uneventful recovery', by describing his pilgrimage as

> a journey, in which one moved, if one moved, stage by stage, or by stations. Every stage, every station, was a completely new advent, requiring a new start, a new birth or beginning. One had to begin, to be born, again and again. Recovery was an exercise in nothing short of birth, for as mortal man grows sick, and dies, by stages, so natal man grows well, and is quickened, by stages – radical stages, existence stages, absolute and new: unexpected, unexpectable, incalculable and surprising.[190]

This description of sequential stages does not contradict the earlier argument that stories represent the fundamental human experience; rather, these 'stations' or 'advents' are defined as rituals or 'acts' defined *within* the frame of an individual's story.[191] Like Erikson's phases of the life cycle, Sacks' stages represent ways in which the recovery process (in the chapters 'Becoming a Patient' and 'Convalescence') can be traced through the breaks or discontinuities in the story (described as 'the unimaginable next step') and, on an existential level, attempts to render the personal significance of the experiences ('Limbo', 'Quickening', 'Solvitur Ambulando').[192] Accordingly, a 'new birth or beginning' becomes a conceptual strategy for refocusing the mind within a narrative of 'healthening'.[193] Such rebirths represent significant moments in a narrative of self-becoming, with self-construction forming a strategy to activate the self to deal with the situation at hand. Instead of moving towards a *telos* where the self is fully realized, the process is revealed as unending: in Sacks' words, a journey leading 'both forwards and backwards'.[194]

---

[189] Ibid., p. 146. The religion that Sacks turns to 'for comfort' in his state of 'limbo' is closer to Jamesian mysticism than any doctrinal belief, for he claims that they do not demand the 'blunt plain commitment that 'religion' involved'; ibid., p. 81. The rituals and evocations of 'art and religion' give him solace and 'hope', in the respect that they 'could communicate, could make sense, make more intelligible – and tolerable'; ibid.

[190] Ibid., pp. 119, 124.

[191] Ibid., pp. 119, 140.

[192] Ibid., p. 119.

[193] Ibid., p. 139.

[194] Ibid., p. 168.

While his description of the story as pilgrimage goes some way to distance the text from the criticism that quest-narratives are merely stories of male conquest, Sacks echoes James' doctrine of the strenuous life by equating activity closely with masculinity. The philosophical emergency into which he is thrown after his operation forces him to seek a redescription of his identity to reconfirm himself as a human being. The enforced passivity of patienthood forces him 'to relinquish all the powers' he normally commanded: in other words, 'the sense and affect of *activity*'.[195] Sacks finds this mortifying because he claims it humbles his 'active, masculine, ordering self', which he equates with 'science', 'self-respect' and 'mind'.[196] Illness and passivity are not only blows to a self defined as human, but also as rational, moral and masculine. If Sacks figures his symbolic story as a 'universal existential experience', following a mythic pattern of descent and ascent (the fall on the mountain into a condition of suffering, then from a state of wilderness and limbo to a re-integration back to health, bringing with it the joy of recovery),[197] but the text actually reveals itself as a male story, then Kusnetz's position is strengthened: in short, the text is a duplicitous one which attempts to hide its own gendered foundations. Furthermore, the fact that it is Sacks' leg that is damaged, an organ commanding mobility (allowing action and movement) and stability (the maintenance of the vertical, or phallic, human form) provides fuel for such an argument. Instead of writing a text which, by interweaving different narrative strands with other generic strains, is open and liberating, Sacks may be seen as merely relying on a narrative framework which conceals a potentially oppressive masculinist myth. One response would be that Sacks has written an example of what a neurological text might be as an encouragement to other patients to write their own texts; he is not so much seeking to overlay every story with a universal symbolic pattern, but outlines some of the investigative and therapeutic possibilities of narrative. Crucially, it is Sacks' body-image (biologically and culturally different from, for example, Ellen West's body-image) which determines the structure of the story.

Jonathan Glover detects three different perspectives for understanding body-image: a visual image of the body, a tactile sensation and an awareness of 'size,

---

[195]   Ibid., p. 79.

[196]   Ibid.

[197]   In a discussion of Coleridge's *Ancient Mariner*, Stanley Cavell quotes Robert Penn Warren who, in his essay 'A Poem of Pure Imagination: An Experiment in Reading (1951), notes two important 'qualities' of the Fall: 'a condition of will . . . "out of time"' and 'the result of no single human motive'; Stanley Cavell, *In Quest of the Ordinary* (Chicago: University of Chicago Press, 1988), p. 57. In line with Sacks' existential story, Cavell's Heideggerian interpretation of Warren suggests that because 'there is no (single) motive' which brings about the Fall, the motive is the existential 'horror of being human itself'; ibid. On this reading, when Sacks stumbles across the bull, figured both as 'magnificent' and 'utterly monstrous', he confronts the 'horror' and 'nausea' of his own existence; Sacks, *A Leg to Stand On*, op. cit., p. 5.

shape and posture'.[198] He argues that usually these three perspectives contribute to create a fully sensuous body-image, although, as James argues in *Principles*, a person is rarely aware of each aspect in isolation. Unless conscious attention is focused explicitly on how the body feels or looks, the body-image retains a background awareness which constitutes, in Israel Rosenfield's phrase, 'the brain's absolute frame of reference': a self-referential space which provides a marker of corporeal and mental continuity.[199] As such, the elusiveness of the pronoun 'I', as an 'indexical' or an empty signifier, is normally given empirical content by the body.[200] However, when this 'frame of reference' of bodily continuity is disturbed due to neurological impairment or radical change in body-image (as it is for Sacks after his operation), Humean attacks on personal identity seem much more convincing. Because of this, Glover claims that the three types of body-images should not be conceived as a 'unity', for 'perhaps some of the distortions involve interference with one [aspect] and not another'.[201] By privileging one of the aspects (for example, vision over tactility or proprioception), it may be possible strategically to reintroduce the memory of a body-image and, as Rosenfield argues, to relocate the familiar self within the strange and estranged.[202]

The neurological damage which Sacks suffers after his operation results not only in a wasting of his leg, but an inability to recognize the leg as belonging to him: 'the day before touching it, I had at least touched *something* – unexpected, unnatural, unlifelike, perhaps but nevertheless something; whereas today, impossibly, I touched nothing at all'.[203] In an attempt to cope with the damage to his nervous tissue, Sacks ceases to recognize the leg as part of himself, jettisoning the leg-image and sealing the body in an adaptive configuration of unity. For Sacks, it is the tactile aspect of the body-image which is most profoundly impaired: he is able to see his leg, but loses the sense that it is connected to the rest of his body. In contrast to the heavy presence of the cast which encases the leg, Sacks dreams that his leg is either inorganic ('chalk or plaster or marble'), 'something friable and incoherent, like sand or cement' lacking 'inner structure or cohesion', or lacking existence (mist, darkness or shadow).[204] He comments

---

[198] Glover, *I*, op. cit., p. 81.

[199] Israel Rosenfeld, *The Strange, The Familiar, and Forgotten* (London: Picador, 1995), p. 45.

[200] Glover, *I*, op. cit., p. 66.

[201] Glover, *I*, op. cit., p. 81.

[202] Rosenfield recalls the 1905 case of Madame I who loses her sense of bodily awareness and capacity to visualize her body as a self-referring entity or to remember her body as a dynamic continuum; see G. Deny and P. Camus, 'Sur une forme d'hypocondrie aberrante due à la perte de la conscience du corps', *Revue Neurologique* (1920), cited in Rosenfield, *The Strange, The Familiar, and Forgotten*, op. cit., pp. 38–44.

[203] Sacks, *A Leg to Stand On*, op. cit., p. 42.

[204] Ibid., pp. 64, 65. When Sacks reflects on his condition, his experiential time, that which consists 'solely of personal moments, life-moments, crucial moments', 'thickens' in the manner of dreams; ibid., p. 8.

that 'none of the dreams seemed to tell any "story": they were fixed and static, like tableaux or dioramas, solely designed, as it were, to exhibit their appalling-boring centre-piece, this nothingness, this phantom, of which nothing could be said'.[205] As such, the fragmentation of a coherent body-image is integrally connected with the (in)ability to tell a cohesive story.[206]

Another example of this fracturing of everyday experience is dramatized in the 'scotoma' which Sacks experiences when he loses the left-side of his visual field: he watches the nurse's gestures but does not see 'her movements as continuous, but, instead, as a succession of "stills" or a static "mosaic world" '.[207] This is closely related to the confusion which accompanies his loss of bodily continuity: 'I could no longer remember having a leg. I could no longer remember how I had ever walked and climbed. I felt inconceivably cut off from the person who had walked and run and climbed just five days before. There was only a "formal" continuity between us.'[208] Sacks, the professional and attentive physician, is aware of the visual connection of the leg to his trunk (the 'formal' sense of bodily continuity), but Sacks, the fearful and experiencing patient, has no sentience of such a connection. Only by continuously casting around for language to describe his loss of feeling, or 'mysterious "abeyance" ', can Sacks, as spectator, affirm this formal continuity and hope that the familiarity of wholeness will follow.[209] But, at this stage, Sacks comments that 'there was no "entering", nor any thought or possibility of entering, these purely sensorial and intellectual phenomena'.[210] The gap between third and first-person enables Sacks to affirm the leg to be 'fine, surgically speaking', but he is nevertheless 'disquieted' and 'shocked' by the loss of feeling.[211] As Rosenfield discerns,

---

[205]   Ibid., p. 65.

[206]   The cohesiveness of the story may be entirely dependent upon the degree of emotional involvement the storyteller, or reader, has with it. Cohesion cannot be understood as a pre-given unity nor is continuity necessarily part of a linear narrative; as Sacks indicates, after the initial mountain confrontation 'there ceased, in a sense, to be any "story", or any particular "mood" to give tension and connection to the days that followed'; ibid., p. 23. He realizes that the story that had been told and the story that is to be told are 'essentially connected', in that Sacks, as narrator, makes these connections and unifies disparate experiences into a narrative whole. As such, the sense of a constructed story echoes the notion of a constructed self as central to the narrative process of 'healthening'.

[207]   Ibid., pp. 67, 70. The description of a 'mosaic world' corresponds to Sacks' discussion of 'mosaic vision' which he detects often accompanies severe migraines. The close overlapping of categories documented in *Migraine*, disorders of visual perception, body-image distortion, speech and language problems, 'states of double or multiple consciousness' and 'elaborate dreamy, nightmarish, trance-like or delirious states', are all dramatized during Sacks' illness in *A Leg To Stand On*; Sacks, *Migraine*, op. cit., pp. 93–6.

[208]   Ibid., p. 58.

[209]   Ibid., p. 93.

[210]   Ibid., p. 111.

[211]   Ibid., p. 91.

'seeing is not by itself "knowing" and that the lack of inner self-reference, together with the incontrovertible sight of the leg, therefore created a paradoxical relation to it'.[212]

Like the post-encephalitic patients of *Awakenings*, Sacks is able to reflect upon an injury which disturbs the 'identity, memory and space' of 'primary consciousness'.[213] The major problem in *Awakenings* and *A Leg to Stand On* is that bodily condition is unrecognizable (and, like James' fringes of experience, unrepresentable in language), or, because it is so changed from that which was previously accepted as familiar, the patient wishes to deny the connection with his or her body. Thus, the activity of narrating the illness can not only connect the outlandish present to a familiar and known past but can clarify disorientating phenomenological experiences. Like Heidegger's attempt to rescue *Dasein* from estrangement, this kind of therapeutic narrative can make the connection which may help recover the 'immediate *thereness* . . . which no thinking could reach'.[214] This 'thereness' constitutes a renewed sense of the self as coextensive with the body and affirms that the self has a place in the world so that, in the utopian words of Sacks, the patient can come to '*know* life, as never before'.[215]

Having insisted on the importance of narrative in making sense of bodily conditions, in which the disconnected and 'fluttering frames' of the body-image can be lent temporal continuity, Sacks emphasizes that neurological damage cannot be repaired in such a way.[216] Autobiographical narration is merely a strategy or a trick, like a mirror, video-film or piece of music used as a '"symbolic" . . . form of self-reference', which interweaves a number of figurative levels.[217] As such, narrative can lend a static image of the body a plasticity and fluidity along its temporal axis. In *A Leg to Stand On* and *Awakenings*, an acknowledgement of the 'uselessness of words', rather than discouraging the patient, may provide the spur to write or tell a revisional narrative, in which the interweaving of different strands may construct a notion of 'healthening' towards which the patient can work.[218] In certain neurological disorders affecting primary consciousness this multi-layered technique of self-creation may help to activate a process of bodily rehabilitation, but, as the next section demonstrates, such narrative techniques do not always work.

---

[212] Rosenfeld, *The Strange, Familiar and Forgotten*, op. cit., p. 53.

[213] Sacks, *A Leg to Stand On*, op. cit., p. 186.

[214] Ibid., p. 118.

[215] This is evident when Sacks comments that he feels that his 'leg had come home, to its home, to me'; ibid, p. 111. This homecoming seems to correspond to Rosenfield's notions of self-reference as well as to the existentially inspired conception of 'thereness'.

[216] Ibid., p. 206.

[217] Rosenfeld, *The Strange, Familiar and Forgotten*, op. cit., p. 58.

[218] Sacks, *A Leg to Stand On*, op. cit., p. 107.

Narrating the Other

In *The Man Who* (1994), the English version of Peter Brook's dramaturgical adaptation of Sacks' work, one of the patients, arising momentarily from his immobility and despair, expresses a realization that he has no habits, but only tricks with which to combat his disorder.[219] Much of the strangeness which Sacks associates with the life-world of those suffering neurological impairments results from the disruption of habitual responses to the environment, previously either innately present (such as reflexive responses) or acquired subconsciously and gradually from an early age. When these background abilities are affected, the patient's phenomenological and physiological worlds are thrown into crisis. In an attempt to recover these lost abilities, not only do the patients have to reorient themselves in their world (through a process of redescription), but they also have to learn different practical methods for compensating for impairments. If habit is associated closely with a place, or a home, in the world, then the arduous task of the patient is one of more than just metaphorical relocation.[220] As Joseph Thomas comments his article 'Figures of Habit in William James' (1993), habit 'is meant to perform a kind of midwifery, easing entry into a more secure, less "uncanny" world'.[221] In cases when these habitual responses are undermined, the resulting uncanny world is usually one of self-estrangement.

This section outlines some of these tricks and techniques deployed to combat self-estrangement as they arise in what has become Sacks' most popular book, *The Man Who Mistook His Wife for a Hat* (1985). Many of these tricks, although ostensibly compensating for the loss of narrative coherence, often serve to recoup the importance of telling stories. By developing his clinical tales, Sacks moves away from the institutionalized case history towards a symbolic structure in which he attempts to rescue the personalities of his patients. Unlike Jacqueline Rose's argument that *The Man Who Mistook* represents an 'unending joke', or Dominique Goy-Blanquet's assertion that the text 'arouse[s] more laughter than compassion', it is more appropriate to argue that if laughter is evoked from the reading of these tales, it better serves to engage the reader's compassion and, often, the discomfort of identification, than arouse mockery or denigration.[222]

---

[219]   Peter Brooks, *The Man Who*, performed on UK tour in 1994.

[220]   For instance, James writes: Tolstoy's 'crisis was the getting of his soul in order, the discovery of its genuine habitat and vocation, the escape from falsehoods into what for him were ways of truth. It was a case of heterogeneous personality tardily and slowly finding its unity and level'; James, *Varieties*, op. cit., p. 186.

[221]   Joseph M. Thomas, 'Figures of Habit in William James', *New England Quarterly*, 66(1) (1993), 15.

[222]   Jacqueline Rose, '*The Man Who Mistook His Wife for a Hat* or *A Wife Is Like an Umbrella* – Fantasies of the Modern and Postmodern', in ed. Andrew Ross, *Universal Abandon?: The Politics of Postmodernism* (Edinburgh: Edinburgh U.P., 1988), p. 238; Dominique Goy-Blanquet, 'Slides of the Cortex', *Times Literary Supplement* (14 May 1993), p. 20.

It is helpful to view Sacks' moral scheme as a response to the model set up by Erving Goffman in *Stigma* (1963), where the outlandish appearance and behaviour of the severely disabled (defined as outlandish through normative social codes which disqualify the individual 'from full social acceptance') marks them out, or stigmatizes them, as Other.[223] In Goffman's view, these differences, whether 'abominations of the body' or 'blemishes of individual character', reduce the stigmatized person 'from a whole and usual person to a tainted, discounted one'.[224] Instead of affirming such stigma, Sacks wishes to depict each patient in a 'fabulous' world which conveys the idiosyncratic impressions of their particular neurological disorder. However, as a caveat, Sacks encounters a double bind. By framing his sketches with sensationalist titles and the rhetoric of fantasy (as can be detected in the title of *An Anthropologist on Mars*), Sacks is in danger of creating a series of grotesque freak-shows which invite a voyeuristic curiosity, instead of encouraging pathos. But, perhaps this is a danger it would be hard to avoid in his engagement with the disorientating experiences of neurological impairment. Against the assertion that Sacks succeeds only in stigmatizing his patients in order to make them interesting, it is more profitable to view such a technique as a creative wager, by which he risks the apparent integrity of the text in order to illuminate his interest in identity and selfhood.

Many of the tales are concerned centrally with the problems of recognition and the disturbance of basic cognitive faculties. The title case study in *The Man Who Mistook* introduces concerns running throughout the first section of the book on neurological losses. Sacks' patient, a musician, Dr P., has 'wholly lost the emotional, the concrete, the personal, the "real" ', although his abstract and musical skills remain unaffected.[225] In their initial encounter, Sacks reports, the patient

> instead of looking, gazing, at me, 'taking me in', in the normal way, made sudden strange fixations – on my nose, on my right ear, down to my chin, up to my right eye – as if noting (even studying) these individual features, but not seeing my whole face, its changing expressions, 'me', as a whole . . . there was just a teasing strangeness, some failure in the normal interplay of gaze and expression. He saw me, he scanned me, and yet . . . [226]

Dr P. suffers no impairment to his optical apparatus, but he incurs damage in the parietal and occipital lobes at the back of his brain which render him unable to relate details and 'tiny features' to 'the picture as a whole'.[227] Dr P. thus

---

[223] Erving Goffman, *Stigma: Notes on the Management of Spoiled Identity* (Harmondsworth: Penguin, 1968), p. 9.

[224] Ibid., p. 12.

[225] Sacks, *The Man Who Mistook*, op. cit., p. 5.

[226] Ibid, p. 8.

[227] Ibid, pp. 8–9.

'constructs the world as a computer construes it, by means of key features and schematic relationships', but he fails to 'behold' or make any sense 'of a landscape or scene'.[228] Because 'he saw nothing as familiar', Dr P. approaches objects and faces as 'if they were abstract puzzles or tests' to painstakingly decipher and, while he can easily recognize abstract shapes, he tends to explain a more complex object like a flower by reducing it to its constituent parts.[229] But, while Dr P. continues to mistake people and objects, Sacks encourages him to nurture a kind of 'inner music' to supply himself with a sense of bodily continuity.[230] By focusing on his sensitivity to musical rhythm, Dr P. is able to provide an accompaniment to activities such as dressing and washing.

Another example of confused cognition is found in the case of Christina, 'The Disembodied Lady', who has lost her 'sixth sense' of proprioception, and with it the automatic 'sensory flow' and 'body tone' which lends a sense of three-dimensionality to the body.[231] As a result she feels 'disembodied', 'pithed'; in Sacks' words, 'she has lost . . . the fundamental, organic mooring of . . . corporeal identity'.[232] As a consequence, Christina has to continually stimulate her body to recognize it and to 'know' that it is there. Similarly in 'Hands', Madelaine J., congenitally blind and suffering from cerebral palsy, cannot 'recognise or identify anything' that is placed in her hands; yet, while her hands are '*mildly* spastic and athetotic' and felt like 'lumps of dough', Sacks detects her 'sensory capacities . . . were completely intact'.[233] She is unable to recognize objects until, stimulated by hunger, she reaches out for a bagel and later learns 'to explore or touch the whole world . . . by a curiously roundabout sort of inference', as she begins to refamiliarize herself with the spatial properties of simple shapes.[234]

These three examples reveal Sacks' ongoing concern with recognition and perception: how his patients conceive of themselves and their location in the world. He concentrates on the most basic of faculties and the most common of activities in order to unearth the confusions which characterize the recurring existential issues of romantic science. Sacks choice of figurative language enables him to focus upon the manifold problems of recognition and acknowledgement. As such, his clinical tales can be read as versions of the

---

[228] Ibid., pp. 14, 12, 9. Dr P.'s story can be read as a parable of reductive interpretation. Just as the 'key features and schematic relationships' which Dr P. is able to identify cannot account for faces or pictures, so too the identification of particular neurological mechanisms cannot do justice to a patient's experiences. Sacks hints at this interpretation when he claims Dr P.'s agnosia is similar to 'a science which eschews the judgmental, the particular, the personal, and becomes entirely abstract and computational'; ibid., p. 19.

[229] Ibid., pp. 13, 12.

[230] Ibid., p. 17.

[231] Ibid., p. 42.

[232] Ibid., p. 50.

[233] Ibid., p. 56.

[234] Ibid., p. 59.

uncanny, especially in his focus on the qualities in his patients which exemplify idiosyncrasy and which serve to transform their familiar and prosaic world into one fantastic and strange. Following Rank's consideration of the mysterious double or shadow self and Erikson's interest in the mystery of the face, 'the uncanny' is coextensive with the unknowability which lies at the heart of identity, an understanding of which shifts the concerns of romantic science away from questions of epistemology towards forms of pragmatic narrative. Sacks gravitates toward the uncanny in order to convey the coexistence of the familiar and the unfamiliar in the experiences of neurologically impaired patients.

Although Sacks attempts to figure his patients in the mythical and fabulous worlds in which he believes them to reside, there is always 'the inaccessibility of a secret': a level of inexplicability which is evident in his reading of his patients and which is also mirrored in the reader's efforts to read Sacks' tales.[235] This inexplicability turns upon the nature of selfhood and how it might be represented. This notion is contrary to Freud's interpretation of the uncanny which, as Hélène Cixous observes, reduces the ambivalence and suggestiveness of Hoffman's 'The Sandman' to what he perceives to be the 'thematic economy of the story'.[236] Thus, Freud undermines his conception of the uncanny by underwriting the story with a primary reading of the castration complex. Sacks resists falling into this trap of reductionism. Although he utilizes the language of neurology in order to organically root his patient's conditions, their experiences remain fundamentally irreducible. Sacks chooses not to prune what Cixous calls 'all "superfluous" detail' from his text, but lets it float on an indefinite level of signification. He is quick to acknowledge areas where his professional learning can do little to unravel the mysteries of illness or to provide therapeutic aids for his patients. Sacks may possess the interpretative tools which his patients have lost but, like the other romantic scientists, as a committed reader of illness he always acknowledges areas of doubt and uncertainty.[237]

There is one particular strain of the uncanny infusing Sacks' tales which moves beyond a simple admission of their strangeness. Freud understands Ernst Jentsch to be interested in the manifestations of acute illnesses ('epileptic fits, and of manifestations of insanity') because 'these excite in the spectator the impression of automatic, mechanical processes at work behind the ordinary appearance of mental activity'.[238] Although the reader of Hoffman's 'The Sandman' can have little doubt that Olympia is a passive automaton contrasted to

---

[235] Jacques Derrida, *Dissemination*, trans. Barbara Johnson (London: Athlone, 1981), p. 71.

[236] Hélène Cixous, 'Fiction and Its Phantoms: A Reading of Freud's Das Unheimliche', *New Literary History*, 7(3) (1976), 534

[237] Jackson attributes such a scenario where 'the narrator is no clearer than the protagonist about what is going on, nor about interpretation' as another characteristic of the fantastic; Jackson, *Fantasy*, op. cit., p. 32.

[238] Freud, 'The Uncanny', op. cit., p. 347.

the strength of Clara's character (despite Nathaniel's confusion), there is an uncertainty as to what is the symbolic significance of the characters. Similarly, although Sacks maintains his faith that on a hidden level all of his patients remain animate, many of the illnesses he describes not only attack the personality of the patient, but question the very substance of the self. As such, the loss of a feeling of selfhood constitutes a crucial disengagement with a 'healthy' life. One of the central themes of *The Man Who Mistook* is the way in which the limits of selfhood problematizes the boundaries between life and death. The insistent question of how far the life of a person can be stripped away before he or she ceases to be sentient or alive haunts the reading of the book.

This kind of hesitancy conforms to Cixous' description of uncanniness: 'the phantasm of the man buried alive represents the confusion of life and death: death within life, life in death, nonlife in nondeath . . . Hence, the horror: you could be dead while living, you can be in a dubious state.'[239] Such horror spills over into a reading of *The Man Who Mistook* and tempers any laughter evoked in the portrayal of stigmatized patients. Sacks' depiction of selfhood as friable and elusive serves to implicate the reader in an identification with the philosophical dilemmas of the patients and neurologist. This identification is, however, always incomplete. The scientific filter of classical neurology reverberates through the text to distance the reader; at times moving beyond the personalities and the experiences of the patients to issues of neural organization. In order to shuttle between these levels of intimacy and explanation, Sacks devises a literary method which encourages the reader to focus upon the peculiar coexistence of familiarity and strangeness. Viewed as a Rankian study of the 'phenomenon of the "double"', Sacks develops the themes of the animate and the inanimate interwoven with those of the habitual and the strange.[240]

He invites the reader to view the tales from a dual perspective, through which he filters their subjects. He claims that he is committed to conceiving humans both as a 'what' and a 'who': as both machine and person.[241] Thus, classical neurology should be complemented by an existential science which can account for the 'representations . . . which are the very thread and stuff of life'.[242] As such, neurological patients should be understood in the abstract, as many interlinked but, sometimes, malfunctioning parts and also as an integrated person who is undergoing profound disorientation and forced to adjust in the face of perceptual or motor disorganization. By concentrating on disorders affecting the right cerebral hemisphere, Sacks focuses upon illnesses which disrupt the 'crucial powers of recognizing reality': powers which are both perceptual and hermeneutic.[243] Closely linked to the preoccupation with selfhood (and its loss) is

---

[239]   Cixous, 'Fiction and Its Phantoms', op. cit., p. 545.

[240]   Sacks, *The Man Who Mistook*, op. cit., p. 234.

[241]   Ibid., p. x.

[242]   Ibid., pp. 142, 140.

[243]   Ibid., p. 1.

a series of vital concerns with reading: how the patient reads his or her condition; how Sacks represents the patient; and how the reader interprets the text. The interpretative oscillation between the levels of the mechanical (inanimate) and the personal (animate) furnishes his clinical tales with uncanny qualities.

Sacks articulates his interest in 'neurological disorders affecting the self' by focusing on right-hemisphere disorder and change in the patient's views of his or her body-image. He claims that 'there is always a reaction, on the part of the affected organism or individual, to restore, to replace, to compensate for and to preserve its identity'.[244] This insistence on recuperation confirms Sacks' belief that there is a fundamental regulative structure which maintains a balance (often a compensatory one) even in individuals suffering from the most profound of illnesses. In *The Man Who Mistook* and *An Anthropologist on Mars* Sacks presents a variety of sketches in which he seeks to describe the many dimensions germane to an understanding of illness. However, if a patient is to invest emotionally in a narrative, then he or she must be able to recognize and connect phenomenological states with one another (by maintaining basic levels of cognition and memory) in order for narration to be beneficial or, indeed, possible. In other words, the patient has to maintain an ability to 'read' his or her condition in order to represent it in a significant form to themselves.

Awakenings and *A Leg to Stand On* present narrative and storytelling as having distinct therapeutic possibilities. However, the worlds of many of the patients portrayed in *The Man Who Mistook* are distorted to such a degree that storytelling cannot adequately connect the perceptual fragments. In cases of profound neurological damages, the kind of redescription of the self which Sacks achieves in *A Leg to Stand On* is barred, resulting in a retardation of the ongoing process of aesthetic self-creation as described by Nietzsche and Foucault. For example, one of the patients in *Awakenings*, Rose R., reappears in *The Man Who Mistook* in the short study, 'Incontinent Nostalgia'.[245] After the privation of her torpor, Rose's awakening is characterized by an increase of liveliness and exuberance, an 'increased libido', 'nostalgia, joyful identification with a youthful self' and the singing of 'innumerable songs of astonishing lewdness'.[246] Despite the return of Rose's life-force, Sacks comments that she appears entirely transfixed in her youthful world of the mid-1920s (at the outbreak of the

---

[244] Ibid., p. 4.

[245] It is notable that in Rose's case history both she and Sacks use the term 'uncanny', but in different senses. Firstly, in describing her transfixed torpor, Rose says, "'It was uncanny . . . My eyes were spellbound. I felt I was bewitched or something, like a rabbit or a snake'"; Sacks, *Awakenings*, op. cit., p. 81. Although she only uses a simile (rather than a transformative metaphor), her unstable condition brings feelings of uncertainty into Rose's expression. Sacks is surprised with the vivid nature of Rose's memory 'considering she is speaking of so long ago'; here, he calls it 'uncanny' because it seems to run against rational expectations; ibid., p. 82.

[246] Sacks, *The Man Who Mistook*, op. cit., p. 144; Sacks, *Awakenings*, op. cit., p. 82.

encephalitis lethargica epidemic). Sacks conjectures that Rose experiences 'her "past" as present . . . perhaps it has never felt "past" for her'.[247] Although Rose acknowledges a formal present through which she is passing, her feeling for life corresponds to her 'ontological age', as a twenty-one year old.[248] One explanation may be that Rose cannot bear to identify herself with her chronological age and so seeks to immerse herself in an idealized world of her past. Sacks suggests that the 'uncontrollable upsurge of remote sexual memories and allusions' which Rose experiences may be due to an increase in nervous excitation induced by L-DOPA, or the provocation of 'disinhibition'.[249] He contrasts her 'incontinent nostalgia' (or 'fossilised memory sequences') with 'forced reminiscence' which mnemonic stimuli may bring about.[250] Sacks concludes that 'an almost infinite number of "dormant" memory-traces' remain as 'subcortical imprints' and 'archaic traces' which are released when the self is disinhibited.[251]

His short commentary on Rose R. provides a touchstone for the whole book: the patient's subjective realities (marked variously by the loss of proprioception or of memory) blur with the categories of time and space. On one level, Rose is a living being who expresses her disinhibition through the language of her youth, but, on another level, she remains asleep or dead to the biological and neurological changes in her life. She seems to recognize her condition, but chooses to immerse herself in another, more vivid, psychological world in which she can express her exuberance. While she is able to interpret her condition, because she has no powers to compensate for the time lost during her catatonia, she chooses to erase it (although it is doubtful as to whether this is a conscious decision). Rose's story not only exemplifies the inability to cope with present realities, but also dramatizes the problems of reading illness.

'The Lost Mariner', Jimmie G., suffers from a severe loss of episodic and short-term memory 'due to alcoholic degeneration of mammillary bodies' characteristic of Korsakov's syndrome.[252] While Jimmie finds no difficulty with abstract problems that can be quickly resolved, he cannot retain the continuity of a conversation with Sacks for longer than two minutes. He occasionally has inklings that he is forgetful ('I do find myself forgetting things, once in a while – things that have just happened'), but he is largely unaware that his participation in the present (that is, Sacks' present) is minimal: 'Jimmie both was and wasn't aware of this deep, tragic loss in himself, loss *of* himself.'[253] Moreover, Jimmie is

---

[247] Ibid, p. 83.

[248] Ibid., p. 87.

[249] Sacks, *The Man Who Mistook*, op. cit., pp. 144, 145.

[250] Ibid., p. 144.

[251] Ibid., p. 145.

[252] Ibid., p. 28. Another study on profound damage to memory is to be found in 'The Last Hippie', one of the 'paradoxical tales' in *An Anthropologist On Mars*, op. cit., pp. 39–72.

[253] Sacks, *The Man Who Mistook*, op. cit., pp. 25, 34.

unable to read (or write) his condition in any conventional way. When Sacks suggests that he keeps a diary he not only continually misplaces it, but he 'could not recognise his earlier entries':

> his entries remained unconnected and unconnecting and had no power to provide any sense of time or continuity. Moreover, they were trivial – 'Eggs for breakfast', 'Watched ballgame on TV' – and never touched the depths. But were there depths in this unmemoried man, depths of an abiding feeling and thinking or had he been reduced to a sort of Humean drivel, a mere succession of unrelated impressions and events?[254]

These comments display those qualities of uncertainty and hesitancy which are characteristic of the uncanny, provoking the question: does Jimmie just function as a neurological machine, or does he continue to exist as a sentient being?

Like Rose R., Jimmie's pre-1945 past is coherent and vivid, but he struggles to acknowledge recent historical events like the filming of the earth from the moon or the building of new aircraft. Sacks suggests he is entirely disconnected from a continuous sense of selfhood which would enable him to root fleeting moments in his own autobiography. However, while Sacks acknowledges the crucial importance of memory, instead of resorting to the model of the 'Humean "being"', he maintains his belief that Jimmie consists of more than just memory: for example, ingrained practical skills (morse-code and touch typing) and qualities of 'feeling, will, sensibilities, moral being – matters of which neuropsychology cannot speak'.[255] He encourages Jimmie in his typing, employing him for 'real work' in the hospital, but he goes on to claim that 'this was superficial tapping and typing; it was trivial, it did not reach the depths'.[256] Sacks emphasizes his romantic sensibilities as he observes a calmness and attentiveness in Jimmie when he sits in the chapel or in the garden: he 'was absorbed in an act, an act of his whole being, which carried feeling and meaning in an organic continuity and unity, a continuity and unity so seamless it could not permit any break'.[257] Although Jimmie cannot express himself through phonetic or graphic means, Sacks detects that this incommunicable communion suggests the survival of Jimmie's integral self.

But the question remains how this integral self can be known or represented without wandering back into the vague metaphysical territory which twentieth-century romantic science wishes to leave behind. In neo-pragmatic vein, Richard Rorty comments in *Philosophy and the Mirror of Nature* (1979) that 'the way things are said' or expressed is more 'essential' (that is, useful) as a method 'of

---

[254] Ibid., p. 34.

[255] Ibid., p. 32.

[256] Ibid., p. 35. Here Sacks seems to underestimate the application and cultivation of skills for renewing a sense of health and well-being in his patients.

[257] Ibid., p. 36.

coping with the world' than talking about a world of factuality.[258] In this sense, the patient does not necessarily have to grasp the technicalities of their disorder to be able to reaffirm themselves by means of expression, whether in words, pictures, music, dance, touch or gestures. However, Rorty privileges rational thought at the expense of other modes of expression: 'the events which make us able to say new and interesting things about ourselves are, in this nonmetaphysical sense, more "essential" to us . . . than the events which change our shapes or our standards of living'.[259] Here, Rorty addresses his comments to his fellow 'relatively leisured intellectuals' and not to people whose very lives (both existential and physical) are determined by the 'change' in their 'shape' and constitution.[260] Rather than asserting a continuous memory or a notion of rational expression as the essential, or primary, constituent of selfhood, it is Sacks' hope that other less orthodox modes of expression can be found to encircle a sense of selfhood. This returns his attention to an emphasis on reading and narration: what it means to tell and retell the self without relying upon the fully functioning capacities of cognition and memory. This line of thought may provide fuel for his critics, but he retains a faith that expression and the barest hint of narrative (fractured and barely coherent) may be detectable in even the smallest gesture.

In an article on the failure of narrative, Richard Allen suggests that the condition of William Thompson, the patient presented in the case 'A Matter of Identity' in *The Man Who Mistook*, exemplifies such a disassociation from the structuring potential of narrative. This loss has little to do with any moral failing on the subject's behalf, but is a direct result of the neurological damage caused by Korsakov's syndrome. Like 'The Lost Mariner', his syndrome causes extreme loss of short-term memory, but leaves archaic memories unimpaired. William's syndrome appears to be more severe than Jimmie's, for he cannot maintain his attention for more than a few seconds at a time. Thus, William lives in a world of simultaneity as a continual present. Like Jimmie, because he has lost connection with an immediate past, William also loses any sense of a coherent identity. For Sacks, it is this very lack of self-awareness that serves to wholly implicate William in his illness. The reader is told that William frenetically and 'continually' creates a series of worlds and selves through the telling of 'ceaseless tales, his confabulations, his mythomania'.[261]

Sacks begins the tale with fragments of a dialogue in which William continually 'misidentifies' Sacks. William projects himself into a habitual, but fictitious, role of a grocer and mistakes Sacks for a customer, a butcher and a mechanic within a few moments. As a reader of William's world, Sacks claims:

---

[258] Richard Rorty, *Philosophy and the Mirror of Nature* (Oxford: Blackwell, 1990), p. 359.

[259] Ibid.

[260] Ibid.

[261] Sacks, *The Man Who Mistook*, op. cit., pp. 105, 106.

he was continually disorientated. Abysses of amnesia continually opened beneath him, but he would bridge them, nimbly, by fluent confabulations and fictions of all kinds. For him they were not fictions, but how he suddenly saw, or interpreted, the world . . . .there was . . . this strange, delirious, quasi-coherence, as Mr Thompson, with his ceaseless, unconscious, quick-fire inventions continually improvised a world around him.[262]

Sacks shows an understanding of how William might experience his world: 'for Mr Thompson . . . it was not a tissue of ever-changing, evanescent fancies and illusion, but a wholly normal, stable and factual world. So far as *he* was concerned, there was nothing the matter.'[263] Unlike the illnesses presented in Sacks' earlier books, severe Korsakov's syndrome strips away *all* sense of self-consciousness, or the interpretative faculty through which it is possible to read the illness experience. Like Rose R.'s retreat into a resplendent past, William's 'shimmering, iridescent, ever-changing' world is a simulacrum of life: a storyteller who is unaware how fractured his narratives have become. There is no stable world (nor home) in which to locate the self because 'the world keeps disappearing, losing meaning, vanishing'.[264] His continual reinvention in scraps of conversation is almost a caricature of the myth of self-becoming: William 'ceaselessly invents and reinvents his self and world in a parody of narrative'.[265] There is no development, no sense of growth, nor an imaginative movement towards a redescription of health. Indeed, for William, there appears to be no consciousness of his predicament: he is totally consumed by the illness. However, Sacks detects a pervading anxiety and a look of 'ceaseless inner pressure' which suggests that on some level William is aware of the 'meaninglessness, the chaos that yawns continually beneath him'.[266]

Part way through William's story, Sacks turns his attention to a brief consideration of the importance of storytelling and recollection:

If we wish to know about a man, we ask 'what is his story – his real, inmost story?' – for each of us is a biography, a story . . . To be ourselves we must *have* ourselves – possess, if need be re-possess, our life-stories. We must 'recollect' ourselves, recollect the inner drama, the narrative, of ourselves. A man *needs* such a narrative, a continuous inner narrative, to maintain his identity, his self.[267]

---

[262] Ibid., p. 104.
[263] Ibid.
[264] Ibid., p. 106.
[265] Allen, 'When Narrative Fails', *Journal of Religious Ethics*, 21(1) (1993), 27.
[266] Sacks, *The Man Who Mistook*, op. cit., p. 106.
[267] Ibid., p. 106.

It is possible to interpret this passage either as a therapeutic homily or as an affirmation of a type of reflective narrative in which the self (situated in a privileged position outside the finished narrative) retrospectively narrates itself. However, Allen detects two other important aspects at work here. Firstly, he understands Sacks to mean 'there is no separation or distance between self and narrative. A person's identity is the narrative she or he embodies.'[268] In other words, the self is written *by* the narrative as a point of reference: a space, a home, an indexical. Secondly, Allen comments:

> Sacks recognizes that one's life narrative – that is, the narrative one is – is more than the enactment of a script . . . a life narrative is more than a structured sequence of words; it is told in part 'unconsciously', and it collects within itself both the nonverbal and the verbal – image, emotion, act and word.[269]

Both insights move away from the type of omnipotent narrative of redescription that Rorty appears to advocate (and which is evident in Binswanger's and, to a certain extent, Erikson's work), and moves towards a fragmentary, hesitant and uncertain narration over which the narrator only ever has partial control. However, this notion of a 'living' narrative leaves two important issues unresolved. Firstly, although authorial control is unnecessary and sometimes restrictive, the patient has to have some capacity to reflect upon the narrative in order to tell it anew. Secondly, if the narrative is to be therapeutic and lead away from a condition of illness towards a different home (tracing the route rather than the goal), then the narrator must be able to restrict the range and complexity of the narrative in order to follow it.

These concerns indicate that the self (that which is written or told) is also the writer or teller who exists outside the frame of the story. Narration taken to the limit, so that the writer of one story becomes the written of another, leads either to the kind of restless metamorphosis characteristic of William's proliferation of narratives, or to an infinite regress where the 'real' self is lost from view in a constantly diminishing perspective. In this sense the recognition of the self is always an uncanny one because it is split: at one and the same time, it is the very point of locational reference (the 'I' which is read) and a perspective lying outside the narrative frame (the 'I' who reads). As the most well-known of current romantic scientists, Sacks' neurological tales indicate the imaginative possibilities and the actual limitations of a tradition of inquiry into selfhood. Indeed, his tales reveal both the proximity of self ('I' can only ever be 'myself')

---

[268] Allen, 'When Narrative Fails', op. cit., p. 28.
[269] Ibid.

and the emptiness and vacuity that lies behind such tautology.[270] Nevertheless, James, Rank, Binswanger, Erikson and Sacks, as practitioners of the ongoing project of romantic science, indicate that when illness strips away the familiarity from the comforting structures of selfhood, the resultant identity crisis (for Erikson) or philosophical emergency (for Sacks) demand that the assumptions which underlie the known and the habitual are radically reoriented.

---

[270] It may be this tendency to become solipsistic that stimulated Sacks to study a community of sufferers in his most recent book-length study *The Island of the Colour-blind* (1996).

Conclusion
The Challenge of Romantic
Science

In the first chapter of *New Maladies of the Soul* (1995), Julia Kristeva traces the history of mind-body dualism from antiquity through to the beginnings of modern psychiatry. Although monistic thought usually triumphs over dualism in the philosophical sphere because it does not need to explain how two distinct realms interact, Kristeva argues that 'the psyche', as a generator of meanings in the experiential sphere, 'shields us from biological fatalism and constitutes us as speaking entities'.[1] For monistic thinkers, dualism offers nothing but 'troublesome contradictions', but it provides the 'complementary dynamics of flux' at the heart of romantic science.[2] Just as early twentieth-century romantic science attempted to dispense with metaphysics, later practitioners, such as Erikson and Sacks, redescribe potentially contradictory elements as a symbiotic cluster of values, which cannot easily be translated into substances. It is this shift from 'substance' to 'meaning' which provides the twentieth-century tradition of romantic science with the impetus to address questions relating to experience, selfhood and health without being seduced at the outset into anti-empirical musings. Not only do alternative languages of the self enable romantic scientists to redescribe 'experience' in ways which free it from the spectre of metaphysics, but they embed the self firmly in the world of others: in Kristeva's words, they provide a 'bond between the speaking being and the other, a bond that endows it with a therapeutic and moral value'.[3]

Kristeva questions whether we still need the 'age-old chimera' of the 'soul' in an age in which 'we are continually decoding the secrets of neurons'.[4] It may be possible to retain an understanding of personal experience within the philosophical frame of neuroscience without reverting to the language of 'spirit' or 'soul' to explain the irreducible presence of Being. To resurrect this depth model of the self (as Fredric Jameson would term it) would be to regress to a coherent meta-narrative which reinstates all the problems of metaphysics through the back door. However, the fear for Kristeva is that, without a language of the soul, the self is 'swept away by insignificant and valueless objects' that erode any notion of depth, leaving a 'false' or 'amphibian' self in its wake who stumbles

---

[1]   Julia Kristeva, *New Maladies of the Soul*, trans. Ross Guberman (New York: Columbia U.P., 1995), p. 4.
[2]   Ibid.
[3]   Ibid.
[4]   Ibid., p. 7.

around in the 'darkroom' of contemporary culture without direction or purpose.[5] Left only with the empty shells of bodies, Kristeva argues that we no longer have the capacity, nor the desire, to turn on the light in the darkroom for fear of metaphysical apparitions which the modern self has supposed to abandon for good. We might be able to claim the conquest of language over metaphysics, but with no inner agency to articulate words in meaningful ways, any vestigial traces of the soul are left for the popular media to dilute into platitudes of display or entertainment (a condition into which Stanley Cavell fears moral perfectionism will degenerate). In the face of this danger, the challenge for romantic science is twofold: firstly, to rescue a sense of self, without reverting to nineteenth-century liberal models of sovereignty or autonomy and, secondly, to discuss the 'experience of self' in potentially therapeutic ways, in order to counteract the death of 'intimacy' which Kristeva claims to be endemic in late twentieth-century culture.[6]

Each of the five figures discussed in this book indicate new ways to redescribe selfhood using a variety of theoretical models, but they encounter numerous methodological problems and institutional barriers which threaten to undermine their projects. For example, Erikson worried that the creative phase of psychoanalysis had expired as early as the 1950s and the resulting professional orthodoxy had actually become an 'instrument of a new kind of anxiety'.[7] The Jamesian notion of 'creative risk' provides a conceptual tool for understanding the achievements, as well as the weaknesses, of twentieth-century romantic science, but does not indicate where speculative experiment should end and theoretical stability begin. Indeed, if romantic science is destined to be a counter-tradition to 'normal science' (in itself a shifting construct) then its practitioners must self-consciously prevent it from achieving any kind of orthodoxy. Its speculative impetus must emphasize the dangers of creating therapeutic narratives as means of self-justification, as much as the possibilities of alleviating adverse conditions for the self. All five figures reinject the language of 'soul' into scientific debate (although they distance themselves from its explicitly religious connotations), but are wary about subduing theoretical speculation or overlooking the broader cultural implications of the term. The language of the soul may enable theorists like Kristeva to switch on the light in the cultural darkroom, but it cannot resolve the conceptual problems of the self without threatening the *raison d'être* of romantic science.

Oliver Sacks continues to explore the limits of narrative and what lies beyond the common-sense beliefs and definitions of 'the human', emphasizing intersubjective communication and cultural transit as two of the defining

---

[5]   Ibid., pp. 7–8.

[6]   Ibid., p. 26.

[7]   Quoted from a 1968 interview with Erikson in Robert Coles, 'Psychoanalysis: The American Experience', in Michael Roth, ed., *Freud: Conflict and Culture*, op. cit., p. 148.

parameters of romantic science. In *The Island of the Colour-blind*, for example, his neuroanthropological explorations in the Pacific islands of Micronesia display his openness to and empathy for its congenitally colour-blind inhabitants ('my visits . . . were . . . not intended to prove or disprove any thesis, but simply to observe'), but also reveal 'primeval' genetic currents which propel his study far beyond the straightforward observation of human activity.[8] This trajectory in Sacks' work is a development of James' interest in the fringes of experience, Rank's emphasis on the influence of intrauterine impulses on human behaviour, Binswanger's conception of individual world-designs and Erikson's epigenetic model of the human life cycle, in addition to their shared interest in what Kristeva calls the 'foreigner [who] lives within us . . . the hidden face of our identity that wrecks our abode'.[9] By combining a child's-eye view of the islands ('a place that was not just botanical for me, but had an element of the mystical'), an aesthetic awareness of his environment ('my senses were actually enlarging, as if a new sense, a time sense, was opening within me') and the scientist's interest in the 'beginnings and originations of all things', Sacks best epitomizes the hybrid nature of romantic science in contemporary thought.[10] This fusion of perspectives both enables him to address what Hannah Arendt calls the indeterminacies of 'the human condition' more intimately than an orthodox scientific approach would permit, and allows him to explore 'realms' which are ostensibly 'remote from the moral or the human' but actually hide archaic forces which continue to impact on human behaviour.[11] Sacks' exemplary study of what can be described as 'enthnographic poetics' is one area of development for romantic science in the early twenty-first century.[12]

Adam Phillips, a Principal Child Psychotherapist in the Wolverton Gardens Child and Family Consultation Centre, London, is another contemporary thinker who bridges the divide between cultural criticism and human science. The subtitle of his volume of essays, *On Kissing, Tickling and Being Bored: Psychoanalytic Essays on the Unexamined Life* (1993), suggests that the uncertainty and unknowability at the heart of romantic science are still productive areas of exploration for practising analysts. Phillips engages with orthodox Freudian ideas throughout his work, but the imaginative exuberance of his essays reveal a romantic attachment to his subject of study. He shares with other contemporaries like Roy Schafer an insistence that the narrative and

---

[8]   Sacks, *The Island of the Colour-blind* (London: Picador, 1996), pp. viii, 224.

[9]   Kristeva, *Strangers to Ourselves*, trans. Leon S. Roudiez (New York: Columbia U.P., 1991), p. 1.

[10]   Sacks, *The Island of the Colour-blind*, op. cit., p. 224.

[11]   See Hannah Arendt, *The Human Condition* (Chicago: University of Chicago Press, 1958); Sacks, *The Island of the Colour-blind*, op. cit., p. 224.

[12]   See George Marcus and Michael Fischer, *Anthropology as Cultural Critique: An Experimental Moment in the Human Sciences* (Chicago: University of Chicago Press, 1986), pp. 73–6.

conversational elements of psychoanalysis (rather than its scientific base) are vital if the analyst is 'to return the reader to his own thoughts' and 'to evoke by provocation'.[13] His attempt to make analytic insights available for a general reading public is central to the extended essay *Terrors and Experts* (1995), in which he argues that the language of expertise can only be productive if analysts eschew the idolatry of the 'religious cult' and understand that 'they are only telling stories about stories; and that all stories are subject to an unknowable multiplicity of interpretations'.[14] Like the five romantic scientists considered here, Phillips argues that current analytic practice should emphasize more than ever 'the fascination of fictions, and the morals of words'.[15]

Another important dimension of romantic science of growing interest is the emphasis on narrative self-creation. For example, the emerging field of 'critical psychology' has begun to absorb recent work in the humanities to develop constructivist notions of the self, without abandoning personal identity entirely to social and cultural forces.[16] One recent essay by Arthur Frank, 'Stories of Illness As Care of the Self' (1998), applies the Foucauldian notion of the 'aesthetics of self' to the practical problems of patient care, with the claim that 'ethics may be no more optional in life than stories are'.[17] Frank does not attempt to circumvent the problem of power within institutional contexts (which is central to Foucault's work and often overlooked by the five romantic scientists considered here), but interrogates the way in which '"therapeutic" and "liberating" speech' is often devised 'as a ruse to control the subject . . . by convincing us of the selves we want and need to become, in order to be "true" to ourselves'.[18] In this sense, the language of authenticity is always in danger of disempowering the self or becoming trivialized, unless 'a shared horizon of moral significance' can be established.[19] The tricky balancing act between attending to the specificities of an individual life and maintaining 'this shared horizon' is central to romantic science and, as Frank claims, may challenge both the 'reader' and the 'teller' of these stories and facilitate 'new kinds of relationships between them'.[20] All five of the

---

[13]   Adam Phillips, *On Kissing, Tickling and Being Bored* (London: Faber & Faber, 1994), p. xx.

[14]   Phillips, *Terrors and Experts*, op. cit., p. xvi. Phillips argues that there are two figures – the 'Enlightenment Freud' and the 'post-Freudian Freud' - at war in Freud's work: the former views the analyst and patient as 'proto-scientists' and the latter understands that they often adopt the roles of 'lover', 'comedian' and 'mystic'; ibid., pp. 7, 9.

[15]   Ibid., p. xvii.

[16]   See, for example, Alan Peterson and Robert Bunton (eds), *Foucault: Health and Medicine* (London: Routledge, 1997).

[17]   Arthur W. Frank, 'Stories of Illness as Care of the Self: A Foucauldian Dialogue', *Health*, 2(3) (1998), 331.

[18]   Ibid., 333.

[19]   Ibid., 343.

[20]   Ibid., 345.

romantic scientists considered here would argue that these new relationships should not involve a retreat from the realities of illness and patient care, but engage with new 'technologies' (in the Foucauldian sense), not as a threat to identity, but as a means for revising or augmenting dominant views of the self.

Bibliography

Abrams, M.H., *The Mirror and the Lamp: Romantic Theory and the Critical Tradition* (New York: Oxford U.P., 1953).
————, *Natural Supernaturalism: Tradition and Revolution in Romantic Literature* (New York: Norton, 1973).
Allen, Richard C., 'When Narrative Fails', *Journal of Religious Ethics*, 21(1) (1993), 27–67.
Allott, R.M. 'Letters', *The Listener* (30 November 1972), 756.
Anderson, Pamela, 'Having It Both Ways: Ricoeur's Hermeneutics of the Self', *Oxford Literary Review*, 15(1–2) (1993), 227–52.
Armstrong, Tim, *Modernism, Technology and the Body: A Cultural Study* (Cambridge, Cambridge U.P., 1998).
Augustine, *Confessions*, trans. Henry Chadwick (Oxford: Oxford U.P., 1992).
Bachelard, Gaston, *The Psychoanalysis of Fire*, trans. Alan Ross (London: Quartet, 1987).
Bain, Alexander, *The Emotions and the Will*, 4th edn (London: Longmans, Green & Co, 1899).
Baldwin, James, *The Fire Next Time* (London: Penguin, 1964).
Bakhtin, Mikhail, *Rabelais and his World*, trans. Helene Iswolsky (Bloomington: Indiana U.P., 1984).
Barzun, Jacques, *A Stroll With William James* (Chicago: Chicago U.P., 1983).
Baudrillard, Jean, *The Ecstasy of Communication*, ed. Sylvère Lotringer, trans. Bernard and Caroline Schutze (New York: Semiotext(e), 1987).
Becker, Ernest, *The Denial of Death* (New York: Free Press, 1973).
Berlin, Isaiah, *Against the Current: Essays in the History of Ideas*, ed. Henry Hardy (London: Hogarth, 1979).
Bermudez, José Luis et al. (eds), *The Body and the Self* (Cambridge, MA: MIT Press, 1996).
Bettelheim, Bruno, *The Uses of Enchantment* (London: Penguin, 1991).
————, *Freud and Man's Soul* (London: Chatto & Windus, 1983).
————, *The Informed Heart* (London: Penguin, 1991).
Binswanger, Ludwig, 'On the Relationship between Husserl's Phenomenology and Psychological Insight', *Philosophy and Phenomenological Research*, 2 (1941), 199–210.
————, 'Existential Analysis and Psychotherapy', in eds Frieda Fromm-Reichmann and J.L. Moreno, *Progress in Psychotherapy* (New York: Grune & Stratton, 1956), pp. 144–8.
————, *Sigmund Freud, Reminiscences of a Friendship*, trans. Nobert Guterman (New York: Grune & Stratton, 1957).
————, 'On a Quote from Hofmannsthal: "What Spirit Is, Only the Oppressed Can Grasp"', trans. Vernon Gras and Irmgard Hobson, *Boundary* 2, 9(2) (1981), 187–95.
————, 'Dream and Existence', trans. Jacob Needleman, in David Hoeller, ed., *Dream and Existence: Michel Foucault and Ludwig Binswanger* (New Jersey: Humanities Press, 1993), pp. 80–105.
Bixler, Julius, 'The Existentialists and William James', *The American Scholar*, 28 (1959), 80–90.
Björkman, Stig et al. (eds), *Bergman on Bergman*, trans. Paul Britten Austin (New York: Simon & Schuster, 1973).
Blackwell, Marilyn, *Gender and Representation in the Films of Ingmar Bergman* (Columbia, SC: Cambden House, 1997).

Blake, William, *William Blake and his Contemporaries* (Cambridge: Fitzwilliam Museum Enterprises, 1986).

Blauner, Jacon, 'Existential Analysis: Ludwig Binswanger's *Daseinanalyse'*, *Psychoanalytic Review*, 44 (1957), 51–64.

Bordo, Susan, 'Anorexia Nervosa: Psychopathology as the Crystallization of Culture', *The Philosophical Forum*, 17(2) (1986), 73–104.

Boss, Medard, *'I dreamt last night . . .'*, trans. Stephen Conway (New York: Gardner Press, 1977).

————, *Existential Foundations of Medicine and Psychology*, trans. Stephen Conway and Anne Cleaves (New York: Aronson, 1979).

————, *Psychoanalysis and Daseinanalysis*, trans. Ludwig B. Lefebre (New York: Da Capo Press, 1982).

Bowie, Andrew, *Aesthetics and Subjectivity: from Kant to Nietzsche* (Manchester: Manchester U.P., 1993).

————, *Schelling and Modern European Philosophy: An Introduction* (London: Routledge, 1993).

————, *From Romanticism to Critical Theory: The Philosophy of German Literary Theory* (London: Routledge, 1997).

Boyle, Nicholas, *Goethe: The Poet and the Age: Volume 1, The Poetry of Desire (1749–1790)* (Oxford: Clarendon, 1991).

Brabant, E. et al. (eds), *The Correspondence of Sigmund Freud and Sandor Ferenczi: Volume 1, 1908–1914*, trans. Peter Hoffer (Cambridge: Harvard U.P., 1993).

Brody, Howard, *Stories of Sickness* (New Haven: Yale U.P., 1987).

Bromberg, Walter, Review of *Childhood and Society*, *Mental Hygiene* (October 1951), 642–4.

Bronowski, Jacob, *Science and Human Values* (London: Hutchinson, 1961).

Brown, Norman O., *Life Against Death: The Psychoanalytical Meaning of History*, 2nd edn (Middletown: Wesleyan U.P., 1985).

Bruns, Gerald L., 'Loose Talk about Religion from William James', *Critical Inquiry*, 11(2) (1984), 299–316.

Buber, Martin, *I and Thou*, trans. R.G. Smith (Edinburgh: T. & T. Clark, 1987).

Buck, Claire, *H.D. and Freud: Bisexuality as a Feminine Discourse* (New York: Harvester Wheatsheaf, 1991).

Buckley, J.H., *The Victorian Temper* (Cambridge: Cambridge U.P., 1981).

Burke, Kenneth, 'Without Benefit of Politics', *The Nation* (18 July 1936), p. 78.

————, *A Grammar of Motives* (Berkeley: University of California Press, 1969).

Burstow, Bonnie, 'A Critique of Binswanger's Existential Analysis', *Review of Existential Psychology and Psychiatry*, 17(2–3) (1981), 245–52.

Caldwell, Patricia, *The Puritan Conversion Narrative: The Beginnings of American Expression* (Cambridge: Cambridge U.P., 1983).

Canguilhem, Georges, *The Normal and the Pathological*, trans. Carolyn R. Fawcett and Robert S. Cohen (New York: Zone Books, 1991).

Carafiol, Peter, *The American Ideal: Literary History as Worldly Activity* (New York: Oxford U.P., 1991)

Carlyle, Thomas, *Selected Writings*, ed. Alan Shelston (London: Penguin, 1971).

————, *Sartor Resartus*, eds Kerry McSweeney and Peter Sabor (Oxford: Oxford U.P., 1991).

Carpenter, William B., *Principles of Mental Physiology* (London: Henry King & Co, 1874).

Cavell, Stanley, *Pursuits of Happiness* (Cambridge, MA: Harvard U.P., 1981).

————, *In Quest of the Ordinary: Lines of Skepticism and Romanticism* (Chicago: University of Chicago Press, 1981).

————, *This New Yet Unapproachable America: Lectures after Emerson after Wittgenstein* (Albuquerque: Living Batch Press, 1989).

————, *Conditions Handsome and Unhandsome: The Constitution of Emersonian Perfectionism* (Chicago: University of Chicago Press, 1990).

————, *The Sense of Walden: An Expanded Edition* (Chicago: University of Chicago Press, 1992).

————, *A Pitch of Philosophy: Autobiographical Exercises* (Cambridge, MA: Harvard U.P., 1994).

————, *Philosophical Passages: Wittgenstein, Emerson, Austin, Derrida* (Oxford: Blackwell, 1995).

Chambers, Iain, *Migrancy, Culture, Identity* (London: Routledge, 1994).

Chapple, Gerald and Hans H. Schute (eds), *The Turn of the Century, German Literature and Art, 1890–1915* (Bonn: McMaster, 1981).

Churchill, Larry R., 'The Human Experience of Dying: The Moral Primacy of Stories over Stages', *Soundings*, 62 (1979), 24–37.

Churchill, Larry R. and Sandra W. Churchill, 'Storytelling in Medical Arenas: The Art of Self-Determination', *Literature and Medicine*, 1 (1982), 73–9.

Cixous, Hélène, 'Fiction and Its Phantoms: A Reading of Freud's Das Unheimliche', *New Literary History*, 7(3) (1976), 525–48.

Clifford, Gay, *The Transformation of Allegory* (London: Routledge & Kegan Paul, 1974).

Coates, Paul, *The Gorgon's Gaze: German Cinema, Expressionism and the Image of Horror* (Cambridge: Cambridge U.P., 1991).

Cohn, Hans, *Existential Thought and Therapeutic Practice* (London, Sage, 1997).

Coleridge, Samuel Taylor, *Biographia Literaria*, ed. George Watson (London: Everyman, 1984).

Coles, Robert, *Erik Erikson: The Growth of His Work* (New York: Da Capo, 1970).

————, *Anna Freud: The Dream of Psychoanalysis* (Reading, MA: Addison-Wesley, 1992).

Copjec, Joan (ed.), *Supposing the Subject* (London: Verso, 1994).

Corlett, William, *Community Without Unity* (Durham, NC: Duke U.P., 1993).

Cotkin, George, 'William James and the "Weightless" Nature of Modern Existence', *San Jose Studies*, 12(2) (1986), 7–19.

————, *William James: Public Philosopher* (Baltimore: Johns Hopkins U.P., 1990).

Critchley, Simon, *Very Little . . . Almost Nothing: Death, Philosophy, Literature* (London: Routledge, 1997).

Culler, Jonathan, *Structuralist Poetics* (London: Routledge, 1975).

————, *The Uses of Uncertainty* (London: Elek, 1984).

Dahlborn, Bo (ed.), *Dennett and His Critics* (Oxford: Blackwell, 1995).

Deleuze, Gilles, *Difference and Repetition*, trans. Paul Patton (London: Athlone, 1994).

Deleuze, Gilles and Félix Guattari, *A Thousand Plateaus: Capitalism and Schizophrenia*, trans. Brian Massumi (London: Athlone, 1992).

de Lauretis, Teresa, 'Rebirth in The Bell Jar', *Women's Studies*, 3(2) (1976), 173–83

de Man, Paul, *Blindness and Insight: Essays in the Rhetoric of Contemporary Criticism* (London: Methuen, 1983).

————, *Romanticism and Contemporary Criticism: The Gauss Seminar and Other Papers*, E.S. Burt et al., eds (Baltimore: Johns Hopkins U.P., 1993).

Dennett, Daniel C., *Content and Consciousness* (London: Routledge & Kegan Paul, 1969).

————, 'Why Everyone is a Novelist', *Times Literary Supplement* (16 September 1988), p. 1016; pp. 1028–9.

————, *The Intentional Stance* (Cambridge, MA: MIT Press, 1990).

————, *Consciousness Explained* (London: Penguin, 1993).

————, 'Our Vegetative Soul', *Times Literary Supplement* (25 August 1995), 3–4.

————, *Darwin's Dangerous Idea: Evolution and the Meanings of Life* (London: Penguin, 1995).

Dennett, Daniel and John Searle, '"The Mystery of Consciousness": An Exchange', *New York Review of Books*, 62(20) (1995), 83–4.

Derrida, Jacques, *Speech and Phenomena and Other Essays on Husserl's Theory of Signs*, trans. David B. Allison (Evanston: Northwestern U.P., 1973).

————, *Writing and Difference*, trans. Alan Bass (London: Routledge, 1978).

————, *Dissemination*, trans. Barbara Johnson (London: Athlone, 1981).

Descartes, René, *A Discourse on Method*, trans. John Veitch (London: Dent, 1986).

Dewey, John, *Problems of Men* (New York: Philosophical Library, 1946).

Dews, Peter, *Logics of Disintegration: Post-Structuralist Thought and the Claims of Critical Theory* (London: Verso, 1987).

————, *The Limits of Disenchantment: Essays on Contemporary European Philosophy* (London: Verso, 1995)

Diggins, John Patrick, *The Promise of Pragmatism: Modernism and the Crisis of Knowledge and Authority* (Chicago: University of Chicago Press, 1994).

Donadio, Stephen, *Nietzsche, Henry James, and the Artistic Will* (New York: Oxford U.P., 1978).

Dreyfus, Hubert, *Being-in-the-World: A Commentary on Heidegger's Being and Time, Division 1* (Cambridge, MA: MIT Press, 1991).

Dreyfus, Hubert and Paul Rabinow, *Michel Foucault: Beyond Structuralism and Hermeneutics* (Chicago: University of Chicago Press, 1982).

DuBow, Wendy, *Conversations with Anaïs Nin* (Jackson: University Press of Mississippi, 1994).

Edelman, Gerald, *The Remembered Present: A Biological Theory of Consciousness* (New York: Basic Books, 1978).

————, *Bright Air, Brilliant Fire* (New York: Penguin, 1994).

Edie, James M., *William James and Phenomenology* (Bloomington: Indiana U.P., 1987).

Edmundson, Mark, *Towards Reading Freud: Self-creation in Milton, Wordsworth, Emerson and Sigmund Freud* (Princeton, NJ: Princeton U.P., 1990).

Eisner, Lotte, *The Haunted Screen*, trans. Roger Greaves (Berkeley: University of California Press, 1969).

Ellenberger, Henri F., *The Discovery of the Unconscious: The History and Evolution of Dynamic Psychiatry* (New York: Basic Books, 1970).

Ellison, Ralph, *Invisible Man* (London: Penguin, 1965).

Emerson, Ralph Waldo, *Essays and Lectures* (New York: Library of America, 1983).

————, *Essays and Poems*, ed. Tony Tanner (London: Everyman, 1992).

Engelhart, Dietrich von, 'Romanticism in Germany', in Roy Porter and Mikulas Teich, eds, *Romanticism in National Context* (Cambridge: Cambridge U.P., 1988), pp. 109–33.

Epstein, Julia, 'Historigraphy, Diagnosis and Poetics', *Literature and Medicine*, 11(1) (1992), 23–44.

Erikson, Erik H., *Young Man Luther: A Study in Psychoanalysis and History* (New York: Norton, 1962).

————, *Gandhi's Truth: On the Origins of Militant Nonviolence* (New York: Norton, 1970).

————, 'Autobiographic Notes on the Identity Crisis', *Daedalus*, 99(4) (1970), 730–59.

————, *Life History and the Historical Moment* (New York: Norton, 1975).

————, 'Reflections on Dr. Borg's Life Cycle', *Daedalus*, 105(2) (1976),1–28.

————, *Dimensions of a New Identity* (New York: Norton, 1979).

————, *A Way of Looking At Things: Selected Papers of Erik H. Erikson 1930–1980*, ed. Stephen Schlein (New York: Norton, 1987).

————, *Identity and the Life Cycle* (New York: Norton, 1994).

————, *Identity: Youth and Crisis* (New York: Norton, 1994).

————, *Insight and Responsibility* (New York: Norton, 1994).

————, *The Life Cycle Completed: A Review* (New York: Norton, 1995).

———— (ed.), *Youth: Change and Challenge* (New York: Basic Books, 1963).

Erikson, Joan, *Wisdom and the Senses: The Way of Creativity* (New York: Norton, 1991).

Faatz, Anita J., 'Individuals in Association', *Journal of the Otto Rank Association*, 1(1) (Fall 1966), 81–8.

————, 'The Summing Up', *Journal of the Otto Rank Association*, 7(1) (June 1972), 41–9.

Feinstein, Howard M., 'The Use and Abuse of Illness in the James Family Circle: A View of Neurasthenia as a Social Phenomenon', *The Psychohistory Review*, 8(1–2) (1979), 6–14.

————, *Becoming William James* (Ithaca: Cornell U.P., 1984).

Ferenczi, Sandor, *The Clinical Diary of Sandor Ferenczi*, ed. J. Dupont, trans. M. Balint and N.Z. Jackson (Cambridge: Harvard U.P., 1988).

Ferenczi, Sandor and Otto Rank, *The Development of Psychoanalysis* (New York: Nervous & Mental Disease Publishing Co, 1925).

Fichtner, Gerhard (ed.), *Sigmund Freud/Ludwig Binswanger: Briefwechsel 1908–1938* (Berlin: Fischer, 1992).

Fischer, Michael, *Stanley Cavell and Literary Skepticism* (Chicago: University of Chicago Press, 1989).

Fisher, Philip, 'The Failure of Habit', in ed. Monroe Engel, *Uses of Literature* (Cambridge, MA: Harvard U.P., 1973), pp. 3–18.

Fitch, Noël Riley, *The Erotic Life of Anaïs Nin* (Boston: Little, Brown, 1993).

Flood, David H. and Rhonda L. Soricelli, 'Development of the Physician's Narrative Voice in the Medical Case History', *Literature and Medicine*, 11(1) (1992), 64–83.

Foster, Hal (ed.) *Postmodern Culture* (London: Pluto, 1983).

Foucault, Michel, *Mental Illness and Psychology*, trans. Alan Sheridan (New York: Harper & Row, 1976).

————, *The Order of Things: An Archeology of the Human Sciences*, trans. unidentified collective (London: Routledge, 1986).

————, *Technologies of the Self: A Seminar with Michel Foucault*, ed. Luther Martin et al. (London: Tavistock, 1988).

————, *The Birth of the Clinic: An Archeology of Medical Perception*, trans. Alan Sheridan (London: Routledge, 1989).

————, *The History of Sexuality Volume 1: An Introduction*, trans. Robert Hurley (London: Penguin, 1990).

————, *The History of Sexuality Volume 2: The Use of Pleasure*, trans. Robert Hurley (London: Penguin, 1990).

————, *The History of Sexuality Volume 3: The Care of the Self*, trans. Robert Hurley (London: Penguin, 1990).

————, *Remarks On Marx: Conversations with Duccio Trombadori*, trans. R. James Goldstein and James Cascaito (New York: Semiotext(e), 1991).

————, 'Dream, Imagination and Existence', in ed. David Hoeller, trans. Forrest Williams, *Dream and Existence: Michel Foucault and Ludwig Binswanger* (New Jersey: Humanities Press, 1993), pp. 31–78.

————, *Ethics: Subjectivity and Truth*, ed. Paul Rabinow, trans. Robert Hurley (London: Penguin, 1997).

Frank, Arthur W., 'The Rhetoric of Self-Change: Illness Experience as Narrative', *The Sociological Quarterly*, 34(1) (1993), 39–52.

Frazer, James, *The Golden Bough* (London: Wordsworth, 1993).

French, Philip and Kersti French, *Wild Strawberries* (London: BFI Publishing, 1995).

Freud, Anna, *Selected Writings*, eds Richard Ekins and Ruth Freeman (London: Penguin, 1998).

Freud, Sigmund, *New Introductory Lectures on Psycho-analysis*, 3rd edn, trans. W.J.H. Sprott (London: Hogarth, 1946).

————, *Introductory Lectures on Psycho-analysis*, 2nd edn, trans. Joan Riviere (London: Allen & Unwin, 1961).

————, *On Metapsychology: The Theory of Psychoanalysis*, ed. Angela Richards, trans. James Strachey (London: Penguin, 1984).

————, *Art and Literature*, ed. Albert Dickson, trans. James Strachey (London: Penguin, 1990).

————, *The Interpretation of Dreams*, ed. Angela Richards, trans. James Strachey (London: Penguin, 1991).

————, *The Psychopathology of Everyday Life*, ed. Angela Richards, trans. Alan Tyson (London: Penguin, 1991).

Freud, Sigmund and Josef Breuer, *Studies on Hysteria*, ed. Angela Richards, trans. James and Alix Strachey (London: Penguin, 1974).

Friedman, Lawrence, *Identity's Architect: A Biography of Erik H. Erikson* (New York: Scribners, 1999).

Fromm, Eric, *Fear of Freedom* (London: Ark, 1984).

————, *The Crisis of Psychoanalysis* (New York: Holt & Co., 1991).

Frye, Northrop, 'Levels of Meaning in Literature', *Kenyon Review*, 12 (1950), 246–62.

————, *Anatomy of Criticism* (London: Penguin, 1957).

Funkenstein, Amos, 'The Polytheism of William James', *Journal of the History of Ideas*, 55(1) (1994), 99–111.

Gay, Peter, *Freud: A Life for our Time* (London: Macmillan, 1989).

Gilligan, Carol, *In a Different Voice: Psychological Theory and Women's Development* (Cambridge, MA: Harvard U.P., 1993).

Gilman, Sander L., *Disease and Representation: Images of Illness from Madness to AIDS* (Ithaca: Cornell U.P., 1988).

Gilroy, Paul, *The Black Atlantic: Modernity and Double Consciousness* (London: Verso, 1993).

Glover, Jonathan, *I: The Philosophy and Psychology of Personal Identity* (London: Penguin, 1991).

Goethe, Johann Wolfgang von, *Essays on Art and Literature*, John Geary ed., *The Collected Works*, vol.e 3 (New Jersey: Princeton U.P., 1980).

Goffman, Erving, *The Presentation of the Self in Everyday Life* (London: Penguin, 1961).

————, *Stigma: Notes On the Management of Spoiled Identity* (Harmondsworth: Penguin, 1968).

————, *Asylums* (London: Penguin, 1970).

Goodman, Russell, *American Philosophy and the Romantic Tradition* (Cambridge: Cambridge U.P., 1990).

Goy-Blanquet, Dominique, 'Slides of the Cortex', *Times Literary Supplement* (14 May 1993), 20.

Graham, David T., 'Health, Disease, and the Mind-Body Problem: Linguistic Parallelism', *Psychosomatic Medicine*, 29(1) (1967), 52–71.

Gras, Vernon W., *European Literary Theory and Practice: From Existential Phenomenology to Structuralism* (New York: Dell, 1973).

Greenblatt, Stephen (ed.), *Allegory and Representation* (Baltimore: Johns Hopkins U.P., 1984).

Grotjahn, Martin, 'Otto Rank on Homer and Two Unknown Letters from Freud to Rank in 1916', *Journal of the Otto Rank Association*, 4(1) (1969), 77.

Hale, Nathan G., Jr, *The Rise and Crisis of Psychoanalysis in the United States: Freud and the Americans, 1917–1985* (New York: Oxford U.P., 1995).

Halliwell, Martin, 'The Trembling Nether Lip: Some Modern Versions of Idiocy', unpublished MA dissertation, University of Exeter (1993).

———, 'Freud at Coney Island: The European Imagination in America', *Over Here: A European Journal of American Culture*, 17(1) (1997), 53–66.

Hampshire, Stuart, 'Amiable Genius', *New York Review of Books*, 35(6) (1988), 17–20.

Harris, Kenneth, *Carlyle and Emerson: Their Long Debate* (Cambridge, MA: Harvard U.P., 1978).

Haughton, Hugh, 'House-calls to Space Cadets', *The Independent on Sunday* (5 February 1995), p. 31.

Havens, Leston L., 'The Development of Existential Psychiatry', *Journal of Nervous and Mental Disease*, 154 (1972), 309–31.

Heidegger, Martin, *Being and Time*, trans. John Macquarrie and Edward Robinson (Oxford: Blackwell, 1962).

———, *Basic Writings*, ed. David Farrel Krell (London: Routledge, 1978).

———, *Basic Problems of Phenomenology*, trans. Albert Hofstadter (Bloomington: Indiana U.P., 1982).

Heller, Erich, *The Disinherited Mind* (Cambridge: Bowes & Bowes, 1952).

Heuscher, Julius E., 'Dread and Authenticity', *The American Journal of Psychoanalysis*, 49(2) (1989), 139–57.

Higham, John, *Writing American History* (Bloomington: Indiana U.P., 1970).

Hinderev, Drew, 'William James on Mystical Epistemology', *Michigan Academician*, 25(1) (1992), 1–7.

Hoare, Rupert, 'A Comparison of the Work of Bultmann and Binswanger', unpublished PhD thesis, University of Birmingham (1973).

Hodgson, Shadworth H., *The Philosophy of Reflection*, vols 1 and 2 (London: Longmans, Green & Co, 1878).

Hoeller, Keith (ed.), *Heidegger and Psychology*, special issue of *Review of Existential Psychology and Psychiatry*, 16(1–3) (1979).

———, *Readings in Existential Psychology and Psychiatry*, special issue of *Review of Existential Psychology and Psychiatry*, 20(1–3) (1987).

———, *Dream and Existence: Michel Foucault and Ludwig Binswanger* (New Jersey: Humanities Press, 1993).

Hoffman, E.T.A., 'The Sandman', *The Golden Pot and Other Tales*, trans. Ritchie Robertson (Oxford: Oxford U.P, 1992), pp. 85–118.

Hofstadter, Douglas and Daniel Dennett, *The Mind's I: Fantasies and Reflections on Self and Soul* (London: Penguin, 1981).

Hofstadter, Richard, *Social Darwinism in American Thought* (Boston: Beacon Press, 1992).

Howarth, William, 'Oliver Sacks: the Ecology of Writing Science', *Modern Language Studies*, 20(4) (1990), 103–20.

Howe, Irving, *The American Newness: Culture and Politics in the Age of Emerson* (Cambridge, MA: Harvard U.P., 1986).

Hume, David, *A Treatise of Human Nature*, ed. L.A. Selby-Bigge (Oxford: Clarendon, 1967).

Hummel, Hermann, 'Emerson and Nietzsche', *New England Quarterly*, 19 (1940), 63–84.

Hunter, Kathryn Montgomery, 'Remaking the Case', *Literature and Medicine*, 11(1) (1992), 163–79.

Husserl Edmund, *The Paris Lectures*, trans. Peter Koestenbaum (The Hague: Martinus Nijhoff, 1964)

———, *Experience and Judgment*, ed. Ludwig Landgrebe, trans. James S. Churchill (London: Routledge & Kegan Paul, 1973).

———, *Cartesian Meditations*, trans. Dorion Cairns (The Hague: Martinus Nijhoff, 1977).

Huxley, Aldous, *Literature and Science* (London: Chatto & Windus, 1963).

———, *The Doors of Perception and Heaven and Hell* (London: Flamingo, 1994).

Ignatieff, Michael, 'Life at Degree Zero', *New Society* (20 January 1983), pp. 95–7.

Ihde, Don, *Hermeneutic Phenomenology: The Philosophy of Paul Ricoeur* (Evanston: Northwestern U.P., 1971).

Irigaray, Luce, *The Irigaray Reader*, ed. Margaret Whitford (Oxford: Blackwell, 1991).

Izenberg, Gerald N., *The Existentialist Critique of Freud: The Crisis of Autonomy* (Princeton, NJ: Princeton U.P., 1976).

Jack, Homer A. (ed.), *The Gandhi Reader* (New York: Grove Press, 1956).

Jackson, Rosemary, *Fantasy: The Literature of Subversion* (London: Methuen, 1981).

Jacobsen, Jens Peter, *Niels Lyhne*, trans. Hanna Astrup Larsen (London: Oxford U.P., 1919).

James, Alice, *The Diary of Alice James*, ed. Leon Edel (Harmondsworth: Penguin, 1964).

James, William, 'Address' on *The Centenary of the Birth of Ralph Waldo Emerson, As Observed in Concord May 25 1903* (Concord: Riverside Press, 1903), pp. 67–77.

———, 'William James's Remarkable Plea for Medical Freedom', *Medical Freedom*, 1(9) (1912), 6–9.

———, *Letters of William James*, vols 1 and 2, ed. Henry James (Boston: Atlantic Press Monthly, 1920).

———, *The Principles of Psychology*, vols 1 and 2 (New York: Dover, 1950).

———, *The Will to Believe and Human Immortality* (New York: Dover, 1956).

———, *Essays in Radical Empiricism and A Pluralistic Universe* (Gloucester, MA: Peter Smith, 1967).

———, *Collected Essays and Reviews* (New York: Russell & Russell, 1969)

———, *The Moral Equivalent of War and Other Essays*, ed. John K. Roth (New York: Harper & Row, 1971).

———, *Pragmatism*, ed. Bruce Kuklick (Indianapolis: Hackett, 1981).

———, *The Varieties of Religious Experience: A Study in Human Nature*, ed. Martin E. Marty (London: Penguin, 1985).

———, *Manuscript Essays and Notes, The Works of William James*, ed. Frederick H. Burkhardt (Cambridge, MA: Harvard U.P., 1988).

Jameson, Fredric, *Postmodernism, Or the Cultural Logic of Late Capitalism* (New York: Verso, 1991).

Jaspers, Karl, *General Psychopathology*, 7th edn, trans. J. Hoening and Marian W. Hamilton (Manchester: Manchester U.P., 1962).

Jentsch, Ernst, 'On the Psychology of the Uncanny', trans. Roy Sellars, *Angelaki*, 2(1) (1995), 7–16.

Johnson, Michael G. and Tracy B. Henley (eds), *Reflections on The Principles of Psychology: William James after a Century* (Hillsdale, NJ: Lawrence Erlbaum, 1990).

Jones, Ernest, *The Life and Work of Sigmund Freud*, eds Lionel Trilling and Steven Marcus (London: Penguin, 1964).

Jung, Carl G., *Memories, Dreams, Reflections* (London: Fontana, 1967).

————, *Dream Analysis: Notes of the Seminar given in 1928–1930*, ed. William McGuire (London: Routledge & Kegan Paul, 1984).

Kakar, Sudhir, 'The Logic of Psychohistory', *The Journal of Interdisiplinary History*, 1(1), (1970), 194.

Kant, Immanuel, *Critique of Pure Reason*, ed. Vasilis Politis (London: Everyman, 1993).

Karpf, Fay B., *American Social Psychology* (New York: McGraw Hill, 1932).

————, *The Psychology and Psychotherapy of Otto Rank: An Historical and Comparative Introduction* (Westport: Greenwood Press, 1970).

Kearney, Richard, *Dialogues with Contemporary Continental Thinkers: The Phenomenological Heritage* (Manchester: Manchester U.P., 1984).

Kemp, T. Peter and David Rasmussen, *The Narrative Path: The Later Works of Paul Ricoeur* (London: MIT Press, 1989).

Kersnowski, Frank L. and Alice Hughes (eds), *Conversations with Henry Miller* (Jackson: University Press of Mississippi, 1994).

Kierkegaard, Søren, *Fear and Trembling/Repetition*, eds and trans. Howard V. Hong and Edna H. Hong (Princeton, NJ: Princeton U.P., 1983).

————, *The Sickness Unto Death*, trans. Alaistair Hannay (London: Penguin, 1989).

King, Richard H., *The Party of Eros: Radical Social Thought and the Realm of Freedom* (Chapel Hill: University of North Carolina Press, 1972).

Kirchner, John H., 'Ellen West, Existential Analysis and the Faustian Motif', *Review of Existential Psychology and Psychiatry*, 7(2) (1966), 119–27.

Kirmayer, Laurence J. (1992) 'The Body's Insistence on Meaning: Metaphor as Presentation and Representation in Illness Experience', *Medical Anthropology Quarterly*, 6(4) (1992), 323–46.

Kirschenbaum, Howard and Valerie Land Henderson (eds), *Carl Rogers: Dialogues* (London: Constable, 1990).

Klein, Dennis B., *The Jewish Origins of Psychoanalysis* (New York: Praegen, 1981).

Kling, Merle, 'The Intellectual: Will he Whither Away?', *New Republic* (8 April 1957), pp. 14–15.

Kolocotroni, Vassiliki, Jane Goldman and Olga Taxidou (eds), *Modernism: An Anthology of Sources and Documents* (Edinburgh: Edinburgh U.P., 1998).

Kracauer, Siegfried, *From Caligari to Hitler: A Psychological History of the German Film* (Princeton: Princeton U.P., 1947).

Kristeva, Julia, *Black Sun: Depression and Melancholia*, trans. Leon S. Roudiez (New York: Columbia U.P., 1989)

————, *Strangers to Ourselves*, trans. Leon S. Roudiez (London: Harvester Press, 1991).

————, *New Maladies of the Soul*, trans. Ross Guberman (New York: Columbia U.P., 1995).

Kuhn, Thomas, *The Structure of Scientific Revolutions* (Chicago: University of Chicago Press, 1970).

Kuklick, Bruce, *The Rise of American Philosophy, Cambridge, Massachusetts, 1860–1930* (New Haven: Yale U.P., 1977).

Kushner, Howard I., 'Pathology and Adjustment in Psychohistory: A Critique of the Erikson Model', *Psychocultural Review*, 1(4) (1977), 493–501.

Kusnetz, Ella, 'The Soul of Oliver Sacks', *Massachusetts Review*, 33(2) (1992), 175–98.

Lacan, Jacques, *Écrits: A Selection*, trans. Alan Sheridan (London: Norton, 1977).

Laing, R.D., *The Voice of Experience* (London: Allen Lane, 1982).

Lang, Candace, 'Autobiography in the Aftermath of Romanticism', *Diacritics*, 12(4) (1982), 2–16.

Lears, T.J. Jackson, *No Place of Grace: Antimodernism and the Transformation of American Culture, 1880–1920* (Chicago: University of Chicago Press, 1984).

Lentricchia, Frank, 'The Return of William James', *Cultural Critique*, 4 (1986), 5–31.

Lester, David, 'Ellen West's Suicide as a Case of Psychic Homicide', *Psychoanalytic Review*, 58 (1971), 251–63.

Lévi-Strauss, Claude, *Structural Anthropology, Volume 1*, trans. Claire Jacobson and Brooke Grundfest Schoef (New York: Basic Books, 1963).

Levin, Jonathan, 'The Esthetics of Pragmatism', *American Literary History*, 6(4) (1994), 658–83.

Levinas, Emmanuel, *The Levinas Reader*, ed. Seán Hand (Oxford: Blackwell, 1989).

Levinson, Henry Samuel, *The Religious Investigations of William James* (Chapel Hill: University of North Carolina Press, 1981).

Lewis, R.W.B., *The James's: A Family Narrative* (London: André Deutsch, 1991).

Lieberman, E. James, *Acts of Will: The Life and Work of Otto Rank* (New York: Free Press, 1985).

Lifton, Robert Jay, 'Protean Man', *Archives of General Psychiatry*, 24 (1971), 298–304.

———, *The Protean Self: Human Resilience in an Age of Fragmentation* (New York: Basic Books, 1993).

Linschoten, Hans, *On the Way to a Phenomenological Psychology, The Psychology of William James* (Pittsburgh: Duquesne U.P., 1968).

Lloyd, G.E.R., *Hippocratic Writings*, trans. J. Chadwick and E.T. Withington (London: Penguin, 1978).

Luria, Alexander, *The Man with a Shattered World: A History of a Brain Wound*, trans. Lynn Solotaroff (Harmondsworth: Penguin, 1975).

———, *The Making of Mind: A Personal Account of Soviet Psychology*, eds Michael Cole and Sheila Cole (Cambridge, MA: Harvard U.P., 1977).

Luther, Martin, *Basic Luther* (Springfield, IL: Templegate, 1954).

Lyotard, Jean-François, *The Postmodern Condition: The Power of Knowledge*, trans. Geoff Bennington and Brian Massumi (Manchester: Manchester U.P., 1984).

———, *The Lyotard Reader*, ed. Andrew Benjamin (Oxford: Blackwell, 1989).

———, *The Inhuman: Reflections on Time*, trans. Geoffrey Bennington and Rachel Bowlby (London: Polity, 1993).

Macey, David, 'Foucault before Foucault', *Radical Philosophy*, 68 (1994), 58–9.

MacCannell, Juliet Flower and Laura Zakarin (eds), *Thinking Bodies* (California: Stanford U.P., 1994).

MacLean, Paul D., 'On the Evolution of Three Mentalities', *Man-Environment Systems*, 5(4) (1975), 213–24.

McLeod, John, *Narrative and Psychotherapy* (London: Sage, 1997).

McLynn, Frank, *Carl Gustav Jung* (London: Bantam, 1996).

Marcuse, Ludwig, 'Nietzsche in America', trans. James C. Fleming, *South Atlantic Quarterly*, 50 (1951), 330–9.

Masson, J.M., *The Assault on Truth: Freud's Suppression of the Seduction Theory* (New York: Farrar, Straus & Giroux, 1984).

May, Rollo, *The Courage to Create* (London: Collins, 1976).

———, *The Discovery of Being* (New York: Norton, 1986).

May, Rollo et al. (eds), *Existence: A New Dimension in Psychiatry and Psychology* (New York: Basic Books, 1958).

McCurdy, Harold, 'Luther: Psychoanalyst', *Contemporary Psychology*, 4(7) (1959), 201–2.

McGuire, William, 'Jung's Complex Reactions (1907): Word Association Experiments Performed by Binswanger', *Spring* (1984), 1–34.

McGuire, William (ed.), *The Freud/Jung Letters*, trans. Ralph Manheim and R.F.C. Hull (London: Penguin, 1991).

McRae, Murdo M., 'Oliver Sacks's Neurology of Identity', in ed. Murdo McRae, *The Literature of Science Perspectives on Popular Scientific Writing* (Athens: University of Georgia Press, 1993), pp. 97–110.

Mead, Margaret, 'The Life Cycle and Its Variations: The Division of Roles', *Daedalus*, 96(3) (Summer 1967), 871–5.

Menaker, Esther, *Otto Rank: A Rediscovered Legacy* (New York: Columbia U.P., 1982).

Menand, Louis 'William James and the Case of the Epileptic Patient', *New York Review of Books*, 45(20) (17 December 1998), 81–93.

Merleau-Ponty, Maurice, *Phenomenology of Perception*, trans. Colin Smith (London: Routledge, 1962).

————, *The Primacy of Perception and Other Essays*, ed. James M. Edie (Evanston: Northwestern U.P., 1964).

————, *Sense and Nonsense*, trans. Hubert L. Dreyfus and Patricia Allen Dreyfus (Evanston: Northwestern U.P., 1964).

Meyer, Donald, *The Positive Thinkers* (New York: Doubleday, 1966).

Mill, John Stuart, *On Liberty* (London: Penguin, 1974).

————, *Autobiography and Literary Essays* (Toronto: University of Toronto Press, 1981).

Mill, John Stuart and Jeremy Bentham, *Utilitarianism and Other Essays* (London: Penguin, 1987).

Miller, David L., 'The "Stone" which is not a Stone: C.G. Jung and the Postmodern Meaning of "Meaning" ', *Spring*, 49 (1989), 110–22.

Miller, Henry, 'Encounter with Rank', *Journal of the Otto Rank Association*, 1(1) (Fall 1966), 57–65.

————, *Tropic of Cancer* (London: Flamingo, 1993).

————, *Tropic of Capricorn* (London: Flamingo, 1993).

Miller, James, *The Passion of Michel Foucault* (London: Flamingo, 1994).

Miller, J. Hillis, 'The Literary Criticism of Georges Poulet', *Modern Language Notes*, 78(5) (1963), 471–88.

————, 'The Geneva School', *Critical Quarterly*, 8(4) (1966), 305–21.

————, 'Geneva or Paris: the Recent Work of Georges Poulet', *University of Toronto Quarterly*, 39(3) (1970), 212–28.

————, *The Ethics of Reading: Kant, de Man, Eliot, Trollope, James, Benjamin* (New York: Columbia U.P., 1987).

Mishara, A.L., 'The Problem of the Unconscious in the Later Thought of L. Binswanger: A Phenomenological Approach to Delusion in Perception and Communication', *Analecta Husserliana*, 31 (1990), 247–78.

Mitchell, Solace and Michael Rosen (eds), *The Need for Interpretation* (London: Athlone, 1983).

Modell, Arnold H., *The Private Self* (Cambridge, MA: Harvard U.P., 1993).

Monroe, William Frank et al., ' "Is There a Person in This Case?" ', *Literature and Medicine*, 11(1) (1992), 45–63.

Morison, Samuel Eliot (ed.), *The Development of Harvard University* (Cambridge, MA: Harvard U.P., 1930).

Mulhall, Stephen, *On Being in the World: Wittgenstein and Heidegger on Seeing Aspects* (London: Routledge, 1993).

Murray, David (ed.), *American Cultural Critics* (Exeter: University of Exeter Press, 1995).

Myers, Gerald E., *William James: His Life and Thought* (New Haven: Yale U.P., 1986).

Nadelman, Heather L., 'Creating an Immortal Life: A Consideration of the Autobiography of Henry James, Sr.', *The New England Quarterly*, 66(2) (1993), 247–68.

Nagel, Thomas, *The View from Nowhere* (Oxford: Oxford U.P., 1986)

Nash, Christopher (ed.), *Narrative in Culture: The Uses of Storytelling in the Sciences, Philosophy, and Literature* (London: Routledge, 1990).

Needleman, Jacob (ed.), *Being-in-the-World: Selected Papers of Ludwig Binswanger*, trans. Jacob Needleman (New York: Basic Books, 1963).

Needleman, Jacob, 'Ludwig Binswanger: 1881–1966', *Review of Existential Psychology and Psychiatry*, 6(2) (1966), 168–70.

Nichols, Ashton, *The Poetics of Epiphany: Nineteenth-Century Origins of the Modern Literary Moment* (Tuscaloosa: University of Alabama Press, 1987).

Nietzsche, Friedrich, *The Will To Power*, trans. Walter Kaufman and R.J. Hollingdale (New York: Random House, 1968).

————, *The Gay Science*, trans. Walter Kaufmann (New York: Random House, 1974).

————, *Twilight of the Idols/The Anti-Christ*, trans. R.J. Hollingdale (London: Penguin, 1990).

————, *Human, All Too Human*, trans. Marion Faber and Stephen Lehmann, (London: Penguin, 1994).

Nin, Anaïs, *The Journals of Anaïs Nin, 1931–1934*, vol. 1, ed. Gunther Stuhlmann (London: Peter Owen, 1966).

————, *The Journals of Anaïs Nin, 1934–1939*, vol. 2, ed. Gunther Stuhlmann (London: Peter Owen, 1967).

————, *The Journals of Anaïs Nin, 1939–1944*, vol. 3, ed. Gunther Stuhlmann (London: Harcourt Brace Jovanovich, 1969).

————, *Anaïs Nin Reader*, ed. Philip K. Jason (London: Peter Owen, 1973).

————, *Delta of Venus* (London: Penguin, 1990).

————, *Winter of Artifice* (London: Peter Owen, 1991).

Nisbet, H.B., *Goethe and the Scientific Tradition* (London: University of London Press, 1972).

Noll, Richard, *The Encyclopedia of Schizophrenic and Psychiatric Disorders* (New York: Facts on File, 1992).

Noske, Barbara, *Humans and Other Animals: Beyond the Boundaries of Anthropology* (London: Pluto Press, 1989).

Novalis, *Pollen and Fragments: Selected Poetry and Prose of Novalis*, trans. Arthur Versluis (New York: Phane Press, 1989).

Norris, Christopher, *Deconstruction: Theory and Practice* (London: Methuen, 1982).

O'Donnell, Patrick, 'Ludwig Binswanger and the Poetics of Compromise', *Review of Existential Psychology and Psychiatry*, 17(2–3) (1981), 235–43.

Olin, Doris (ed.), *William James: Pragmatism in Focus* (London: Routledge, 1992).

Olney, James, *Metaphors of Self: The Meaning of Autobiography* (New Jersey: Princeton U.P., 1972).

Osborne, Peter, *The Politics of Time: Modernity and Avant-Garde* (London: Verso, 1995).

Parker, Gail Thain, *Mind Cure in New England: From the Civil War to World War I* (Hanover, NH: University Press of New England, 1973).

Pater, Walter, *The Renaissance: Studies in Art and Poetry*, ed. Adam Phillips (Oxford: Oxford U.P., 1986).

Pease, Donald, *Visionary Compacts: American Renaissance Writings in Cultural Context* (Madison: University of Wisconsin Press, 1987).

Pellegrino, Edmund, *Humanism and the Physician* (Knoxville: University of Tennessee Press, 1979).

Perry, Ralph Barton, *The Thought and Character of William James*, vols 1 and 2 (Boston: Little, Brown, 1935).

Peterson, Alan and Robert Bunton (eds), *Foucault: Health and Medicine* (London: Routledge, 1997).

Phillips, Adam, *On Kissing, Tickling and Being Bored* (London: Faber & Faber, 1994).

————, *Terrors and Experts* (London: Faber & Faber, 1995).

Poirier, Richard, *Poetry and Pragmatism* (London: Faber & Faber, 1992).

Poirier, Suzanne et al., 'Charting the Chart – An Exercise in Interpretation(s)', *Literature and Medicine*, 11(1) (1992), 1–22.

Posnock, Ross, *The Trial of Curiosity: Henry James, William James, and the Challenge of Modernity* (New York: Oxford U.P., 1991).

Poulet, Georges, *Studies in Human Time*, trans. Elliott Coleman (New York: Harper Torchbook, 1956).

————, 'The Self and the Other in Critical Consciousness', *Diacritics*, 2 (1972), 46–50.

Puffer, Ethel Dench, 'The Loss of Personality', *Atlantic Monthly*, 85 (1900), 195–204.

Putnam, Ruth Anna (ed.), *The Cambridge Companion to William James* (Cambridge: Cambridge U.P., 1997).

Rae, Patricia, 'William James and Contemporary Criticism', *Southern Review*, 21(3) (1988), 307–14.

Rajchman, John and Cornel West (eds), *Post-Analytic Philosophy* (New York: Columbia U.P., 1985).

Ramsay, Bennett, *Submitting to Freedom: The Religious Vision of William James* (New York: Oxford U.P., 1993).

Rank, Otto, 'Psychoanalysis of Organic Conditions', *Medical Journal and Record*, 120(3) (1924), xxxiii–xxxv.

————, 'Self-Inflicted Illness', *Proceedings of the California Academy of Medicine* (February 1935), 8–18.

————, *Beyond Psychology* (New York: Dover, 1941).

————, *The Myth of the Birth of the Hero and Other Writings*, ed. Philip Freund (New York: Vintage, 1959).

————, 'The Psychological Approach to Personal Problems', *Journal of the Otto Rank Association*, 1(1) (1966), 12–23.

————, *The Don Juan Legend*, ed. and trans. David G. Winter (Princeton, NJ: Princeton U.P., 1975).

————, *Will Therapy*, trans. Jessie Taft (New York: Norton, 1978).

————, *Truth and Reality*, trans. Jessie Taft (New York: Norton, 1978).

————, 'The Artist', trans. Eva Salomon, *Journal of Otto Rank Association*, 15(1), 1980, 1–63.

————, 'Literary Autobiography', trans. Jessie Taft, *Journal of Otto Rank Association*, 16(30) (1981), 3–38.

————, *Art and Artist: Creative Urge and Personality Development*, trans. Charles Francis Atkinson (New York: Norton, 1989).

————, *The Double: A Psychoanalytic Study*, ed. and trans. Harry Tucker, Jr. (London: Karnac, 1989).

————, *The Incest Theme in Literature and Legend: Fundamentals of a Psychology of Literary Creation*, trans. Gregory Richter (Baltimore: Johns Hopkins U.P., 1992).

————, *The Trauma of Birth*, trans. E. James Lieberman (New York: Dover, 1993).

————, *A Psychology of Difference: The American Lectures*, ed. Robert Kramer (Princeton, NJ: Princeton U.P., 1996).

————, *Psychology and the Soul*, trans. Gregory C. Richter and E. James Lieberman (Baltimore: Johns Hopkins U.P., 1998).

Rapaport, Herman, *Heidegger and Derrida: Reflections on Time and Language* (Lincoln: University of Nebraska Press, 1989).

Reck, Andrew J., *Introduction to William James* (Bloomington: Indiana U.P., 1967).

Rice, James L., *Dostoevsky and the Healing Art: An Essay in Literary and Medical History* (Ardis: Ann Arbor, 1985).

Ricoeur, Paul, *Freud and Philosophy: An Essay on Interpretation* (New Haven: Yale U.P., 1970).

————, 'Narrative Time', *Critical Inquiry*, 7(1) (1980), 169–90.

Rieff, Philip, *Freud: The Mind of the Moralist* (Chicago: University of Chicago Press, 1979).

————, *The Triumph of the Therapeutic: Uses of Faith After Freud* (Chicago: University of Chicago Press, 1987).

Riese, Walther, 'The Impact of Romanticism on the Experimental Method', *Studies in Romanticism*, 2 (1962–63), 12–22.

Rilke, Rainer Maria, *Selected Poems* (London: Penguin, 1964).

Roazen, Paul, *Erik H. Erikson: The Power and Limits of a Vision* (New York: Free Press, 1970).

————, *Freud and His Followers* (New York: Da Capo Press, 1992).

Robinson, Virginia, 'The Double Soul: Mark Twain and Otto Rank', *Journal of the Otto Rank Association*, 6(1) (1971), 32–9.

Rogers, Carl, 'The Loneliness of Contemporary Man', *Review of Existential Psychology and Psychiatry*, 1(2) (1961), 94–101.

Roosevelt, Theodore, *The Strenuous Life: Essays and Addresses* (London: Thomas Nelson & Sons, 1900).

Rorty, A.O. (ed.), *The Identities of Persons* (California: University of California Press, 1976).

Rorty, Richard, *Consequences of Pragmatism: Essays 1972–1980* (Brighton: Harvester, 1982).

————, *Contingency, Irony, and Solidarity* (Cambridge: Cambridge U.P., 1989).

————, *Philosophy and the Mirror of Nature* (Oxford: Blackwell, 1990).

————, *Essays on Heidegger and Others: Philosophical Papers Volume 2* (Cambridge: Cambridge U.P., 1991).

Rose, Jacqueline, '*The Man Who Mistook His Wife for a Hat* or *A Wife Is Like an Umbrella* – Fantasies of the Modern and Postmodern', in Andrew Ross, ed., *Universal Abandon?: The Politics of Postmodernism* (Edinburgh: Edinburgh U.P., 1988), pp. 237–50.

Rosenfield, Israel, *The Strange, Familiar, and Forgotten: An Anatomy of Consciousness* (London: Picador, 1995).

Rosenzweig, Saul, *The Historic Expedition to America (1909): Freud, Jung and Hall the King-maker* (St Louis: Rana House, 1992).

Ross, Dorothy (ed.), *Modernist Impulses in the Human Sciences 1870–1930* (Baltimore: Johns Hopkins U.P., 1994).

Roth, Michael (ed.) *Freud: Conflict and Culture: Essays on His Life, Work and Legacy* (New York: Knopf, 1998).

Roudinesco, Elisabeth, *Jacques Lacan and Co.: A History of Psychoanalysis in France, 1925–1985*, trans. Jeffrey Mehlman (Chicago: University of Chicago Press, 1990).

Ruddick, Lisa, 'Fluid Symbols in American Modernism: William James, Gertrude Stein, George Santayana, and Wallace Stevens', in ed. Martin Bloomfield, *Allegory, Myth and Symbol* (Cambridge, MA: Harvard U.P., 1981).

————, 'William James and the Modernism of Gertrude Stein', in ed. Robert Kiely, *Modernism Reconsidered* (Cambridge, MA: Harvard U.P., 1983).

Rudnytsky, *The Psychoanalytic Vocation: Rank, Winnicott, and the Legacy of Freud* (New Haven: Yale U.P., 1991).

Ruf, Frederick J., *The Creation of Chaos: William James and the Stylisitic Making of a Disorderly World* (New York: State University of New York Press, 1991).

Ryan, Judith, *The Vanishing Subject: Early Psychology and Literary Modernism* (Chicago: University of Chicago Press, 1991).

Sacks, Oliver, *Migraine: The Evolution of a Common Disorder* (London: Faber, 1970).

————, 'The Great Awakening', *The Listener* (26 December 1972), pp. 521–4.

————, *The Man Who Mistook His Wife For A Hat* (London: Picador, 1986).

————, 'Clinical Tales', *Literature and Medicine*, 5 (1986), 16–23.

————, 'Travel Happy (1961)', *Antæus*, 61 (1988), 391–406.

————, 'Neurology and the Soul', *New York Review of Books* (22 November 1990), 44–50.

————, 'Luria and "Romantic Science"', in Elkonon Goldberg, ed., *Contemporary Neuropsychology and the Legacy of Luria* (Hillsdale, NJ: Erlbaum, 1990).

————, *Awakenings* (London: Picador, 1991).

————, *A Leg To Stand On* (London: Picador, 1991).

————, *Migraine* (London: Faber & Faber, 1991).

————, *Seeing Voices* (London: Picador, 1991).

————, 'The Poet of Chemistry', *New York Review of Books* (4 November 1993), 50–6.

————, *An Anthropologist on Mars: Seven Paradoxical Tales* (London: Picador, 1995).

————, *The Island of the Colour-blind* (London: Picador, 1996).

Sadler, William A., Jr, *Existence and Love: A New Approach in Existential Phenomenology* (New York: Scribners, 1969).

Said, Edward, *Beginnings: Intention and Method* (New York: Basic Books, 1975).

Sallis, John, *Delimitations: Phenomenology and the End of Metaphysics*, 2nd edn, (Bloomington: Indiana U.P., 1995).

Santayana, George, 'What is Aesthetics?', in eds Justus Buchler and Benjamon Schwartz, *Obiter Scripta: Lectures, Essays and Reviews* (New York: Scribners, 1936), pp. 30–40

————, *Santayana on America: Essays, Notes, and Letters on American Life, Literature, and Philosophy*, ed. Richard Colton (New York: Lyon, 1968).

Schafer, Roy, *A New Language for Psychoanalysis* (New Haven: Yale U.P., 1976).

————, *The Analytic Attitude* (New York: Basic Books, 1983).

————, *Retelling a Life: Narration and Dialogue in Psychoanalysis* (New York: Basic Books, 1992).

————, 'Reading Freud's Legacies', in Joseph H. Smith and Humphrey Morris, eds, *Telling Facts: History and Narration in Psycho-analysis* (Baltimore: Johns Hopkins U.P., 1992), pp. 1–20.

Schamdasani, Sonu and Michael Münchow (eds), *Speculations after Freud: Psychoanalysis, Philosophy and Culture* (London: Routledge, 1994).

Scheler, Max, *The Nature of Sympathy*, trans. Peter Heath (London: Routledge & Kegan Paul, 1954).

Schelling, F.W.J. von, *On the History of Modern Philosophy*, trans. Andrew Bowie (Cambridge: Cambridge U.P., 1994).

Segal, Robert A. (ed.), *In Quest of the Hero* (Princeton, NJ: Princeton U.P., 1990).

Shechner, Mark, *After the Revolution: Studies in the Contemporary Jewish-American Imagination* (Bloomington: Indiana U.P., 1987).

Shklovsky, Victor, 'Art as Technique', in eds and trans. Lee T. Lemon and Marion J. Reis, *Russian Formalist Criticism: Four Essays* (Lincoln: University of Nebraska Press, 1965), pp. 5–22.

Schlüpmann, Heide, 'The First German Film: Rye's *The Student of Prague* (1913)', in Eric Rentschler, ed., *German Film and Literature: Adaptations and Transformations* (New York: Methuen, 1986), pp. 9–24.

Schmidl, Fritz, 'Sigmund Freud and Ludwig Binswanger', *Psychoanalytic Quarterly*, 28 (1959), 40–58.

Schopenhauer, Arthur, *The Word as Will and Idea*, vol. 1, 6th edn, trans. R.B. Haldane and J. Kemp (London: Kegan Paul, Trench, Trübner & Co, 1907).

————, *Essays and Aphorisms*, trans. R.J. Hollingdale (Harmondsworth: Penguin, 1970).

Schrader, George A., 'Existential Psychoanalysis and Metaphysics', *The Review of Metaphysics*, 13(1) (1959), 139–64.

Schwartz, Michael and Osborne Wiggins, 'Science, Humanism and the Nature of Medical Practice: A Phenomenological View', *Perspectives in Biology and Medicine*, 28(3) (1985), 331–61.

Scudder, Vita, 'The Moral Dangers of Musical Devotees', *Andover Review*, 7 (1887), 46–53.

Searle, John R., *The Rediscovery of the Mind* (Cambridge, MA: MIT Press, 1992).

————, 'The Mystery of Consciousness', *New York Review of Books*, 62(17) (1995), 60–6 and 62(18) (1995), 54–61.

Seidman, Bradley, *Absent at the Creation: The Existential Psychiatry of Ludwig Binswanger* (Roslyn Heights: Libra, 1983).

Seigel, Jerrold, 'Avoiding the Subject: A Foucaultian Itinerary', *Journal of the History of Ideas*, 51(2) (1990), 273–99.

Shaw, Charles Gray, 'Emerson the Nihilist', *International Journal of Ethics*, 25 (1914), 68–86.

Sidney, Philip, *A Defence of Poetry*, ed. Jan van Dorsten (Oxford: Oxford U.P., 1989).

Siegfried, Charlene Haddock, 'Vagueness and the Adequacy of Concepts', *Philosophy Today*, 26(4) (1982), 357–67.

Silverman, Hugh (ed.), *Gadamer and Hermeneutics* (New York: Routledge, 1991).

Simmel, Georg, 'The Metropolis and Mental Life', *The Sociology of Georg Simmel*, ed. Kurt Wolff (New York: Free Press, 1950).

Simons, Jon, *Foucault and the Political* (London: Routledge, 1995).

Simonton, Dean, *Greatness: Who Makes History and Why* (New York: The Guilford Press, 1994).

Sinfield, Alan, *Faultlines: Cultural Materialism and the Politics of Dissident Reading* (Oxford: Clarendon, 1992).

Skinner, Quentin (ed.), *The Return of Grand Theory in the Human Sciences* (Cambridge: Cambridge U.P., 1990).

Snelders, T.A.M., 'Romanticism and Naturphilosophie and the Inorganic Natural Sciences, 1797–1840: An Introductory Survey', *Studies in Romanticism*, 9(4) (1970), 193–215.

Snow, C.P., *The Two Cultures, The Rede Lecture, 1959* (Cambridge: Cambridge U.P., 1965).

Sonnemann, Ulrich, *Existence and Therapy: An Introduction to Phenomenological Psychology and Existential Analysis* (New York: Grune & Stratton, 1954).

Spanos, William V., 'Breaking the Circle: Hermeneutics as Dis-closure', *Boundary* 2, 5(2) (1977), 421–57

————, 'The Indifference of Différance: Retrieving Heidegger's Destruction', *Annals of Scholarship*, 2 (1981), 109–29.

Spector, Jack, *The Aesthetics of Freud: A Study on Psychoanalysis and Art* (New York: McGraw-Hill, 1974).

Spencer, Herbert, *First Principles of Psychology* (New York: Appleton & Co, 1870).

Spiegelberg, Herbert, *Phenomenology in Psychology and Psychiatry: A Historical Introduction* (The Hague: Martinus Nijhoff, 1969).

Sprigge, T.L.S., *James and Bradley: American Truth and British Reality* (Chicago & La Salle: Open Court, 1993).

Stallybrass, Peter and Allon White, 'Bourgeois Hysteria and the Carnivalesque', in Simon During, ed., *The Cultural Studies Reader* (London: Routledge, 1993), 284–92.

Stevens, Richard, *Erik Erikson: An Introduction* (Milton Keynes: The Open University Press, 1983).

Stoichita, Victor I., *A Short History of the Shadow* (London: Reaktion, 1997).

Strachey, James and Anna Freud (eds), *The Standard Edition of the Complete Psychological Works of Sigmund Freud* (London: Hogarth Press, 1961).

Strout, Cushing, 'William James and the Twice-Born Sick Soul', *Daedalus*, 97(3) (1968), 1062–82.

———, 'The Pluralistic Identity of William James: A Psycho-historical Reading of The Varieties of Religious Experience', *American Quarterly*, 23(2) (1971), 135–52.

Sturrock, John, *The Language of Autobiography: Studies in the First Person Singular* (Cambridge: Cambridge U.P., 1993).

Swales, Martin, *The German Bildungsroman from Wieland to Hesse* (Princeton, NJ: Princeton U.P., 1978).

Taft, Jessie, *Otto Rank* (New York: Julian Press, 1958).

Tallack, Douglas, 'Transcendentalism and Pragmatism', in ed. Mick Gidley, *Modern American Culture: An Introduction* (London: Longman, 1993), pp. 68–93.

Tanner, Tony, *The Reign of Wonder* (New York: Harper & Row, 1965).

Taylor, Eugene, *William James on Exceptional Mental States: The 1896 Lowell Lectures* (Amherst: University of Massachusetts Press, 1984).

Taylor, Eugene and Robert H. Wozniak (eds), *Pure Experience: The Response to William James* (Bristol: Thoemmes Press, 1996).

Taylor, Charles, 'Interpretation and the Sciences of Man', *Review of Metaphysics*, 25(1) (1971), 3–51.

———, *Hegel* (Cambridge: Cambridge U.P., 1975).

———, *Sources of the Self* (Cambridge, MA: Harvard U.P., 1989).

———, *Philosophical Arguments* (Cambridge, MA: Harvard U.P., 1995).

Taylor, Paul, 'A Study in the Right Frame of Mind', *The Independent* (28 April 1994), II, p. 23.

Thayer, H.S. (ed.), *Pragmatism: The Classic Writings* (New York: New American Library, 1970).

Thiele, Leslie, *Timely Meditations: Martin Heidegger and Postmodern Politics* (Princeton, NJ: Princeton U.P., 1995).

Thiselton, Anthony C., *New Horizons in Hermeneutics* (London: HarperCollins, 1992).

Thomas, Joseph M., 'Figures of Habit in William James', *New England Quarterly*, 66(1) (1993), 3–26.

Todorov, Tzvetan, *The Fantastic: A Structural Approach to a Literary Genre*, trans. Richard Howard (Cleveland: The Press of Cape Western Reserve University, 1973).

Townsend, Kim, *Manhood at Harvard: William James and Others* (Cambridge, MA: Harvard U.P., 1996).

Tricomi, Albert H., 'The Rhetoric of Aspiring Circularity in Emerson's "Circles" ', *ESQ: A Journal of the American Renaissance*, 18(4) (1972), 271–83.

Tymms, Ralph, *German Romantic Literature* (London: Methuen, 1955).

Valdés, Mario J. (ed.), *A Ricoeur Reader: Reflection and Imagination* (New York: Harvester Wheatsheaf, 1991).

Varela, Francisco J. et al. (eds), *The Embodied Mind: Cognitive Science and Human Experience* (Cambridge, MA: MIT Press, 1993).

Wachterhauser, Brice R. (ed.), *Hermeneutics and Modern Philosophy* (Albany, NY: State University of New York Press, 1986).

————, *Hermeneutics and Truth* (Evanston, IL: Northwestern University Press, 1994).

Wall, Patrick, 'Is there a Higher Nervous System?', *The Listener* (3 August 1972), 139–41.

Wallenstein, Robert S. and Leo Goldberger (eds), *Ideas and Identities: The Life and Work of Erik Erikson* (Madison: International Universities Press, 1998).

Webber, Andrew, *The Doppelgänger: Double Visions in German Literature* (Oxford: Clarendon, 1996).

Weber, Max, *The Protestant Ethic and the Spirit of Capitalism*, trans. Talcott Parsons (London: Routledge, 1992).

Werth, Lee F., 'The Banks of the Stream of Consciousness', *History of Philosophical Quarterly*, 3(1) (1986), 89–105.

West, Cornel, *The American Evasion of Philosophy: A Genealogy of Pragmatism* (Madison: University of Wisconsin, 1989).

Wetzels, Walter D., 'Aspects of Natural Science in German Romanticism', *Studies in Romanticism*, 10 (1970–71), 44–59.

Wheeler, Kathleen, *Romanticism, Pragmatism and Deconstruction* (Oxford: Blackwell, 1993).

White, Hayden, *The Content of the Form: Narrative Discourse and Historical Representation* (Baltimore: Johns Hopkins U.P., 1992).

White, Morton, *Pragmatism and the American Mind* (New York: Oxford U.P., 1973).

Whitehead, Alfred North, *Science and the Modern World, The Lowell Lectures, 1925* (Cambridge: Cambridge U.P., 1929).

Wild, John, 'On the Nature and Aims of Phenomenology', *Philosophy and Phenomenological Research*, 3 (1943), 85–95.

————, 'Contemporary Phenomenology and the Problem of Existence', *Philosophy and Phenomenological Research*, 20 (1959), 166–80.

————, *The Radical Empiricism of William James* (New York: Doubleday, 1970).

Willcox, Louise Collier, 'Nietzsche: A Doctor for Sick Souls', *North American Review*, 194 (1911), 765–74.

Wilshire, Bruce, *William James and Phenomenology: A Study of The Principles of Psychology* (Bloomington: Indiana U.P., 1968).

————, 'William James's Theory of Truth Phenomenologically Considered', in ed. Peter Caws, *Two Centuries of Philosophy in America* (London: Blackwell, 1980).

Wilson, Daniel J., 'Science and the Crisis of Confidence in American Philosophy, 1870–1930', *Transactions of the Charles S. Peirce Society*, 23(2) (1987), 235–62.

Wing, J. K., 'Distorting Mirrors', *Times Literary Supplement* (7 February 1986), 146.

Wood, David (ed.), *On Paul Ricoeur: Narrative and Interpretation* (London: Routledge, 1992).

Wordsworth, William and Samuel Taylor Coleridge, *Lyrical Ballads*, eds R.L. Brett and A.R. Jones (London: Routledge, 1988).

Wulff, Henrik et al., *Philosophy of Medicine: An Introduction* (Oxford: Blackwell, 1986).

Wygal, Winnifred, 'The Political Inertia of the Intellectual', *Radical Religion*, 1(3) (1936), 31–4.

Yalom, Irvin D., *Existential Psychotherapy* (New York: Basic Books, 1980).

Žižek, Slavoj, *The Sublime Object of Ideology* (London: Verso, 1989).

Index